From Perturbative to Constructive Renormalization

Princeton Series in Physics

edited by Philip W. Anderson, Arthur S. Wightman, and Sam B. Treiman

Quantum Mechanics for Hamiltonians Defined as Quadratic Forms *by Barry Simon*

The Many-Worlds Interpretation of Quantum Mechanics *edited by B. S. DeWitt and N. Graham*

Homogeneous Relativistic Cosmologies *by Michael P. Ryan, Jr., and Lawrence C. Shepley*

The $P(\Phi)_2$ Euclidean (Quantum) Field Theory *by Barry Simon*

Studies in Mathematical Physics: *Essays in Honor of Valentine Bargmann edited by Elliott H. Lieb, B. Simon, and A. S. Wightman*

Convexity in the Theory of Lattice Gases *by Robert B. Israel*

Works on the Foundations of Statistical Physics *by N. S. Krylov*

Surprises in Theoretical Physics *by Rudolf Peierls*

The Large-Scale Structure of the Universe *by P. J. E. Peebles*

Statistical Physics and the Atomic Theory of Matter, From Boyle and Newton to Landau and Onsager *by Stephen G. Brush*

Quantum Theory and Measurement *edited by John Archibald Wheeler and Wojciech Hubert Zurek*

Current Algebra and Anomalies *by Sam B. Treiman, Roman Jackiw, Bruno Zumino, and Edward Witten*

Quantum Fluctuations *by E. Nelson*

Spin Glasses and Other Frustrated Systems *by Debashish Chowdhury* (*Spin Glasses and Other Frustrated Systems* is published in co-operation with World Scientific Publishing Co. Pte. Ltd., Singapore.)

Weak Interactions in Nuclei *by Barry R. Holstein*

Large-Scale Motions in the Universe: A Vatican Study Week *edited by Vera C. Rubin and George V. Coyne, S.J.*

Instabilities and Fronts in Extended Systems *by Pierre Collet and Jean-Pierre Eckmann*

More Surprises in Theoretical Physics *by Rudolf Peierls*

From Perturbative to Constructive Renormalization *by Vincent Rivasseau*

From Perturbative to Constructive Renormalization

Vincent Rivasseau

Princeton Series in Physics

PRINCETON UNIVERSITY PRESS
PRINCETON, NEW JERSEY

Published by Princeton University Press, 41 William Street,
Princeton, New Jersey 08540
In the United Kingdom: Princeton University Press, Oxford

Library of Congress Cataloging-in-Publication Data

Rivasseau, Vincent, 1955–
From perturbative to constructive renormalization / Vincent Rivasseau.
p. cm. — (Princeton series in physics)
Includes bibliographical references and index.
ISBN 978-0-691-63682-5
1. Renormalization (Physics) 2. Quantum field theory.
3. Perturbation (Mathematics) 4. Constructive mathematics.
I. Title. II. Series.
QC174.17.R46R58 1991
530.1′43—dc20 90-49514

This book has been composed in Times Roman

Princeton University Press books are printed on acid-free paper, and meet the guidelines for permanence and durability of the Committee on Production Guidelines for Book Longevity of the Council on Library Resources

Printed in the United States of America by Princeton University Press, Princeton, New Jersey

10 9 8 7 6 5 4 3 2 1

A l' unique étoile

Contents

Acknowledgements

The material presented in this book grew out of common work with C. de Calan on uniform bounds for renormalized Feynman amplitudes and with J. Feldman, J. Magnen and R. Sénéor on phase space analysis applied to perturbative and constructive renormalization theory. They are in this sense coauthors of these notes. I thank them deeply, together with all my other collaborators and colleagues, for the joy, stimulation and sharings of day to day work. I am particularly indebted to J. Lascoux, A. Wightman and E. Speer for guiding my first steps in this domain.

For many years I have benefited from the friendly encouragement of Professors A. Wightman and F. Dyson. I hope to thank them by this small testimony to the inspiring beauty of the subject that, in different ways, they pioneered.

I thank D. Ruelle for the first suggestion to write such a book, and I apologize to him and many others for taking such a long time to complete it.

I thank the Universities of Princeton and of Belo-Horizonte for support and for allowing me to experiment with the teaching of the material of this book (in this respect I thank more particularly Profs. A. Wightman, M. O'Carroll and R. Schorr, and all the students which attended the corresponding courses for their help). Support of the Institute for Advanced Study and of my home institutions, the CNRS and the Ecole Polytechnique, is also gratefully acknowledged.

I thank C. Kopper and A. Wightman for their careful critical reading of the manuscript.

Finally I thank Marie-France, Jean-Noël, Christian, and Marie, Annie and Jacques, François and Thérèse; their affection was also essential for this book to exist.

PART I

INTRODUCTION TO EUCLIDEAN FIELD THEORY

Chapter I.1

The Ultraviolet Problem

Quantum field theory is an attempt to describe the properties of elementary "point-like" particles in terms of relativistic quantum fields. It is now widely believed to offer a coherent mathematical framework for relativistic models (like the "standard $U(1) \times SU(2) \times SU(3)$ model"). These models include all the particles and interactions observed up to now except gravity. Therefore, together with general relativity, field theory is the backbone of our current understanding of the physical world. In the future a new, more unifying framework may be adopted, like the currently promising superstring theory, which is a relativistic and quantum description of extended one dimensional objects instead of point-like particles; nevertheless even in this case it is extremely likely that field theory will remain important in many situations, just as classical mechanics is still today.

This situation is relatively recent. Until the 70's the very statement that quantum field theory might provide a coherent mathematical framework at all was not widely accepted. The main doubts on the mathematical consistency of quantum field theory were due to the persistence of ultraviolet problems (and to the lack of successful models for strong interactions: QCD, the present field theory of strong interactions, did not exist). Let us sketch what these ultraviolet problems are and why they are important.

An ultraviolet problem is one which is due to the existence of arbitrarily small length scales, or equivalently of arbitrarily large frequencies in the Fourier analysis of a theory. Such problems are inherent to the formalism of quantum field theory, because it is a crucial assumption that the fields live on a continuous space time. One might wonder whether this continuity condition has anything to do with physics and whether the whole problem is not a mathematical artifact. After all it is reasonable to expect that space time will conserve its smoothness only until the Planck scale, where quantum aspects of gravitation might distort it significantly. This Planck scale might provide a physical ultraviolet cutoff; this is what seems to occur in superstring theories where it is conjectured that at least in perturbation theory, there are no ultraviolet divergences. However the Planck scale is much higher than the typical scales that field theory tries to describe, and is completely inaccessible to direct experiments.

In fact the most compelling reason for which we are interested in the continuous formulation of field theory is the same for which we are interested in the thermodynamic limit of statistical systems. In statistical mechanics this limit corresponds to systems of infinite volume. We know that in nature macroscopic systems are in fact finite, not infinite, but they are huge with respect to the atomic scale. The thermodynamic limit is an adequate simplification in this case, since it allows one to give a precise mathematical content to the physically relevant questions (like dependence of the limit on boundary conditions, existence of phase transitions etc....). Since a limit has been taken, the power of classical analysis may be applied to these questions. It would be much harder and less natural to try to define the analogous notions for a large finite system, just as it is difficult and often inappropriate to make discrete approximations to some typically continuous mathematics like topology.

From this point of view the ultraviolet problem appears central and inescapable in field theory. A limit has to be performed; if this limit were not to exist, the corresponding mathematical formalism would be of little interest.

Historically quantum field theory was plagued by two successive ultraviolet "diseases" which raised doubts on the existence or consistency of the ultraviolet limit. In both cases the situation looked bad for many years until a way out of the crisis was found. The first and most famous ultraviolet disease has been recognized almost since the birth of quantum field theory. It is the presence of divergences due to the integration over high momenta in the loops of Feynman integrals. In the φ_4^4 theory which will be discussed soon, one of the simplest of these divergences is the divergence of the second order graph which we call the "bubble" (see Fig. I.1.1).

By momentum conservation the amplitude for this graph is only a function of $k = k_1 + k_2$. In Euclidean space, this amplitude (apart from combinatoric coefficients discussed later) is given by the integral:

$$\int \frac{d^4p}{(p^2 + m^2)[(p+k)^2 + m^2]}, \tag{I.1.1}$$

which diverges logarithmically for large values of p. (Similar divergences occur of course in the more physical theory of quantum electrodynamics).

Around 1950, this disease was cured by the invention of perturbative renormalization by Feynman, Schwinger, Tomonaga, Dyson and others (see e.g., [Dy1]). Basically it amounts to a redefinition of the physically observable parameters of the theory which pushes the infinities into unobservable "bare" parameters. It took more than a decade to put this perturbative theory of renormalization on a completely firm mathematical basis. Roughly speaking, the main result states that theories which are renormalizable from the naive power counting point of view can indeed be renormalized without changing the formal structure of the Lagrangian. More precisely one can replace the bare parameters of the Lagrangian by formal power series in the renormalized parameters (usually the coupling constant), so that the resulting perturbative expansion in the renormalized coupling is finite to all orders, as a formal power series. We call this important theorem the BPH theorem (Bogoliubov–Parasiuk–Hepp [BP][He1]) although as usual it incorporates a lot of former work and was followed by important extensions and refinements. It was somehow a surprise to discover that this theorem, developed for quantum electrodynamics or the φ_4^4 model, remained also true for non-Abelian gauge theories ['tH1][LZ], in which case it is highly non trivial to check that the counterterms

Figure I.1.1 (Left) The bubble graph. (Right) A chain of bubbles.

required do not break gauge invariance, i.e., can be absorbed in a redefinition of the field strength and of the coupling constant. By power counting analysis alone, this would not be true: one has to incorporate additional information coming from Slavnov identities or BRS invariance [Tay][Sla][BRS]. The traditional proof (see [IZ]) relies on a dimensional regularization in which gauge invariance is maintained, but this approach has some drawbacks: the dimensional regularization is complicated and cannot be used up to now in a constructive (non-perturbative) program.

But quantum electrodynamics, the only firmly established field theory has been plagued by another ultraviolet problem, raised in particular by L. Landau and other physicists of the Russian school. We call it the "renormalon" problem, although this name was introduced much later. It does not occur any more at the level of individual Feynman graphs, but it affects the perturbative series as a whole. In φ_4^4 (which is similar to electrodynamics in this respect), there are several ways to discover the problem. One of them is to consider the leading-log behavior of the renormalized 4 point function at n-th order, $S_n^{4,R}$, and large external momenta. Let m be the mass of the particles. One finds (in Euclidean space):

$$S_n^{4,R}(k) \simeq_{k\to\infty} g_R^n[\beta_2 \log(k/m)]^{n-1}, \qquad \text{(I.1.2)}$$

where g_R is the renormalized coupling constant and β_2 is a numerical coefficient (for single component φ_4^4 it is $9/2\pi^2$ or $3/16\pi^2$ if one writes, as usual, the interaction as $\varphi^4/4!$). Various ways of playing with formula (I.1.2) give rise to various troubles, all related.

When the 4 point function is inserted into a convergent loop like the triangular 6-point graph of Fig. I.1.2, at 0 external momenta, one obtains contributions to the n-th order of perturbation theory for the 6 point function proportional to

$$g^n \int_{|k|\geq m} \frac{d^4k}{(k^2+m^2)^3}[\beta_2 \log(k/m)]^{n-3} \simeq (n-3)!g^n\beta_2^{n-3}. \qquad \text{(I.1.3)}$$

Figure I.1.2 (Left) The triangular graph. (Right) The triangular graph with a chain of bubbles attached.

These contributions are not summable over n (since they add up with the same sign they are also not Borel summable). Therefore the renormalon problem appears as a difficulty in summing perturbation theory. However one might also consider the asymptotic behavior in k of the bare (unrenormalized) 4-point function:

$$S^{4,B}_{n,k}(0) \simeq_{k\to\infty} (-g_B)^n[\beta_2 \log(k/m)]^{n-1}, \tag{I.1.4}$$

where g_B is the bare coupling constant. Since in the theory of perturbative renormalization this function should be the counterterm of the theory, i.e., the difference between the renormalized coupling g_R and the bare coupling g_B, one gets the formula:

$$g_R = g_B - \sum_{n=2}^{\infty}(-g_B)^n[\beta_2 \log(k/m)]^{n-1} = \frac{g_B}{1+g_B\beta_2 \log(k/m)}. \tag{I.1.5}$$

Therefore if $g_B > 0$, $\lim_{k\to\infty} g_R = 0$ (no matter how g_B is chosen as a function of k). This is the phenomenon of "charge screening" called also the triviality problem for φ_4^4 or QED. In the ultraviolet limit, only the trivial noninteracting theory seems to exist. Still another possibility is to invert formula (I.1.5) to obtain

$$g_B(k) \simeq \frac{g_R}{1-g_R\beta_2 \log(k/m)}. \tag{I.1.6}$$

Keeping g_R fixed, g_B becomes negative at an energy of order $me^{+1/\beta_2 g_R}$, therefore the theory should become unstable or inconsistent at this "Landau" energy scale. The singularity in (I.1.6) is sometimes called a Landau ghost. But of course one should not really trust (I.1.6). One should rather remark that as $k \to \infty$, g_B increases and becomes of order unity around the Landau energy, after which perturbation theory itself and in particular (I.1.2)–(I.1.6) should no more be valid. The behavior of g_B is turned into a strong coupling problem, intractable up to now, except perhaps by Monte Carlo simulations.

Let us summarize the characteristic features of the renormalon problem. As remarked already, it affects the summability of perturbation theory as a whole (therefore physicists might call it a non-perturbative problem, although we would prefer to consider it as a "strong coupling" problem). It is truly an ultraviolet disease, because it does not occur in the theory with fixed ultraviolet cutoff, no matter how large. It is easy for instance to check that the $n!$ behavior of (I.1.3) appears neither in the bare nor in the renormalized perturbative series with fixed ultraviolet cutoff. Finally the name "renormalon problem" although perhaps awkward, is justified because the disease arises from the introduction of counterterms, hence from the use of perturbative renormalization. As will indeed be discussed at length in this book, the perturbative theory of renormalization cures the first problem of infinities too well. It introduces both some pieces of counterterms which we call "useful" because they make the renormalized Feynman amplitudes finite, but also some pieces of counterterms which we call "useless." These useless counterterms do not cure any divergence. Furthermore they are the ones which are responsible for the renormalon problem!

To observe the distinction between useful and useless counterterms, one needs some detailed Fourier analysis such as the one provided by phase space decomposition. Then it appears that the sole reason for introducing useless counterterms is the locality requirement and the insistence upon writing the renormalized series in terms of fixed renormalized constants which do not depend on the energy scale. In this book we will argue that one should drop this restriction, and adopt a more effective perturbation theory with an infinite number of scale dependent "effective constants." In this way we rediscover that the renormalization group point of view is the right way to investigate the renormalon problem.

Historically the renormalon disease and its investigation by the invention of the renormalization group was not discussed exactly in these terms; the emphasis was on the invariance of the theory under changes of the (arbitrary) subtraction scale (from this invariance came the very name "renormalization group") and on high energy asymptotic behavior, not on $n!$ behavior at large order and the problem of summing up perturbation theory. Of course both points of view are closely related. Anyway the problem was serious enough to raise again doubts on the consistency of quantum field theory through the 60's.

As a reaction the constructive program was launched, and in the early 70's a major milestone was reached when superrenormalizable theories of various types were built and checked to be free of inconsistencies. Reviews or books on this first period of constructive theory are [Er1][Si1][GJ2]. Although the results and techniques used have proved very influential in many areas of physics and mathematics, this success was nevertheless not of specific relevance to the renormalon problem. For theoretical physicists, a convincing way to escape this problem was really found with the major discovery of asymptotic freedom in non-Abelian gauge theories [Po][GW]. This occurred just at the right time to complete the spectacular rebirth of quantum field theory: non-Abelian gauge theories had been developed as realistic models for the electroweak interaction and had been shown to be perturbatively renormalizable. It gave in turn a major impetus to adopt them to describe strong interactions as well.

Asymptotic freedom occurs when the coefficient β_2 in (I.1.2) is negative. As a result equations (I.1.5)–(I.1.6) are inverted. It is now the bare charge which is screened. At large energies the particles behave like free point-like objects (hence the name of asymptotic freedom). Looking at (I.1.3) and changing the sign of β_2 we see that the renormalon problem still prevents "ordinary" summation, but the corresponding contributions now become alternate; therefore an other type of summability, like Borel summability, becomes possible.

The discovery of asymptotic freedom in non-Abelian gauge theories convinced the theoretical physics community that these quantum field theories are indeed mathematically consistent. Many physicists believe that there is no longer any surprise to be expected in the ultraviolet problem for gauge theories (see however ['tH4] for an exception). But this belief has yet to be substantiated by a non-perturbative, mathematically rigorous analysis.

To understand rigorously the concept of asymptotic freedom beyond perturbation theory, one way is to construct first some consistent models of renormalizable theories with such a behavior. There exist models of this kind simpler than the non-Abelian gauge theories in 4 dimensions, namely fermionic models in two dimensions with many components and a quartic interaction [MW][GrNe]. Also the φ_4^4 model, although not asymptotically free in the ultraviolet direction, is asymptotically free in the infrared direction. Although the corresponding constructive problem is a problem of statistical mechanics rather than field theory (the ultraviolet cutoff is not removed), it is very similar in mathematical structure. This road has been followed by K. Gawedzki and A. Kupiainen [GK2-3-4] and by J. Feldman, J. Magnen, R. Sénéor and the author [FMRS3-4-5], who, with somewhat different technical tools, succeeded in building these models. In fact it is the main goal of this book to present in a more systematic and accessible form than the original papers the technique of multiscale or phase space expansion, as developed and applied in our collaboration with J. Feldman, J. Magnen and R. Sénéor. This technique originated in the constructive work of Glimm and Jaffe [GJ1].

To extend our rigorous understanding of asymptotic freedom to non-Abelian gauge theories in a finite volume is the next natural challenge. (The large volume problem is indeed a different one, where one has to deal with a strong coupling problem and physical issues like quark confinement which seem still much farther from a rigorous analysis). This ultraviolet consistency of non-Abelian gauge theories on compact manifolds is not only a key issue in theoretical physics, but is also becoming one in geometry, in particular since E. Witten related Donaldson's invariants to (still formal) functional integrals of some (supersymmetric) Yang–Mills theory [Wit]. Various versions of Yang–Mills theories now seem to be the link between the most fascinating problems of geometry (homotopy and knots in three dimensions, symplectic geometry and conformal theories in two dimensions and differential geometry in four dimensions). A constructive understanding of the corresponding functional integrals would be therefore a major progress in pure mathematics.

The problem has been attacked first by T. Balaban [Ba2-9]. In an impressive sequence of papers, completed recently, he establishes an ultraviolet stability bound for the effective action of a lattice gauge theory after iterating a large number of clever block-spin transformations. From this result it is expected that the continuum limit of gauge invariant observables like Wilson loops can be constructed. Hence at least these observables should be free from inconsistencies and this is a very important result. This work is now followed by a related program of P. Federbush [Fe2-7][FedW], still in progress, which uses also a lattice regularization and phase space cells.

But even after completion of these works, many questions remain in particular because the expectation values of the fields of the theory (in a particular gauge) are not built in the lattice-based approach. It is important to know whether such expectation values, which are not gauge invariant, can also be built or, if they cannot, to understand better why. The main difficulty for this other program lies in a lack of positivity of the gauge-fixed functional integrals, related to the Gribov problem [Gri]. Progress in understanding this difficult problem has been slow up to now, but interesting as well as surprising results may lie ahead [Zw1-4][DeZw1-2].

After this quick historical overview of the ultraviolet problem in field theory, let us describe the structure of this book.

The first part is devoted to some introductory material on field theory, on the φ^4 model which is the training ground for most of the book, and on perturbation theory and Feynman graphs. We keep this part very brief, with some particular emphasis on the aspects which will be most useful for the rest of the books, namely the Euclidean formulation of field theory, Feynman graphs and amplitudes, Gaussian functional measures, ultraviolet cutoffs, and the Nevanlinna–Sokal version of Borel summability which is very convenient for mathematical physics. Although we try to be reasonably self-contained, we assume some familiarity of the reader with quantum field theory, for instance with [IZ].

Then in the second part we apply the idea of "phase space chopping" to the study of perturbation theory, focusing on renormalization and using the φ_4^4 model for simplicity. We derive a uniform BPH theorem which adds reasonable estimates to the finiteness content of the original BPH theorem; the first theorem of this kind appeared in [dCR1]. The version worked out in this book contains some improvements other previous results, in particular concerning the external momentum dependence of the bounds. Then we show that the analysis of the bounds obtained leads one naturally to reshuffle the bare or renormalized perturbation theory into a better form, effective perturbation theory. In the case of simplified asymptotically free models for which the number of graphs at order n is not too large, like the wrong sign planar φ_4^4 model ['tH5-7][Ri1], this reshuffling is even sufficient to construct the model. This part is closed by a discussion of the large order *behavior* of perturbation theory, a problem truly at the border between perturbative and constructive techniques.

In the third part, devoted to constructive theory, we start with an introduction to the key techniques of cluster and Mayer expansion; these techniques extend in a natural way to phase space under the name of "multiscale expansions." Our goal is to apply them first to the infrared limit of critical φ_4^4; we show how positivity of the interaction can be used to handle the so called "domination" or large field problem, using first the case of convergent power counting as a warm-up; the construction of the full model is then explained. Although it is not yet fully formalized as a neat chain of definitions and lemmas, we tried our best to improve the original construction [FMRS5], both by simplifying many technical aspects and by making some others more explicit, in particular the definition of counterterms and the mechanism by which "constructive renormalization" works. We continue with the construction of the "Gross Neveu" model in two dimensions, which is a genuine renormalizable field theory, for which we do not try to perform an extensive rewriting of the initial constructions because we think they are as a whole somewhat more satisfying. The construction of the non renormalizable Gross–Neveu model in three dimensions, which requires interesting additional techniques, is also sketched.

We conclude with a chapter which describes a tentative approach to the problem considered above, namely the construction of the continuum limit for the gauge-fixed non-Abelian gauge theories YM_4 in a finite (small) volume. This chapter contains some previously unpublished material based on a collaboration with J. Feldman, J. Magnen and R. Sénéor, and it assumes some familiarity with the standard perturbative approach to gauge theories (chapter 12 in [IZ]). We restrict for simplicity to the pure $SU(2)$ theory. We want to construct the continuum limit of the gauge fixed theory in a regular gauge, because this theory has a regular propagator which is compatible with phase space chopping. Although the phase space chopping does not respect gauge invariance, it should nevertheless be possible to recover gauge invariance in the limit, at least in the form of Slavnov identities which express invariance under gauge transformations continuously connected to the identity. This is because only the relevant or marginal effects of gauge breaking cutoffs matter for the final theory, and these effects can be compensated by means of a finite number of appropriate non gauge invariant counterterms. There are some stability requirements for these counterterms, which as we show can be met.

This approach has led only to modest results up to now, but at least as a by-product we sketch how it should give a control of perturbative renormalization of non-Abelian gauge theories with reasonable bounds which avoids the traditional use of dimensional regularization (which is not adapted to constructive purposes).

As will be discussed in the last section of this chapter, the difficulties that we met in our program are presumably an other aspect of the famous "Gribov" problem [Gri], a problem which does not show up in the standard perturbative renormalization group analysis of non-Abelian gauge theories. We explain the problem of Gribov "copies" and give some explicit proofs that with a good infrared cutoff Gribov copies of the origin are absent. This shows that there is a certain amount of strict positivity in the usual gauge fixed action, but unfortunately this positivity seems up to now too weak for constructive purposes. It might be necessary to supplement standard gauge conditions such as the Landau gauge with additional conditions which resolve the ambiguities associated to Gribov copies. This is an approach advocated since many years by Zwanziger and collaborators; we conclude by a brief review of this point of view, and of the possibility that some vacuum expectation values may not follow the standard ultraviolet behavior expected from the perturbative renormalization group analysis [DeZw1]. This would be a very surprising effect which clearly calls for a rigorous clarification.

To conclude, let us apologize sincerely to the many experts in all these areas whose work is not properly cited or accounted for in this book. Neither on perturbative nor on constructive theory, neither at the level of the subjects treated nor at the level of the references does this book intend to be an extensive or even a fair review. The main reason is that we do not feel able to report properly on something other than the particular techniques and point of view that we have personally used, which we know may not be always the most elegant or the most powerful ones. As a result, this book may be considered more as a "guided tour" of the author's favorite subjects than as an exhaustive review on perturbative or constructive field theory.

Nevertheless we want to mention at this stage that concerning the part of this book on perturbative renormalization, another elegant formalism using phase space chopping was developed in parallel to ours by Gallavotti and Nicolò, and later by Feldman, Hurd, Rosen and Wright. This formalism is very closely related to the one presented here, the main difference being that it avoids the use of Feynman graphs, making in a sense the combinatorics of cancellations more transparent. Here we stick to the point of view of Feynman graphs, which we think are presumably familiar to the potential reader, and refer for this other approach to [Gal][GaNi][FHRW].

Concerning the third part of this book, we mentioned already that a renormalization group solution to many of the same problems was worked out simultaneously by K. Gawedzki and A. Kupiainen [GK2-4], using the formalism of block spin transformations. These beautiful works and many others related will not be described at all, again for lack of competence but not of interest!

A last word is in order on the mathematical level of rigor. We intend to describe only rigorous results, but we do not always provide complete proofs of all the statements. Also we always prefer the particular to the general, for pedagogical reasons. We try to convey better the underlying ideas by phrasing as much as possible our arguments in ordinary words rather than in fully formalized equations. We know that this is dangerous and that the ratio of equations to words in scientific papers has been sometimes compared to that of signal to noise. In any case we did honestly our best to mention all the difficulties, even when they are apparently purely technical. In conclusion we hope that this book can be useful, in particular to the beginner, by explaining the natural link between perturbation theory, which is conceptually simpler, and constructive field theory, which has had for too long the reputation of being a much harder subject.

Chapter I.2

Euclidean Field Theory. The O.S. Axioms

From the classical work of the founding fathers of axiomatic field theory, we learn that the minimal mathematical requirements or axioms for a field theory are conveniently expressed in terms of the vacuum expectation values of products of the field operators, the "Wightman functions" [SW]. From these quantities one could also, at least in principle, compute more physical quantities like the S matrix. We could start directly from the Wightman axioms or their Euclidean counterpart, the Osterwalder–Schrader axioms, but we prefer to motivate them first with a brief sketch of their relation to the S matrix formalism, without any attempt to mathematical rigor. For this sketch, we follow [dC]; for a more complete study on general aspects of quantum field theory, we refer to [BS][BD][IZ].

In ordinary (Minkowski) quantum field theory, the fields are technically "operator valued distributions," i.e., they take their values in some set of operators on the Hilbert space of physical states, H. The main physical properties that are required for field theory are relativistic covariance and microcausality. Therefore H should bear a unitary representation of the Poincaré group, for which the only invariant state is the vacuum. The generators of the Poincaré group, the momentum operators P should have positive norm: $P^2 \geq 0$. And since signals do not propagate faster than light, operators which are smeared with test functions whose supports are "space like" separated should commute (or anticommute for fermions).

Since field theory is a second quantized formalism, the Hilbert space is a Fock space:

$$H = \bigoplus_{n=0}^{\infty} H_n. \tag{I.2.1}$$

H_0, of dimension 1, is generated by a particular vector ψ_0, called the vacuum. H_1 is the space of 1-particle states, and H_n, the space of n-particle states, is the n-th tensor product of H_1, symmetrized for bosons and antisymmetrized for fermions. For instance for free massive scalar bosons, H_1 is generated by a complete orthonormal set f_i of positive energy solutions of the Klein–Gordon equation:

$$(\partial_\mu \partial^\mu + m^2)f = 0. \tag{I.2.2}$$

The scalar product in this case is

$$\delta_{ij} = \langle f_i, f_j \rangle = i \int d^3x f_i^* \overleftrightarrow{\partial_0} f_j = \int d\mu_p \tilde{f}_i^* \tilde{f}_j, \tag{I.2.3}$$

where $f_i = \int d\mu_p \tilde{f}_i(p) e^{-ip\cdot x}$; $d\mu_p$ is the Lorentz invariant measure:

$$d\mu_p = \frac{d^3p}{(2\pi)^3 2p_0} = \frac{d^4p}{(2\pi)^4} 2\pi\delta(p^2 - m^2)\theta(p_0), \tag{I.2.4}$$

and we use the convention $u\overleftrightarrow{\partial_0}v = u(\partial_0 v) - (\partial_0 u)v$. The free scalar bosonic field φ also satisfies the Klein Gordon equation (I.2.2), which derives from the free Lagrangian:

$$L_0(\varphi) = \frac{1}{2}\left[\partial_\mu\varphi\partial^\mu\varphi - m^2\varphi^2\right], \tag{I.2.5}$$

and it can be developed into creation and annihilation operators which satisfy canonical commutation relations:

$$\varphi(x) = \int d\mu_p [a(p)e^{-ip\cdot x} + a^+(p)e^{ip\cdot x}], \tag{I.2.6}$$

$$[a(p), a(p')] = [a^+(p), a^+(p')] = 0;$$
$$[a(p), a^+(p')] = 2p_0(2\pi)^3\delta_3(\vec{p} - \vec{p}\,'). \tag{I.2.7}$$

The action of the field on the Hilbert space is then best described by smearing these operators with solutions f of the Klein–Gordon equation. We can define

$$a_f = \int d\mu_p \tilde{f}(p) a(p), \qquad a_f^+ = \int d\mu_p \tilde{f}(p) a^+(p). \tag{I.2.8}$$

The vacuum is then annihilated by any annihilation operator a_f, and a generating set for the full Hilbert space H is obtained by the action of a finite number of creation operators a_f^+ on the vacuum (which is therefore called a cyclic vector). The action of a_f and a_f^+ is then determined by the commutation relations coming from (I.2.7):

$$[a_f, a_{f'}] = [a_f^+, a_{f'}^+] = 0; \qquad [a_f, a_{f'}^+] = \langle f, f' \rangle . \tag{I.2.9}$$

To solve the Klein–Gordon equation one can introduce Green's functions (advanced, retarded or the symmetric one called Feynman's propagator), and the vacuum expectation value of a time ordered product of field operators (or generalized, N-point Green's functions) is given by a sum over pairings of these fields of the corresponding product of propagators. Such pairings are called Wick contractions. The result is called Wick's theorem. We will find this rule again in the context of Gaussian integration.

The theory of the free field is therefore completely explicit. But we are in fact in search of an interacting quantum field theory. We look for field theories which admit a simple Lagrangian, polynomial in the fields and their derivatives. For scalar fields this Lagrangian will be decomposed into the free piece L_0 given by (I.2.5) and a higher order polynomial L_i which is the interaction. Let us use Φ for the interacting field, to distinguish it from the free field φ. For instance in the Φ^n theory the interaction is $g\Phi^n$ and the corresponding field equation is a non linear generalization of the Klein–Gordon equation:

$$(\partial_\mu \partial^\mu + m^2)\Phi = ng\Phi^{n-1} . \tag{I.2.10}$$

The traditional approach to a collision process is to start from a system of free particles at time $-\infty$ and to end up also with a system of free particles at time $+\infty$. The corresponding asymptotic spaces H_{in} and H_{out} should be therefore isomorphic to H and the collision process should be represented by a unitary matrix called the S matrix mapping H_{in} to H_{out}. Cross sections are then obtained directly from the matrix elements of S in a suitable basis. Unitarity of S is necessary so that probabilities of outgoing states add up to 1, no matter what the incoming state is. One requires also invariance of S under the Poincaré group, and stability of the vacuum (which should therefore be also invariant by S). However many mathematical problems arise with this approach. First (I.2.10) contains a multiplication of distributions, which is generally an ill-defined operation. Secondly, solutions of the non-linear equation (I.2.10) cannot be asymptotic to free fields hence to solutions of the linear one, except in a certain weak sense. The result of [LSZ] is that under such a suitable asymptotic condition, there are "reduction formulae" which express the matrix elements of S in terms of the Green's functions G_N (or time ordered vacuum expectation values) of the interpolating (interacting) field Φ:

$$G_N(z_1, \ldots, z_N) = \langle \psi_0, T(\Phi(z_1), \ldots, \Phi(z_N)\psi_0\rangle . \qquad (I.2.11)$$

The Gell-Mann–Low formula gives in turn these functions as vacuum expectation values of a similar product of free fields with e^{iL_I} inserted:

$$G_N(z_1, \ldots, z_N) = \frac{\left\langle \psi_0, T\left[\varphi(z_1), \ldots, \varphi(z_N) e^{i\int dx L_I(\varphi(x))}\right] \psi_0 \right\rangle}{\left\langle \psi_0, T\left(e^{i\int dx L_I(\varphi(x))}\right) \psi_0 \right\rangle} . \qquad (I.2.12)$$

This formula is difficult to justify because the usual argument, based on the so called "interaction picture" is wrong: by a theorem of Haag, there is no way to relate the free field φ to the interacting one Φ by a unitary operator [SW]. Nevertheless the Gell-Mann–Low formula can be rigorously justified at least at the level of perturbation theory (in the sense of formal power series in the coupling g appearing in front of L_i). Branching the coupling in an adiabatic way one finds (I.2.12) up to renormalization ambiguities, which means that a certain finite part has to be taken in (I.2.12) [EG].

In the functional integral formalism proposed by Feynman [FH], the Gell-Mann–Low formula is itself replaced by a functional integral in terms of an (ill-defined) "integral over histories" which is formally the product of Lebesgue measures over all space time. It is interesting to notice that the integrand appearing in this formalism contains the full Lagrangian $L = L_0 + L_I$, not only the interacting one. The corresponding formula is the Feynman–Kac formula:

$$G_N(z_1, \ldots, z_N) = \frac{\int \prod_j \varphi(z_j) e^{i\int L(\varphi(x))dx} D\varphi}{\int e^{i\int L(\varphi(x))dx} D\varphi} . \qquad (I.2.13)$$

This formula may actually have first appeared in print in [MaS]. It was found in a heuristic way by Feynman [Fey1][FH], and Kac related it, at least in one space time dimension, to the rigorous work of Wiener and Ornstein–Uhlenbeck on path integrals [Kac1-2].

The functional integral has potentially many advantages. First it relates Wick's theorem to the rules of Gaussian integration, and therefore makes perturbation theory very transparent. The fact that the full Lagrangian appears in (I.2.13) is interesting when symmetries of the theory are present which are not separate symmetries of the free and interacting Lagrangians, as is the case for non-Abelian gauge theories. It is also well adapted to constrained quantization, and to the study of non-perturbative effects. But for a long time nobody knew whether some day some rigorous meaning could be attached to it.

In fact the difficulty of making rigorous sense out of either the Gell-Mann–Low or the Feynman–Kac formula lead to the investigation of what minimal properties should be expected from the Green's functions of a field theory, no matter from which Lagrangian or formula they may come from. Several axiomatic schemes were developed. The time ordered Green's functions can also be computed from ordinary (not time ordered) vacuum expectation values of products of the fields operators, called Wightman functions. It is for these functions that the most popular axiomatic scheme, the Wightman axioms was formulated. It includes a regularity assumption (temperedness), a relativistic transformation law, spectral conditions, hermiticity, local commutativity, a positiveness condition and a cluster property. The "Wightman reconstruction theorem" [Wig1] ensures that from such properties one can reconstruct in a unique way a Hilbert space with a unitary representation of the Poincaré group, positive spectrum condition for the momentum operators, the correct properties for the domain, regularity and transformation law under Poincaré of the field operators, local commutativity (or microscopic causality) and uniqueness and cyclicity of the vacuum. This Hilbert space formulation is called the Gårding–Wightman axiomatic scheme. But in the absence of interacting models, all these axioms could be checked only for free fields. In the late sixties, the constructive program was launched in order to provide at least some non trivial models for these axioms, and the simplest thing to do was to return to Lagrangians and to formulas (I.2.12) or (I.2.13).

There is a deep analogy between the Feynman–Kac formula and the formula which expresses correlation functions in classical statistical mechanics. For instance, the correlation functions for a lattice Ising model are given by

$$\left\langle \prod_{i=1}^{n} \sigma_{x_i} \right\rangle = \frac{\sum_{\{\sigma_x = \pm 1\}} e^{L(\sigma)} \prod_i \sigma_{x_i}}{\sum_{\{\sigma_x = \pm 1\}} e^{L(\sigma)}}, \tag{I.2.14}$$

where the index x runs over the discrete points of the lattice, and $L(\sigma)$ in the simplest case contains only nearest neighbor interactions and possibly a magnetic field h:

$$L(\sigma) = \sum_{|i-j|=1} J\sigma_i\sigma_j + \sum_i h\sigma_i. \tag{I.2.15}$$

By analytically continuing (I.2.13) in time to the Euclidean points, it is possible to complete the analogy with (I.2.14), hence to establish a firm contact with statistical mechanics. This idea also allows to give a rigorous meaning to the Euclidean path integral, at least for a free bosonic field. Indeed the corresponding Euclidean measure $Z^{-1}e^{-\int L_0(\varphi x)dx}D\varphi$, where Z is a normalization factor, can be defined easily as a Gaussian measure on the Schwartz space S' of rapidly decreasing distributions, using the general theory of such measures and a theorem by Minlos [Er1][Si1]. This is simply because L_0 is a quadratic form of positive type.

The Green's functions continued to Euclidean points are called the Schwinger functions of the model, and they are given by the Euclidean Feynman–Kac formula:

$$S_N(z_1, \ldots, z_N) = Z^{-1} \int \prod_{j=1}^{N} \varphi(z_j) e^{-\int L_I(\varphi(x))dx} d\mu_0(\varphi), \tag{I.2.16}$$

$$Z = \int e^{-\int L_I(\varphi(x))dx} d\mu_0(\varphi). \tag{I.2.17}$$

This formula is still formal, even in a finite volume; for instance the factor $e^{-\int L_I(\varphi(x))}$ might be ill-defined since it involves multiplying distributions, or might not decrease at large φ, in the case of an "unstable" potential. However the gain from an oscillating factor in (I.2.13) to a real one in (I.2.15), turns out to be enormous in practice to make sense out of such a formula. When the potential is stabilizing, we get direct decrease in modulus of the integrand in the functional integral, something much easier to use than a rapidly oscillating factor.

The Euclidean formulation also helps to understand and control the classical limit: classical solutions are the fields for which the action is stationary. In the classical limit in Minkowski space, to show that the functional integral is dominated by the contributions near these classical solutions involves a stationary phase method; but in Euclidean space it is a steepest descent method, usually a much easier problem to control.

But Euclidean field theory would not have attracted so much interest, were it not for theorems which allow to go back from Euclidean to

Minkowski space. There are several axiomatic schemes in Euclidean space which ensure such a recovery of a Wightman theory and the corresponding theorems are discussed quite extensively in [Si1]. We restrict ourselves here to a brief survey of the Osterwalder–Schrader (O.S.) axioms [OS]. These axioms are expressed as properties of the Schwinger functions S_N, not of the field φ. Therefore, strictly speaking, they are a kind of "Euclidean formulation of field theory" rather than "Euclidean field theory." A set of axioms which apply directly to the Euclidean measure, not the Schwinger functions, has been devised by Nelson [Nel][Si1]; the key property is the Markovian character of these measures. There is no doubt that this scheme of axioms is in a sense deeper, but in practice it is more difficult to check Nelson's axioms than Osterwalder–Schrader axioms. This is why we use Osterwalder–Schrader scheme in this book. It is an efficient scheme for constructive purposes and it is convenient in the context of this book, where the Schwinger functions are the natural objects under study, both in the perturbative and constructive parts.

The O.S. axioms include five properties:

- OS1 A regularity property
- OS2 Euclidean covariance
- OS3 O.S. positivity
- OS4 Symmetry
- OS5 Cluster property

OS1 is purely technical; for the analytic continuation to Minkowski space to work one must check that the set of moments S_N does not grow too quickly with N. This axiom is usually very easy to check in constructive theory.

OS2 simply states that the Schwinger functions are invariant under a global Euclidean transformation, in the case of a scalar bosonic field. Of course for theories with spinors, the proper law of transformations of the spinor fields under the Euclidean group must be taken into account. After analytic continuation, this property ensures the proper covariance under the Poincaré group.

OS3 is the most interesting axiom. It states that the expectation value of a function (like a polynomial) of the fields multiplied by the same function after reflection on any hyperplane (and complex conjugation) must be positive. This key property ensures after continuation the locality axiom and positivity of the Hilbert space in the Wightman axiomatic scheme.

OS4 is just full symmetry for the Schwinger functions under permutations of the external arguments, as should be the case for bosons. For fermions, of course, one should replace this rule by antisymmetry.

OS5 states that the Schwinger functions asymptotically factorize when two sets of arguments are taken far apart. This ensures the uniqueness of the vacuum in the Wightman axioms. In theories where all particles are massive, the clustering is exponential with the separation distance. For the two point function, the rate of decay is called the mass gap.

There are in fact several technical possibilities for OS1, depending on whether one requires full equivalence with the Wightman's axioms or simply a reconstruction theorem OS $\to$ W. For a discussion of this and a full mathematical presentation of these axioms we refer to [OS] [Si1].

The important result is:

Osterwalder Schrader reconstruction theorem

Any set of functions satisfying OS1–OS5 determine a unique Wightman theory whose Schwinger functions form precisely that set.

From now on we forget the Minkowski space and all the background briefly reviewed above. We always assume that we are in a d dimensional Euclidean space $\mathbb{R}^d$. Our starting point is the Euclidean Feynman–Kac formula; our goal is to make rigorous sense out of it, and to check the validity of Osterwalder–Schrader axioms for the corresponding Schwinger functions.

Chapter I.3

The Φ^4 Model

A Introduction

The simplest interacting field theory is the theory of a one-component scalar bosonic field φ with quartic interaction $g\varphi^4$ (φ^3 which is simpler looks unstable). In $\mathbb{R}^d$ it is called the φ^4_d model. For $d = 1, 2, 3$ the model is superrenormalizable and has been built by constructive field theory. For $d = 4$ it is renormalizable in perturbation theory. Although a constructive version may not exist [Aiz][Frö], it remains a valuable tool at least for a pedagogical introduction to renormalization theory.

Formally the Schwinger functions of the φ^4_d are the moments of the measure:

$$d\nu = \frac{1}{Z} e^{-(g/4!)\int \varphi^4 - (m^2/2)\int \varphi^2 - (a/2)\int(\partial_\mu\varphi\partial^\mu\varphi)} D\varphi. \qquad \text{(I.3.1)}$$

- g is the coupling constant, usually assumed positive or complex with positive real part;
- m is the mass; it fixes an energy scale for the theory;
- a is the wave function constant. We often assume it to be 1.
- Z is a normalization factor which makes (I.3.1) a probability measure.
- $D\varphi$ is a formal product $\prod_{x \in \mathbb{R}^d} d\varphi(x)$ of Lebesgue measures at every point of $\mathbb{R}^d$.

Remark that the quadratic piece in the exponential of (I.3.1) describes the free propagation of particles with a positive definite propagator $(p^2+m^2)^{-1}$ in Fourier space. This suggests a first improvement of (I.3.1) towards mathematical respectability. Consider the translation invariant propagator $C(x,y) \equiv C(x-y)$ (with slight abuse of notation), whose Fourier transform is

$$\hat{C}(p) = \frac{1}{(2\pi)^d} \frac{1}{p^2+m^2} . \tag{I.3.2}$$

However this propagator is ill-defined in x space for $d \geq 2$ since the corresponding Fourier integral is not absolutely convergent. We can use Minlos theorem and the general theory of Gaussian processes [Er1] [Si1] to define $d\mu(\varphi)$ as the Gaussian measure on $S'(\mathbb{R}^d)$ whose covariance is C. Let us recall how this Gaussian measure is defined, and what are its elementary properties.

B The Gaussian Measure

Let $E = S(\mathbb{R}^d)$ be the Schwarz space of smooth test functions with fast decrease at infinity and $E^* = S'(\mathbb{R}^d)$ be its dual space, the space of tempered distributions. A set A in E^* is called a cylinder set if it is the set of all distributions φ such that $(\varphi(f_1), \ldots, T\varphi(f_n)) \in \Omega$ where $f_1, \ldots, f_n$ are fixed elements of E and Ω is a fixed Borel set in $\mathbb{R}^n$. A cylinder measure is a probability measure (i.e., it is normalized) on the σ-algebra Σ_c generated by the cylinder sets. The corresponding notion of measurability is not very familiar but one can usually check that particular functions are cylinder-measurable by using the decomposition of S' as a countable union of compact subsets. To illustrate this point of view we recall the following lemma which relates the σ-algebra of cylinder sets to the strong topology of S' [CL]:

Lemma I.3.1

S' is a countable union of compact metrizable subsets K_n on which the weak and strong topology of S' coincide. On each K_n the notion of cylinder-measurability coincides with the usual notion of measurability (corresponding to the Borel σ-algebra generated by the open balls for the metric). In particular the σ-algebra Σ_c is the same as the σ-algebra Σ_s generated by the strongly open sets of $S'(\mathbb{R}^d)$.

Proof ([CL]) Obviously $\Sigma_c \subset \Sigma_s$. Using Hermite expansions we can indeed identify S with the space of rapidly decreasing sequences and

S' with the set of polynomially bounded sequences. We can write S' as the union of the sets

$$K_n = \{(a_i) : |a_i| \leq n(1+i^n) \quad \forall i\}.$$

These sets are compact (as product of compact balls) and metrizable. They are also cylinder-measurable as countable intersections of the cylinder sets $K_n^j = \{(a_i) : |a_j| \leq n(1+j^n)\}$, and also measurable with respect to Σ_s (as strongly closed sets). A subset B of S' is measurable with respect to Σ_c or Σ_s if and only if $B \cap K_n$ is measurable with respect to the same σ-algebra for all n. But on K_n, which is metrizable and compact, the strong and weak topologies are the same, hence the restriction to K_n of Σ_c and Σ_s agree. This concludes the proof.

We return now to the definition of Gaussian measures on S'. By Bochner's classic theorem we know that a complex function c is the characteristic function of some probability measure on $\mathbb{R}$ if it is a continuous function of positive type with $c(0) = 1$ (normalization condition). Minlos theorem is a generalization of Bochner's theorem. In general when we have a complex valued function c on some vector space E, which is of positive type, normalized ($c(0) = 1$) and continuous on each one dimensional linear subspace in E, it is the characteristic function of some probability measure on the algebraic dual of E, equipped with the σ-algebra generated by the algebraic cylinder sets. But in infinite dimensional situation we all know that algebraic duals are usually not the right objects to consider. Minlos theorem is interesting because it states that for a nuclear space E (such as Schwarz space) if the function c is continuous not only on every real line of E but also with respect to the topology of E, then the probability measure can be realized on the dual E^* rather than on the algebraic dual.

Let us state Minlos theorem for the particular case of a quadratic form on the Schwarz space of test functions:

Theorem I.3.1

Let $B(f,g)$ be a positive semi-definite bilinear form on $E = S(\mathbb{R}^d)$ which is continuous; then applying Minlos theorem to $c(f) = e^{-(1/2)B(f,f)}$, there exists a Gaussian cylinder measure $d\mu$ on $E^* = S'$ whose covariance is B.

Let us apply this theorem to the particular case $B(f,g) = \langle f, (-\Delta + m^2)^{-1} g\rangle$, where Δ is the Laplacian and the inner product is the L^2 product. This bilinear form is certainly well defined and continuous on the Schwarz space of test functions, (as can be seen by taking

Fourier transforms) hence there is a corresponding cylinder probability measure on $S'(\mathbb{R}^d)$, which from now on we call the free field Gaussian measure $d\mu(\varphi)$ with propagator C, C being defined as in (I.3.2). $C(x,y)$ is the kernel corresponding to the bilinear form B, i.e., $B(f,g) = \int dxdy f(x)g(y)C(x,y)$. As remarked in our case C is translation invariant. By some slight abuse of language, $C(x,y)$ is a function of $x-y$ (which has to be of positive type) which we will also note C, and call the propagator of the theory or the covariance of the Gaussian measure.

This Gaussian measure on $\mathbb{R}^d$ satisfies Nelson's axioms and has moments which satisfy all the Osterwalder–Schrader axioms; it is also called the free Euclidean field of mass m in d dimensions. Using the rules of Gaussian integration one can check that the connected functions (defined below by (I.3.16)) all vanish in this case except the two point function which is equal to the propagator. Theories which satisfy this last property are called generalized free field theories; they correspond to Gaussian measures with propagators which are in general more complicated than $(-\Delta + m^2)^{-1}$. Of course we are in fact interested in studying interacting theories which do not correspond to Gaussian measures, such as the φ^4 theory (I.3.1) and it seems natural to try to understand them as perturbations of the corresponding free field theory. This means that the measure (I.3.1) should be written as

$$\frac{1}{Z} e^{-(g/4!)\int_{\mathbb{R}^d} \varphi^4} d\mu(\varphi). \tag{I.3.3}$$

and the Schwinger functions should be written as

$$S_N(z_1, \ldots, z_N) = \frac{1}{Z} \int \varphi(z_1) \ldots \varphi(z_N) e^{-(g/4!)\int \varphi^4} d\mu(\varphi). \tag{I.3.4}$$

However this is still a formal expression, first of all because the multiplication of distributions is in general an illegal operation, then because the integral over $\mathbb{R}^d$ (infinite volume limit) might also be ill-defined. The first problem is a local regularity problem, hence an ultraviolet problem; the second is a long distance or infrared problem.

To investigate in an informal way how severe is the ultraviolet problem of multiplying the distributions in (I.3.3–4) we may investigate in more detail the local properties of typical sample fields in the support of the measure $d\mu$.

For $d = 1$ $d\mu$ is nothing but the ordinary Wiener measure with exponential killing rate (corresponding to the presence of the mass m^2). It is well known that it is supported by continuous functions (in fact Hölder continuous with exponent¡1/2), not distributions. This reflects the fact that in one dimension there is really no singularity at coinciding points (i.e., for $x = y$) in $C(x, y)$: the Fourier transform of (I.3.2) is a well defined absolutely convergent integral. Therefore in one dimension the fourth power of φ in (I.3.3–4) is well defined. In 2 dimensions, the situation is a bit worse, and with probability one the support of $d\mu$ is made of true distributions which are not functions; however these distributions are still not very rough. After applying a smoothing operator $(-\Delta + m^2)^{-\varepsilon}$ (i.e., multiplying their Fourier transform by $(p^2 + m^2)^{-\varepsilon}$ and Fourier transforming back to direct space, no matter how small ε is, we get locally square integrable functions [Ree] (in fact we get Hölder continuous with exponents smaller than $\varepsilon/2$ or equal to $\varepsilon/2$ with logarithmic corrections; also the smearing could be performed in only one direction [Ree][CL]). This reflects the fact that the coinciding point singularity in the two dimensional propagator is only logarithmic. The situation gets worse in higher dimensions; in dimension d we have to apply a smoothing operator $(-\Delta + m^2)^{(1/2)-(d/4+\varepsilon)}$ to get back to local functions. Therefore for $d \geq 2$ we cannot define directly the fourth power of φ in (I.3.1) and we have to build the interacting measure (I.3.1) by a limiting process. In one way or another this means that we have to introduce cutoffs. Then the exact regularity properties of the sample fields for the Gaussian or the final interacting measure (they may not be the same) will not be very important for the construction of the theory. This is because typically in this book we are going to use very smooth cutoffs which correspond to Gaussian measures with very regular sample fields. Then we study the convergence of the corresponding moments (the Schwinger functions) as the cutoffs are removed and eventually we use Osterwalder–Schrader axioms to check that a sensible theory has been obtained, but for that we do not need support properties of the underlying interacting measure.

C Cutoffs

Let us define a smooth ultraviolet cutoff first. One of the most convenient is the "α-space cutoff." It may also be called a heat kernel regularization of the propagator, or a regularization of the proper time of the path in the Wiener representation of the propagator as an integral over random paths. It suppresses in a smooth way the high frequencies in (I.3.2). To define it we write the α or parametric representation of the propagator:

$$\hat{C}(p) = \frac{1}{(2\pi)^d} \int_0^\infty e^{-\alpha(p^2+m^2)} d\alpha \tag{I.3.5}$$

$$C(x,y) = \frac{1}{(2\pi)^d} \int_0^\infty d\alpha \int e^{ip\cdot(x-y)-\alpha(p^2+m^2)} d^d p$$

$$= \frac{1}{(4\pi)^{d/2}} \int_0^\infty \frac{d\alpha}{\alpha^{d/2}} e^{-\alpha m^2 - |x-y|^2/(4\alpha)} \tag{I.3.6}$$

(remark that (I.3.6) is well defined except at coinciding points $x = y$, where for $d \geq 2$ it is a divergent integral). We suppress the contributions of parameters α less than κ and get:

$$C_\kappa(x,y) = \frac{1}{(4\pi)^{d/2}} \int_\kappa^\infty \frac{d\alpha}{\alpha^{d/2}} e^{-\alpha m^2 - |x-y|^2/(4\alpha)} \tag{I.3.7}$$

$$\hat{C}_\kappa(p) = \frac{1}{(2\pi)^d} \int_\kappa^\infty e^{-\alpha(p^2+m^2)} d\alpha = \frac{1}{(2\pi)^d} \frac{1}{p^2+m^2} e^{-\kappa(p^2+m^2)}. \tag{I.3.8}$$

When $\kappa \to 0$, one recovers the full propagator. In contrast with (I.3.6), (I.3.7) is well defined everywhere. The corresponding bilinear form on $S(\mathbb{R}^d)$ is $B_\kappa(f,g) = \langle \hat{f}, [1/(2\pi)^d][e^{-\kappa(p^2+m^2)}/(p^2+m^2)]\hat{g}\rangle$, where $\hat{f}$ and $\hat{g}$ are the Fourier transform of f and g. It satisfies obviously the hypothesis for Minlos theorem, but the corresponding Gaussian measure $d\mu_\kappa$ has much better support properties than $d\mu$. It is in fact supported with probability one by smooth C^∞ functions. We can get an intuitive understanding of why this is true by remarking that in the naive formulation (I.3.1) the inverse of the covariance appears, so that for functions whose Fourier transform does not decay exponentially fast (at least approximately as $e^{-\kappa p^2/2}$) the corresponding exponential quadratic factor in (I.3.1) blows up and the measure is 0 for such fields. But fields whose Fourier transform decays that fast have infinitely many derivatives. Of course this is not completely correct because (I.3.1) is only a formal guide. To make this idea more precise one can follow the strategy of [CL], and prove:

Theorem I.3.2

The support of $d\mu_\kappa$ is made of distributions which are locally smooth C^∞ functions and which grow at infinity as $\sqrt{\log|x|}$. More precisely

$\mu_\kappa(C^\infty \cap S') = 1$, and if we call also μ the restriction of $d\mu_\kappa$ to $C^\infty \cap S'$, we have $\mu_\kappa(SFC_d) = 1$, where SFC_d, the set of Sample Fields with Cutoff in dimension d, is defined as

$$SFC_d \equiv \left\{ \varphi \in C^\infty(\mathbb{R}^d) \cap S'(\mathbb{R}^d) : \limsup_{|x|\to\infty} \frac{\varphi(x)}{\sqrt{\log|x|}} = \sqrt{2d} \right\}$$

(Remark that SFC_d, our best estimate for the support of $d\mu_\kappa$, is not a vector subspace of C^∞).

Proof In what follows, the value of κ is fixed (we may suppress the corresponding subscript to simplify the notations). In [CL] it is shown that if the propagator $C(x,y)$ is Hölder continuous with some exponent $2a$ then the sample paths are in particular Hölder continuous for all exponents $a' < a$ (in fact logarithmic local behaviors are also studied in [CL]). It is also shown that for a Hölder continuous translation invariant propagator $C(x-y)$ with $C(0) = 1$ and $C(x-y)|x-y|^d$ bounded, then the behavior at infinity of the sample paths is in $\sqrt{\log|x|}$; more precisely it is exactly as specified in the definition of the set SFC_d. Hence the main difference between our Theorem I.3.2 and the results of [CL] is in the local smoothness; we start with a cutoff propagator C_κ which is not only Hölder continuous but in fact smooth, and we want to prove that the resulting sample fields are smooth. We give now an outline of how this can be done using the following theorem of [Ree], which is in fact a corollary of the proof of Minlos theorem [Hid].

Theorem I.3.3

Let E be a nuclear space, $d\mu$ a measure on E^*, c its characteristic function. Let $\langle , \rangle_n$ be a continuous inner product on E and H_{-n} be the completion of E with respect to this scalar product. Suppose that c is continuous on H_{-n}. Let T be a Hilbert Schmidt operator on H_{-n} which is one to one, and such that $E \subset \operatorname{Ran} T$, $T^{-1}(E)$ is dense in H_{-n}, and the map $T^{-1} : E \to H_{-n}$ is continuous. Then the support of $d\mu$ is in fact contained in $(T^{-1})^* H_n \subset E^*$, where $H_n \equiv H^*_{-n}$ is the dual of H_{-n}.

We apply this theorem to the scalar product $\langle f,g\rangle_n \equiv \langle f, P^{-n} g\rangle$, where $\langle , \rangle$ is the regular L^2 product, and $P = -\Delta + m^2$. $P^{n/2}$ is a unitary map from $H_0 \equiv L^2$ into H_n, the completion of S with respect to $\langle , \rangle_n$. Our covariance is continuous on H_{-n} for any n since we have $C_\kappa(f,g) = \int [(d^d p)/(2\pi)^d] \hat{f}(p)[e^{-\kappa(p^2+m^2)}/(p^2+m^2)]\hat{g}(p)$ which is convergent on H_{-n} and bounded by $(n-1)!\kappa^{-n+1}$ times $\langle f,g\rangle_n$. We apply the theorem with

$T = P^{n/2-a}Q^{-a}P^{-n/2}$ where $a = d/4 + \varepsilon$, $\varepsilon > 0$ and $Q = x^2 + 1$. Then T is Hilbert Schmidt on H_{-n} because $T_0 = P^{-a}Q^{-a}$ is Hilbert–Schmidt on H_0, since it is an integral operator with kernel in $L^2(\mathbb{R}^{2d})$. (This is because $(p^2 + m^2)^{-2a}$ is integrable in p and Q^{-2a} is integrable in x). Furthermore T satisfies the hypotheses of the theorem. The dual of H_{-n} is the space H_n of functions which are sent to L^2 by the differential operator $P^{n/2}$. These functions are n times weakly derivable, which means that they are at least of class C^{n-1}. Therefore $d\mu_\kappa$ has support on the set $\{P^{a-n/2}Q^a f, f \in L^2\}$. Since this is true for any n, we conclude that $d\mu_\kappa$ has support on smooth C^∞ functions. This together with the results of [CL] completes the proof of Theorem I.3.2.

There is an other point of view which is even more enlightening. Since the covariance with cutoff is just the multiplication in p space of the standard covariance by the function $F = e^{-\kappa(p^2+m^2)}$ one can consider that the corresponding Gaussian process is the same as the standard one without cutoff but realized on a space which is obtained by convoluting the standard sample distributions with a fixed deterministic kernel, namely the Fourier transform of $\sqrt{F}$. In this point of view this convolution automatically smears out the rough distributions on which the measure with no cutoff was supported and make them C^∞.

By Theorem I.3.2 multiplications are well defined on the support of $d\mu_\kappa$ and derivatives can be taken in the ordinary sense. Every polynomial in the field, like φ^4 has a meaning. However the fields are not bounded and they are not typically in L^4, so that in order to make sense out of (I.3.1) we need still to introduce an infrared or volume cutoff.

The simplest thing is to integrate the interaction only in a finite volume box Λ, usually some d-dimensional cube. In this way we obtain the functional integral

$$\frac{1}{Z(\Lambda,\kappa)} e^{-(g/4!)\int_\Lambda \varphi^4} d\mu_\kappa(\varphi) \tag{I.3.9}$$

This is the φ^4 theory with ultraviolet cutoff and finite volume interaction, and (I.3.9) is now a well defined formula for complex g such that $\mathrm{Re}\, g > 0$. Indeed by Theorem I.3.2 above the exponential is a well defined function on the space of smooth functions on which $d\mu_{X,\kappa}$ is supported; this exponential is bounded by 1, hence it is in $L^1(d\mu_{X,\Lambda,\kappa})$, provided that we can check that it is measurable. For that the essential step is to check that on the support of $d\mu_\kappa$ the evaluation at a point

$\varphi \to \varphi(x)$ is measurable for all x; then any power of φ evaluated at a point will be measurable, and integrals of such powers on a compact domain will be also measurable because they are limits of corresponding Riemann sums and the limit of a sequence of measurable functions is measurable.

To check that the evaluation at a point is a measurable function, we can for instance introduce some countable approximation to $\mathbb{R}^d$, like the n dimensional dyadic numbers D^d considered in [CL], and let X be the space of all real valued functions φ on D^d. We consider the Gaussian measure $d\hat{\mu}$ on X with covariance C defined by (I.3.7) restricted to dyadic values. This is a Gaussian measure on the natural dyadic cylinder σ-algebra (the smallest one for which, for all dyadic x, the evaluation functions $\varphi \to \varphi(x)$ are measurable). Because there is a natural identification between continuous functions on $\mathbb{R}^d$ and their restriction to dyadic values we can find a corresponding isomorphism between the Gaussian measure on X and the true one, and the evaluation at a point in the true problem inherits the measurability of the evaluation at a point for the dyadic problem [CL]. Still an other simpler way of reasoning (already mentioned above) is to consider the measure with cutoff as the measure without cutoff but on the space of distributions convoluted with a smoothing kernel. Then the evaluation at a point, which is the same as applying the field to a Dirac test function, which is normally illegal, becomes the same as applying the unsmeared field to a smeared Dirac function, and this is directly a measurable function from the definition of the cylinder σ-algebra on S'.

The well defined measure (I.3.9) is our usual starting point in this book, but similar measures with different cutoffs such as Pauli–Villars ultraviolet cutoffs of sufficient degree can be defined rigorously in the same way, and may be convenient in some particular situations. Typical Pauli–Villars cutoffs suppress the high momenta in a polynomial way, hence they correspond to sample fields of a given class C^k.

Also there are situations where one may want to restrict not only the interaction but also the Gaussian measure to a finite volume. For this purpose one can still use a finite box Λ, with some set of prescribed boundary conditions X for the Laplace operator. Periodic ($X = p$) or Dirichlet ($X = D$) boundary conditions are the most usual. The corresponding Gaussian measure and normalization factors with ultraviolet and infrared cutoffs are noted $C_{X,\Lambda,\kappa}$, $d\mu_{X,\Lambda,\kappa}$, $Z_{X,\Lambda,\kappa}$ or simply C, $d\mu$, Z, depending on context; we try to forget subscripts or superscripts when it seems harmless. We have to be careful however that these volume cutoffs may spoil theorem I.3.2; for instance if we define a finite covariance $C_{\Lambda,\kappa}(x,y) = \chi_\Delta(x)C_\kappa(x,y)\chi_\Delta(y)$ with $\chi_\Delta(x)$ the characteristic function of Δ it is no longer continuous in x or y, and theorem I.3.2 does not hold. However again the best point of view is to consider the corresponding measure $d\mu_{\Lambda,\kappa}$ as applied to functions $\chi_\Delta\varphi$ where φ is a sample field for $d\mu_\kappa$, hence is smooth. Such products are no longer continuous functions because the characteristic function is not continuous; nevertheless multiplication can still be defined and functions like $\int_\Delta \varphi^4$ are still measurable, so that the corresponding theory is still well defined. It is this theory which is the natural starting point in the Brydges–Battle Federbush cluster expansion scheme considered in section III.1.

Still an other possibility is to use a smooth function with compact support for the volume cutoff instead of a sharp characteristic function; this may be technically useful in some cases. However we must remember that compact support in x space is not compatible with fast decay in p space and conversely, so that ultimately some compromise on the infrared and ultraviolet properties of the sample fields is unavoidable.

Theories with cutoffs such as (I.3.9) violate some of the axioms; for instance the volume cutoff breaks Euclidean invariance and an ultra violet cutoff such as (I.3.7–8) violates reflection positivity (axiom O.S.3 in section I.2). To construct a satisfying theory we have to perform both the ultraviolet limit $\kappa \to 0$ and the infinite volume limit, also called thermodynamic limit. The thermodynamic limit consists in defining in fact a sequence volume cutoffs, typically a sequence of boxes Λ_n such that $\cup_n\Lambda_n = \mathbb{R}^d$, and studying the limit of the theory in Λ_n as $n \to \infty$. In this book we do not pay to much attention to the particular shape of the boxes and the particular boundary conditions used because the models that we are going consider are in their "high-temperature phase," in which case the thermodynamic limit (when it can be constructed) turns out to be independent of the particular sequence of boxes and of boundary conditions chosen.

Let us recall that although (after appropriate renormalization) the normalization Z and the unnormalized Schwinger functions $Z \cdot S_N$ in a fixed box can have separately an ultraviolet limit, the thermodynamic limit makes sense only for so called intensive quantities such as the pressure $(1/|\Lambda|)\log Z(\Lambda)$ or the normalized Schwinger functions S_N (defined in (I.3.4)).

The reader might conclude from this lengthy discussion that even functional integration with cutoffs is something quite complicated. There is however a conceptually simpler regularization scheme, which provides a cheaper road to a well defined functional integral formula for the cutoff φ^4 theory. It is the lattice regularization, which we discuss now. In this scheme the functional space is finite dimensional, and the functional integral reduces therefore to the ordinary Lebesgue integral. However it has also some disadvantages (e.g., it is not Euclidean invariant) and we think that it may be a good idea to emphasize techniques which are not limited to the lattice situation. After all we are interested in performing an ultraviolet limit which should be independent of the particular regularization scheme.

D The Lattice Regularization

In the lattice φ^4 model continuous space time $\mathbb{R}^d$ is replaced by a discrete lattice grid of spacing δ, hence by $\delta\mathbb{Z}^d$. We have to substitute a discrete analogue for the Laplacian. We keep in the Lagrangian the term $\sum_{i=1}^{d}(\partial_i\varphi)^2(x)$ but ∂_i is now the discrete derivative along the unit lattice vector e_i:

$$\partial_i\varphi(x) = \frac{1}{\delta}[\varphi(x+\delta e_i) - \varphi(x)]. \tag{I.3.10}$$

It is a straightforward computation to check that with this choice, the free field theory on the lattice is again given by a Gaussian measure $d\mu_{\text{lattice},\delta}$ with a propagator:

$$C^{\delta}_{\text{lattice}} = \frac{1}{(2\pi)^d}\int_{-\pi/\delta}^{\pi/\delta} d^d p \frac{e^{ip\cdot(x-y)}}{m^2 + 2\delta^{-2}\sum_{i=1}^{d}(1-\cos\delta p_i)}. \tag{I.3.11}$$

With these notations the measure for the lattice φ^4_d theory in a volume Λ is just:

$$\frac{1}{Z(\delta,\Lambda)} e^{-(g/4!)\sum_{x\in\Lambda\cap\delta\mathbb{Z}^d}\varphi^4(x)}\, d\mu_{\text{lattice},\delta}(\varphi). \tag{I.3.12}$$

In a finite volume, since there is only a finite number of lattice points, we can even make sense out of our starting formula (I.3.1). The lattice Laplacian built with lattice derivatives (I.3.10) involves a diagonal piece plus nearest neighbor terms, and we can therefore recast (I.3.12) in the form [GaRi]:

$$\frac{1}{Z(\delta,\Lambda)} e^{-(g/4!)\sum_x\varphi^4(x)-(\mu^2/2)\delta^d\sum_x\varphi^2(x)+\beta\sum_{x,y}\varphi(x)\varphi(y)}\prod_x d\varphi(x) \tag{I.3.13}$$

where the sums $\sum_x$ and the product $\prod_x$ are taken over $\Lambda\cap\delta\mathbb{Z}^d$ and $\sum_{x,y}$ is performed over the pairs of nearest neighbors in $\Lambda\cap\delta\mathbb{Z}^d$.

Furthermore:

$$\beta = \delta^{d-2} \qquad \text{and} \qquad \mu^2 = m^2 + 2d\delta^{-2}. \tag{I.3.14}$$

It is now obvious that such an expression can be taken literally. There is only a finite product of Lebesgue measures in (I.3.13), hence the functional integral is an ordinary finite dimensional integral. Every sum is finite, and for $\operatorname{Re} g \geq 0$ this functional integral is finite, hence normalizable. One interest of (I.3.13) is that it exhibits the lattice φ^4 theory as a finite system of classical continuous "spins," with nearest neighbor interaction, inverse temperature β and single spin measure

$$e^{-(g/4!)\varphi^4 - (\mu^2/2)\delta^d \varphi^2}\, d\varphi. \tag{I.3.15}$$

There are many techniques of statistical mechanics which apply to such a system, in particular correlation inequalities (see e.g. [Si1][Aiz][BFS][Frö]). One drawback, however, is that the form of the propagator (I.3.11) is not so simple. Hence in the next section on perturbative renormalization we will not use this lattice regularization at all.

Nevertheless when turning to constructive theory, it is good to keep several types of cutoffs at hand, and to show uniqueness of the limit obtained with these different cutoffs. This uniqueness may follow from a theorem relating the limit in a unique way to some well defined perturbation expansion, like the Borel summability theorems which will be discussed later. The advantage is that the final theory will typically retain all properties preserved by at least one set of these different cutoffs. This idea was applied to check the axioms for the Gross–Neveu model in [FMRS5]. From this point of view, we remark that the two cutoffs introduced in this section, the α-space and lattice cutoff are complementary, because they violate different axioms. The α-space cutoff, like the Pauli–Villars and other "momentum" space cutoffs, preserves Euclidean invariance but violates O.S. positivity. In contrast the lattice cutoff preserves a discrete version of O.S. positivity, namely reflection positivity with respect to hyperplanes of symmetry of the lattice; in the case of a cubic lattice this means the planes passing through the sites or halfway between. However the lattice is only invariant with respect to a few discrete Euclidean symmetries, hence the lattice regularization is not Euclidean invariant.

E Connected Functions, Vertex Functions

Returning to our basic objects of study, the Schwinger functions, we observe that they still contain some rather trivial substructure which one may want to trim out in order to arrive at more interesting irreducible objects. It is a standard useful construction to build from the Schwinger functions two other classes of functions called respectively the connected Schwinger functions and the one-particle irreducible (in short 1PI) Schwinger functions (in statistical mechanics connected functions are called Ursell functions or cumulants).

The connected Schwinger functions are given by:

$$C_N(z_1,\ldots,z_N) = \sum_{P_1\cup\cdots\cup P_k=[1,N];\, P_i\cap P_j=\emptyset} (-1)^{k+1}\prod_{i=1}^{k} S_{p_i}(z_{j_1},\ldots,z_{j_{p_i}}) \tag{I.3.16}$$

where the sum is performed over all distinct partitions of [1,N] into k subsets P_1, ..., P_k, P_i being made of p_i elements called j_1, ..., j_{p_i}. For instance the connected 4-point function in the high-temperature region of the φ^4 theory, where all odd Schwinger functions vanish due to the unbroken $\varphi \to -\varphi$ symmetry, is simply given by:

$$\begin{aligned} C_4(z_1,\ldots,z_4) = {} & S_4(z_1,\ldots,z_4) - S_2(z_1,z_2)S_2(z_3,z_4) \\ & - S_2(z_1,z_3)S_2(z_2,z_4) - S_2(z_1,z_4)S_2(z_2,z_3). \end{aligned} \tag{I.3.17}$$

The 1PI functions, also called vertex functions, $\Gamma_N(z_1,\ldots,z_N)$ are slightly harder to derive from the Schwinger functions. Nevertheless 1PI functions may be defined rigorously directly from the Schwinger functions, without relying on perturbation theory as their names seem to suggest. This direct construction involves combining the formalism of the first Legendre transform with T. Spencer's idea of testing irreducibility through decoupling surfaces [Sp2]. For a complete presentation of this point of view we refer to [CFR]. Here we will give simply a naive formal definition, and admit that vertex functions can be derived rigorously from Schwinger functions and vice versa. We will also admit that the perturbation theory of these objects coincides with the usual graphical definition of one particle irreducibility.

The generating functional for the connected functions is:

$$C(J) = \log Z(J) = \log \int e^{\varphi\cdot J} d\nu(\varphi) \tag{I.3.18}$$

in the sense that

$$C_N(z_1,\ldots,z_N) = \prod_{i=1}^{N} \frac{\delta}{\delta J(z_i)} C(J)|_{J=0}. \tag{I.3.19}$$

The first Legendre transform $\Gamma(A)$ is defined as

$$\Gamma(A) = C\{J(A)\} - \frac{\delta C}{\delta J}\{J(A)\} \cdot J(A) \qquad \text{(I.3.20)}$$

where the "scalar product" is defined by

$$\frac{\delta C}{\delta J} \cdot J = \int dx \frac{\delta C}{\delta J(x)} J(x),$$

and $J(A)$ is defined by inverting the equation:

$$A(J)(x) = \frac{\delta}{\delta J(x)} C(J) - \frac{\delta}{\delta J(x)} C(J)|_{J=0}. \qquad \text{(I.3.21)}$$

This Legendre transform is the generating functional for the vertex functions, which are therefore given by:

$$\Gamma_N(z_1, \ldots, z_N) = \prod_{i=1}^{N} \frac{\delta}{\delta A(z_i)} \Gamma(A)|_{A=0}. \qquad \text{(I.3.22)}$$

Chapter I.4

Feynman Graphs and Amplitudes

A Graphs

The perturbation expansion is an expansion in powers of the coupling constant. In the case of the Schwinger functions (I.3.4) this means that one writes:

$$S_N(z_1, \dots, z_N) = \frac{1}{Z} \sum_{n=0}^{\infty} \frac{(-g)^n}{n!} \left[\int \frac{\varphi^4(x)}{4!} \right]^n \varphi(z_1) \dots \varphi(z_N) d\mu(\varphi). \quad \text{(I.4.1)}$$

It is possible to perform explicitly the functional integral of a polynomial in the fields with respect to a Gaussian measure. The result, called in physics "Wick's theorem," gives at any order a sum over "Wick contractions," i.e., ways of pairing together the fields in (I.4.1). More precisely, we can label the integration variables in (I.4.1) as $x_1, \dots, x_n$, and the 4 fields in a monomial φ^4 as $\varphi_1\varphi_2\varphi_3\varphi_4$. Then we commute the spatial integral over $x_1, \dots, x_n$ and the functional integral. The result is a Gaussian integral over a polynomial in φ of degree $4n+N$, namely $\prod_{i=1}^{n} \prod_{k=1}^{4} \varphi_k(x_i) \prod_{j=1}^{N} \varphi(z_j)$. By the rules of Gaussian integration, the result is:

$$\sum_{\Pi} \prod_{l \in \Pi} C(x_l, y_l) \quad \text{(I.4.2)}$$

where the sum is over contraction schemes Π namely the partitions of the set of all $4n+N$ fields into $2n+N/2$ pairs l, called lines. For such a line l, x_l and y_l are the arguments of the fields in the pair. (There is no ordering ambiguity in (I.3.2) since $C(x,y)=C(y,x)$). There are exactly $(4n+N-1)(4n+N-3)\dots 5.3.1=(4n+N-1)!!$ such contraction schemes.

Formally at order n the result of perturbation theory is therefore simply the sum over all these schemes Π of the spatial integrals over $x_1,\dots,x_n$ of the integrand (I.3.2) times the factor $(1/n!)(-g/4!)^n$. These integrals are then functions (in fact distributions) of the external positions $z_1,\dots,z_N$ But they may diverge either because they are integrals over all of $\mathbb{R}^4$ (no volume cutoff) or because of the singularities in the propagator C at coinciding points.

For practical computations, it is obviously more convenient to gather all the contractions which lead to the same topological structure, hence the same integral. This leads to the notion of Feynman graphs or diagrams. To any such graph is associated a contribution or amplitude, which is the sum of the contributions associated with the corresponding set of Wick contractions. The Feynman rules summarize how to compute this amplitude with its correct combinatoric factor. In the case of Euclidean φ^4_d these rules are rather simple and we will describe them in detail, introducing some terminology and notations which are used throughout the rest of the book.

Let us start with some elementary abstract graph theory. A labeled unoriented graph is a triple (V,L,I) made of a set of points or vertices V, a set of lines or edges L and a relation between them This relation is a matrix $I=i_{l,v}$, called the incidence matrix of the graph, whose rows are labeled by the lines of L and whose columns are labeled by the points in V. The entries of this matrix are positive integers such that for each line l the relation $\sum_v i_{l,v}=2$ holds. The vertices for which $i_{l,v}\neq 0$ are called the end points of the line l; there are two such end points except for a line l such that $i_{l,v}=2$ for some v; such a line is called in the physics literature a tadpole, and has only one endpoint.

A pictorial representation of the graph is obtained by plotting the vertices and representing the lines as connecting links between their endpoints. For any $v\in V$, the number $\sum_l i_{l,v}$ is called the coordination number of the vertex.

Using the incidence matrix, we can define in a rigorous way all the basic topological notions which are obvious from the pictorial representation of the graph, e.g., the notion connected components of a graph and its number of independent loops.

We can also define a subgraph of a graph as a graph which is a subtriple (V', L', I') with $V' \subset V$, $L' \subset L$, I' being the corresponding restricted matrix. Since the subtriple has to be a graph, remark that lines come always with their end points attached to them. A subgraph which has no connected component reduced to a single vertex is called regular. Remark that there are always $2^{l(G)}$ regular subgraphs in a graph G (if we includes the empty one); they are in one to one correspondence with the subsets of lines of G.

One can also define unlabeled graphs of various types as equivalence classes of labeled graphs under the action of some groups. For instance we can consider the subgroup of the product of the permutation groups of the lines and of the vertices which leaves invariant the incidence matrix; this is called the automorphism group of the labeled graph and its order, $\sigma(G)$ is called the symmetry number of the graph. An unlabeled graph or diagram may be then defined as an equivalence class of labeled graphs under permutation of lines and vertices, and the number of labeled graphs G equivalent to a diagram Γ is $n!.l!/\sigma(\Gamma)$, where $\sigma(\Gamma)$ is the symmetry number of any representative of the class Γ.

Most of the time we will not need considering oriented graphs, but let us remark that starting with an unoriented labeled graph, we can define an orientation as a particular partition of the two end vertices into one initial vertex and one final vertex, for each line of the graph which is not a tadpole. Then the corresponding oriented incidence matrix $\varepsilon_{l,v}$ is defined as $+1$ if v is the final vertex of l, -1 if v is the initial vertex of l and 0 otherwise. This matrix will be useful to write down the momentum representation of Feynman amplitudes.

We are going now to customize these general notions for our particular problem of describing the bare φ^4 perturbative expansion. When studying Schwinger functions like (I.4.1) the external variables play a passive rôle so that it is indeed wise to alter slightly the general definitions of graph theory to take advantage of this.

In a labeled graph, let us call the vertices with coordination number equal to 1 the external vertices; we label them as $V_1, \ldots, V_N$. The other vertices, with coordination number greater than 1 are called internal vertices and labeled as $v_1, \ldots, v_n$. We shall say that a labeled graph is a φ^4 Feynman graph if any such internal vertex has coordination number exactly equal to 4, so that there are exactly four "half-lines" hooked to it; "half-lines" are loosely defined by cutting each line in the middle to get two half-lines. This notion of half-line is useful because we will see that in Feynman graphs half-lines correspond to the fields after Wick's theorem has been applied. Examples of φ^4 graphs are shown in Fig. I.4.1.

Feynman graphs without any labeling of the internal vertices or the internal lines are called Feynman diagrams (remark that external vertices should remain labeled). In this section we will maintain the distinction between Feynman graphs and diagrams in order to discuss the related issue of symmetry factors. Since this issue is the only one for which the distinction is important, we warn the reader that from the next section on, we will use only the word *graph*.

Clearly if we apply Wick's theorem to (I.4.1) we obtain contributions ("Wick contraction schemes") which correspond to φ^4 Feynman graphs or diagrams; their N external vertices correspond to the external fields in (I.4.1), their n internal vertices correspond to the n monomials φ^4 in (I.4.1), and their lines correspond to the propagators. However the contraction schemes sit somewhere in between of the fully labeled graphs and the diagrams; they have labeled vertices and labeled fields or "half-lines." Therefore it is not immediately clear how many contraction schemes correspond to a given Feynman diagram with the above definition. This problem will be discussed below when we define the symmetry factors of diagrams.

We use always the following notations for a Feynman graph G:

- $n(G)$ or simply n is the number of internal vertices of G, or the order of the graph.
- $l(G)$ or l is the number of internal lines of G, i.e., lines hooked at both ends to an internal vertex of G. We use also widely l as an index for lines; in cases where confusion is possible, we try to note always the number of lines as $l(G)$.
- $N(G)$ or N is the number of external vertices of G; it corresponds to the order of the Schwinger function one is looking at. When $N = 0$ the graph is a vacuum graph, otherwise it is called an N-point graph.

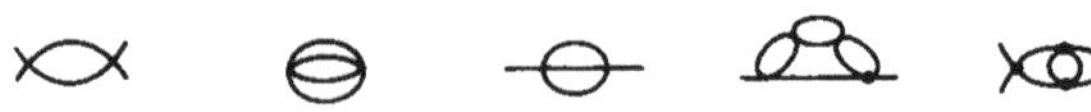

Figure I.4.1 Some Φ^4 graphs.

- $c(G)$ or c is the number of connected components of G,
- $L(G)$ or L is the number of independent loops of G.

For a *regular* φ^4 graph, i.e., a graph which has no line hooked at both ends to external vertices, we have the relations:

$$l(G) = 2n(G) - N(G)/2 \tag{I.4.3}$$

$$L(G) = l(G) - n(G) + c(G) \tag{I.4.4}$$

and, for a connected graph:

$$L(G) = n(G) + 1 - N(G)/2. \tag{I.4.5}$$

We like to define the superficial degree of convergence, which is the opposite of the more standard degree of divergence. For φ^4_d it is:

$$\omega(G) = 2l(G) - dL(G) \tag{I.4.6}$$

so that for a connected graph:

$$\omega(G) = (4-d)n(G) + \frac{d-2}{2}N(G) - d. \tag{I.4.7}$$

It will be important also (and not completely straightforward) to extend the above definitions to subgraphs. In this book subgraphs are important in the context of power counting and for the study of renormalization, but for that purpose we can always restrict ourselves to consider subgraphs which are *regular* (no isolated vertices) and which furthermore correspond to subsets of *internal* lines of G. Hence there are exactly $2^{l(G)}$ such subgraphs in G. We call the lines in the subset defining F the internal lines of F, and their number is simply $l(F)$, as before. Similarly all the vertices of G hooked to at least one of these internal lines of F are called the internal vertices of F and considered to be part of F; their number by definition is $n(F)$. But remark that with our rule no external vertex of G can be of this kind. Precisely for this reason, the notion of external vertices does not generalize in the straightforward way to such subgraphs. Nevertheless for power counting we need at least to define a generalization of the number N for subgraphs. A good convention is to call external half-line of F every half-line of G which is not in F but which is hooked to a vertex of F; it is then the number of such external half-lines which we call $N(F)$. With this convention one has for φ^4 subgraphs the same relation (I.4.3) as for regular φ^4 graphs.

The definitions of c, L and ω for subgraphs are then straightforward, and relations (I.4.4)–(I.4.7) extend to them.

It will be also useful to have a particular name for the vertices of F attached to at least one external half-line of F. We call them the "border" vertices. The internal vertices of F which are not border-vertices are called inside-vertices. We will see that border-vertices partly play the role of external vertices for subgraphs, in the sense that it is sometimes natural to integrate over the position of inside-vertices and to consider the corresponding "subamplitudes" as functions (in fact distributions) of the positions of the border-vertices. Examples of subgraphs are pictured in Fig. I.4.2.

Finally for the renormalization of φ_4^4 it will be sometimes convenient to define generalized φ^4 graphs as graphs with four types of internal vertices, the regular internal vertices with four half lines plus three kinds of special internal vertices for counterterms:

- the coupling constant counterterm, with four half lines,
- the mass counterterms, with two half-lines,
- the wave function counterterms, again with two half-lines, and two arrows to distinguish them from the mass counterterms. We will see that these arrows correspond to derivative acting on the corresponding propagators.

The counterterms are pictured as thicker "blobs" to distinguish them from the ordinary ones. Generalized φ^4 graphs are shown in Fig. I.4.3. Usually it will be clear from context which class of graphs is considered.

B Amplitudes

To compute the amplitude associated to a φ^4 Feynman diagram, we have to add the contributions of the corresponding contraction schemes. This is best done by introducing the intermediate notion

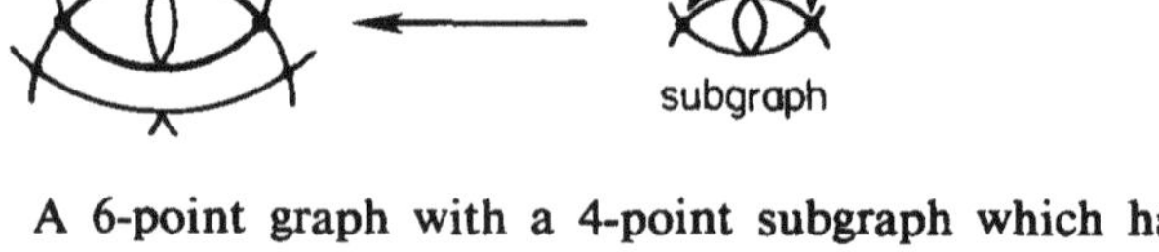

Figure I.4.2 A 6-point graph with a 4-point subgraph which has 2 border vertices.

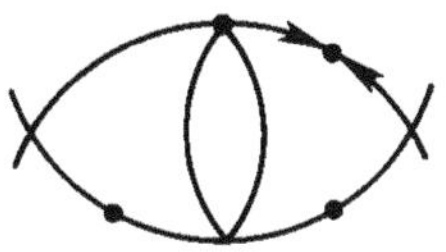

Figure I.4.3 A generalized Φ^4 graph (with one coupling constant insertion, two mass insertions and one wave function insertion).

of a vertex-labeled graph. This is an equivalence class of fully labeled graphs under the action of the permutation group of the lines. For instance in the particular case where no lines have the same incidence relations (i.e., there are no identical rows in the incidence matrix) there are $l(G)!$ fully labeled graphs for one vertex-labeled graph, otherwise there are $l(G)!/\sigma_l(G)$, where $\sigma_l(G)$ is an integer (the order of the corresponding stabilizing subgroup of the permutation of the lines).

Now a contraction scheme is still something different; it is a partition of the $4n+N$ labeled field variables into $2n+N/2$ pairs. For each such contraction scheme, there is an underlying vertex-labeled graph, obtained by erasing the four labels of the half-lines or fields hooked to each internal vertex. More precisely for each of the n internal vertices there is an associated permutation group of the 4 corresponding variables, so we have a group F with $(4!)^{n(G)}$ elements which acts on the contraction scheme. The equivalence classes of contraction schemes under the action of this group can be identified with vertex-labeled Feynman graphs. The order of the corresponding stabilizing subgroup only depends on the equivalence class, hence is a number $S_l(G)$ which is a function of the vertex-labeled graph, called the symmetry factor of the Feynman vertex-labeled graph. Remark that $S_l(G)$ in fact does not depend on the labeling of internal vertices, hence is a function of the underlying diagram.

Then the computation of the amplitude associated to a vertex-labeled Feynman graph is summarized by the rules:

- To each line l_j with end vertices at positions x_j and y_j, associate a propagator $C(x_j, y_j)$.
- To each internal vertex, associate $(-g)/4!$.
- Count all the contraction schemes corresponding to this vertex-labeled graph. The number should be of the form $(4!)^n/S_l(G)$ where $S_l(G)$ is an integer (If it is not the case, try again).

- Multiply all these factors, divide by $n!$ and sum over the position of all internal vertices.

This gives the bare amplitude of a vertex-labeled graph, neglecting possible divergences. To get the bare amplitude for a *diagram*, we count the number of vertex-labeled graphs giving rise to this diagram (by erasing internal labeling). This means again that we have to quotient by some group which is now the permutation group of the internal vertices. The corresponding number is therefore of the form $n!/S_2(G)$, where $S_2(G)$ is an integer, again the order of some automorphism subgroup. $S(G) = S_1(G)S_2(G)$ is called the total symmetry factor for the diagram G. Some of these factors are shown in Fig. I.4.4.

The formula for the bare amplitude of a diagram without taking symmetry factors into account is a distribution in $z_1, \ldots, z_N$:

$$A_G(z_1, \ldots, z_N) \equiv \int \prod_{i=1}^{n} dx_i \prod_{l \in G} C(x_l, y_l). \tag{I.4.8}$$

As stated above, this integral suffers from possible divergences. But the corresponding quantities with both volume cutoff and ultraviolet cutoff κ are well defined. They are:

$$A^{\kappa}_{G,\Lambda}(z_1, \ldots, z_N) \equiv \int_{\Lambda^n} \prod_{i=1}^{n} dx_i \prod_{l \in G} C_\kappa(x_l, y_l). \tag{I.4.9}$$

The integrand is indeed bounded and the integration domain is a compact box.

Returning to (I.4.1) the *unnormalized* Schwinger functions are formally given by the sum over all diagrams with the right number of external lines of the corresponding Feynman amplitudes, taking symmetry factors into account:

$$ZS_N = \sum_{\varphi^4 \text{ diagrams } G \text{ with } N(G)=N} \frac{(-g)^{n(G)}}{S(G)} A_G \equiv \sum_n (-g)^n b_n^N. \tag{I.4.10}$$

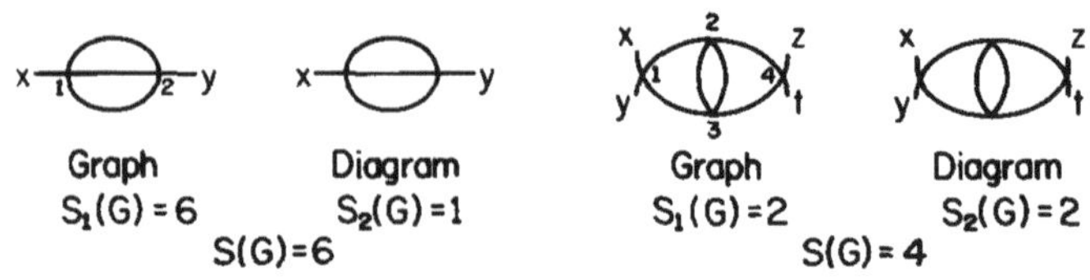

Figure I.4.4 Some symmetry factors.

Z itself, the normalization, is given by the sum of all vacuum amplitudes:

$$ZS_N = \sum_{\varphi^4 \text{ diagrams } G \text{ with } N(G)=0} \frac{(-g)^{n(G)}}{S(G)} A_G \equiv \sum_n (-g)^n b_n^0. \qquad \text{(I.4.11)}$$

Indeed the $n!$ and $4!$ factors in (I.4.1) cancel against the corresponding order of the full group which acts from contraction schemes to Feynman diagrams, and only the non trivial order $S(G)$ of the stabilizer remains in the denominator. It is the reason for which one usually choose an interaction of the form $\varphi^4/4!$†.

From translation invariance, we do not expect $A^{\kappa}_{G,\Lambda}$ to have a limit as $\Lambda \to \infty$ if there are vacuum subgraphs in G. But we can remark that an amplitude factorizes as the product of the amplitudes of its connected components at least as far as the integral is concerned; the only subtle point is the combinatoric factors, which are not easy to disentangle at the level of diagrams because of the factor $S(G)$, but which are reasonably transparent at the level of contraction schemes. Let us consider (I.4.11). Clearly each contraction scheme w creates a partition of the n vertices of (I.4.1) into k subsets $V_1, \ldots, V_k$ of $n_1, \ldots, n_k$ vertices which are the connected components of the contraction scheme. Summing first over the partition and then over the schemes which give rise to this partition, one has to take care of the proper combinatoric factor $n!/\prod_{i=1,\ldots,k} n_i!$ to attribute the vertices to each subset, and of an additional $1/k!$ since partitions are unordered sums; then the initial sum is put in the form $(1/k!)\sum_{V_1,\ldots,V_k} \prod_l A(V_l)$ where $A(V)$ is a power series exactly similar to the initial one, except that when applying Wick's theorem, only Wick's contractions which connect together all the vertices of V are allowed. Therefore one has proved that the pressure $p(\Lambda) = (1/|\Lambda|)\log Z(\Lambda)$ has a perturbation expansion given by the sum of all vacuum *connected* Feynman amplitudes:

$$p(\Lambda) = \sum_{\varphi^4 \text{ connected vacuum diagrams } G} \frac{1}{|\Lambda|} \frac{(-g)^{n(G)}}{S(G)} A_{G,\Lambda} \equiv \sum_n (-g)^n a_n^0. \qquad \text{(I.4.12)}$$

---

†This completes our discussion of the symmetry factors; in practical perturbative computations it is of course important to have the smallest number of integrals to compute, hence to use Feynman diagrams with their symmetry factor, but for abstract combinatoric results such as the ones in this book it is usually wise to remain at the level of contraction schemes.

Later in this book we introduce the Mayer expansion which is a systematic way of computing the logarithm of partition functions, and we invite the reader to prove again equation (I.4.12) using this general method.

When G is a vacuum connected diagram, there is a single overall translation invariance in its amplitude; therefore using the rapid decay of (I.3.6) for large $|x-y|$ we expect $(1/|\Lambda|)A_{G,\Lambda}$ to have a limit as $\Lambda \to \infty$, hence the pressure, not the normalization, is the correct quantity to consider in the thermodynamic limit. For G such a vacuum connected diagram we write:

$$\lim_{\Lambda\to\infty} \frac{1}{|\Lambda|} A_{G,\Lambda} = I_G \equiv \int_{\mathbb{R}^d} \cdots \int_{\mathbb{R}^d} \prod_{i\neq 1} dx_i \prod_l C(x_l, y_l)|_{x_1=0}. \qquad \text{(I.4.13)}$$

With similar simple combinatoric arguments or with the formalism of the Mayer expansion, we can also factorize the vacuum diagrams in the expansion of unnormalized Schwinger functions, and get for the normalized ones:

$$S_N = \sum_{\substack{\varphi^4 \text{ diagrams } G \text{ with } N(G)=N \\ G \text{ without any vacuum subdiagram}}} \frac{(-g)^{n(G)}}{S(G)} A_G \equiv \sum_n (-g)^n a_n^N. \qquad \text{(I.4.14)}$$

Again in (I.4.14) it is possible to pass to the thermodynamic limit (in the sense of formal power series) because using the exponential decrease of the propagator, each individual diagram has a limit (at fixed external arguments). There is of course no need to divide by the volume for that because each connected component in (I.4.14) is tied to at least one external source, and they provide the necessary breaking of translation invariance.

Finally with the help of the Mayer expansion or by direct reasoning one can determine the perturbative expansions for the connected Schwinger functions and the vertex functions. As expected the connected Schwinger functions are given by sums over connected amplitudes:

$$C_N = \sum_{\varphi^4 \text{ connected diagrams } G \text{ with } N(G)=N} \frac{(-g)^{n(G)}}{S(G)} A_G \equiv \sum_n (-g)^n a_{n,c}^0 \qquad \text{(I.4.15)}$$

and the vertex functions are the sums of the *amputated* amplitudes for 1PI or proper diagrams. They are the diagrams which remain connected even after removal of any given internal line. The amputated amplitudes are defined in momentum space by omitting the

Fourier transform of the propagators of the external lines. It is therefore convenient to write these amplitudes in the so called momentum representation:

$$\Gamma_N = \sum_{\varphi^4 \text{ proper diagrams } G \text{ with } N(G)=N} \frac{(-g)^{n(G)}}{S(G)} A_G^T(z_1, \ldots, z_N)$$

$$\equiv \sum_n (-g)^n a_{n,p}^N \qquad \text{(I.4.16)}$$

$$A_G^T(z_1, \ldots, z_N) \equiv \frac{1}{(2\pi)^{dN/2}} \int dp_1 \ldots dp_N e^{i\sum p_i z_i} A_G(p_1, \ldots, p_N), \qquad \text{(I.4.17)}$$

$$A_G(p_1, \ldots, p_N) = \int \prod_{l \in G} \frac{d^d p_l}{p_l^2 + m^2} \prod_{v \in G} \delta\left(\sum_l \varepsilon_{l,v} p_l\right). \qquad \text{(I.4.18)}$$

Remark in (I.4.18) the δ functions which ensure momentum conservation at each internal vertex v; the sum inside is over both internal and external lines; each internal line which is not a tadpole is oriented in an arbitrary way and each external line is oriented towards the inside of the graph; then the incidence matrix $\varepsilon(l, v)$ was defined above as 1 if the line l arrives at v (its final vertex), -1 if it starts from v (its initial vertex) and 0 otherwise. Remark also that there is an overall momentum conservation rule $\delta(p_1 + \cdots + p_N)$ hidden in (I.4.18). By Euclidean invariance it is clear that I_G depends only of the Euclidean invariants built on the external momenta, which are scalars; this remark is important for interpolation of amplitudes to non-integer dimension [Er2]. Notice that in the φ^4 theory a connected vacuum diagram is always proper, and indeed for $N = 0$ the definition (I.4.18) coincides with (I.4.13). We may define P_v as the sum of all external momenta entering vertex v, in which case the momentum conservation function at each vertex becomes $\delta(P_v + \sum_l \varepsilon_{l,v} p_l)$, where the sum runs now over internal lines l only.

The drawback of the momentum representation lies in the necessity for practical applications to eliminate the δ functions by a "momentum routing" prescription, and there is no canonical choice for that. Another interesting representation is the parametric representation which is obtained after the x space or p space integration has been explicitly performed, using respectively representations (I.3.6) or (I.3.5) of the propagator. This is possible because these integrations are quadratic. The result is a compact formula with one scalar integration over a parameter α for each internal line of the diagram. To write down explicitly this formula requires (for the first time in this book) the key combinatoric notion of a tree, so we pause for a while to give some corresponding definitions and notations.

C Trees

A tree can be defined as a graph which is connected and without loops. If the constraint of connectedness is removed we have a forest, which is therefore nothing more than a finite set of trees. As in the distinction between graphs and diagrams, we have to decide whether the vertices of the tree will be labeled or not. One of the most useful aspects of trees lies in the partial ordering relations they provide. More precisely, to any particular vertex of a tree is associated a natural partial ordering obtained by drawing the tree with that particular vertex as its root. This partial ordering means that a vertex is "higher" than an other if the particular unique path connecting it to the root passes through the other. Maximal vertices may be called the "leaves" and the tree together with its particular vertex a "rooted tree."

Up to now this description is rather intuitive; sometimes the problem of counting trees arise and the reader may like to have a more set-theoretic definition. We propose to define both unordered and ordered trees. A labeled unordered tree (or simply a tree) with n vertices is simply a set T of $n-1$ unordered pairs or links $\{l_1, \ldots, l_{n-1}\}$ among n elements (vertices) which can be represented as $\{1, \ldots, n\}$, with the property (P):

$$\text{For any couple of vertices } (i,j), \text{ there is a } \textit{unique} \text{ path from } i \text{ to } j \text{ in } T. \tag{P}$$

A rooted tree is a tree plus the choice of a particular vertex; hence there are n rooted trees with n vertices for any single tree.

One may call "tree shapes" the quotient of trees by the equivalence relation of erasing the labels of vertices. For $n = 3, 4, 5$, there are respectively 1, 2 and 3 tree shapes, but unfortunately such a simple series does not continue and we do not know whether there is a simple formula for the number of tree shapes at any order.

The tree shape is somewhat probed by the coordination numbers of the tree, which are the numbers d_i of pairs containing i. It is for the number of labeled unordered trees, with or without fixed coordination numbers, that Cayley's theorem gives a simple formula:

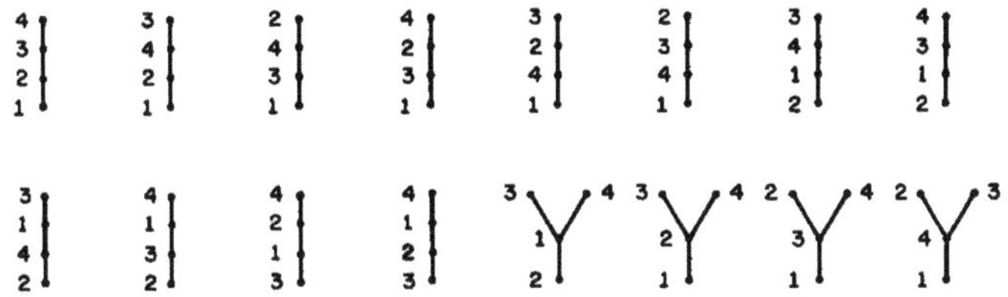

Figure I.4.5 The 16 trees of Cayley's theorem for $n = 4$.

Theorem I.4.1: Cayley's theorem

The number of labeled unordered trees with n vertices is n^{n-2}. The number of such trees with fixed coordination numbers d_i is $(n-2)!/\prod_i (d_i - 1)!$.

Proof This famous combinatoric result can be proved by an induction on n. To compute the number $t(n, \{d_i\})$ of trees with n vertices and fixed coordination numbers we can (up to a relabeling of vertices) assume that $d_1 \geq d_2 \geq \cdots d_n = 1$. Then considering that vertex n has to be joined to some particular vertex $i \leq n-1$ in the tree, we derive the recursion relation

$$t(n, \{d_1, d_2, \cdots d_n\}) = \sum_{i=1}^{n-1} t(n-1, \{d_1, d_2, \cdots, d_i - 1, \cdots, d_{n-1}\})$$

But this recursion relation is similar to the recursion relation for the multinomial coefficients which generalize Pascal's rule for binomial coefficients. Identifying the first terms in the induction leads to the second piece of the theorem, which states that $t(n, \{d_i\}) = (n-2)!/\prod_i (d_i - 1)!$. Then the standard multinomial theorem leads easily to the first piece of the theorem, which states that the total number of trees is n^{n-2}.

Figure I.4.6 An ordered tree with the corresponding "ancestor function" a.

For instance for $n=4$, there are sixteen trees, depicted in Fig. I.4.5. For coordination numbers (3,1,1,1) there is a single possibility, because the tree shape is the first one and the vertex with $n_i = 3$ must be at the center; for coordination numbers (2,2,1,1), there are 2 possibilities.

It is also sometimes useful to define a "labeled ordered rooted tree" or in short an ordered tree as a rooted tree plus a total ordering which is compatible with the partial ordering relation of the rooted tree, and such that the root has label 1 (see Fig. I.4.6). Hence it may be viewed as a permutation j of $\{1,\ldots,n\}$ satisfying $j(1)=1$, plus an "ancestor" function a from $\{1,\ldots,n-1\}$ to $\{1,\ldots,n-1\}$ which satisfies, instead of (P), the property:

$$a(k) \leq k \quad \forall k. \tag{P'}$$

The links of the trees are the pairs $(i(k), j(k+1))$, $k = 1, \ldots, n-1$, where $i(k) \equiv j(a(k))$. Such a structure appears naturally in the cluster expansion scheme of section III.1. There are at most $(n-1)!$ ordered trees for an unordered one; there are for instance 176 ordered trees for $n=4$, but no simple general formula.

Inside a graph one defines a spanning tree (or in short, a tree) as a subset of internal lines which is a tree and which connects together all the internal vertices of the graph (hence is maximal for the number of lines); a two tree is a spanning tree minus one line, hence it splits the internal vertices into two connected subsets. Remark that graphically a two tree is either a forest of two trees or a single tree plus an isolated vertex (if the line removed from the spanning tree was an extremal one).

D Parametric Representation

With these notions, let us return to the parametric representation and prove:

$$A_G(p_1,\ldots,p_N) = \delta\left(\sum_v P_v\right)\cdots$$
$$\int_0^\infty \prod_l d\alpha_l e^{-\sum_l \alpha_l m^2 - V_G(\alpha,p)/U_G(\alpha)} \frac{1}{[U_G(\alpha)]^{d/2}}, \tag{I.4.19}$$

where U_G and V_G are polynomials in α depending on the particular topology of the graph, called the Symanzik polynomials. Their explicit expression is:

$$U_G = \sum_S \prod_{l \text{ not in } T} a_l, \tag{I.4.20}$$

$$V_G(p, a) = \left(\sum_T \prod_{l \text{ not in } T} a_l \right) \left(\sum_{a \in T_1} p_a \right)^2 . \tag{I.4.21}$$

In (I.4.20) the sum runs over the spanning trees S of G, and in (I.4.21) over the two-trees T of G which separate G into two connected components, each containing a non empty set of external lines, one of which is T_1 (by overall momentum conservation, (I.4.21) does not change if T_1 is replaced by the set of external lines of the other connected component, which is the complementary of T_1).

Proof Following [IZ], we rewrite the δ functions expressing momentum conservation as:

$$(2\pi)^d \delta(P_v + \sum_l \varepsilon_{l,v} p_l) = \int dy_v e^{-iy_v \cdot \left(P_v + \sum_l \varepsilon_{l,v} p_l\right)}. \tag{I.4.22}$$

Then we exchange the order of integration and integrate over each p_l (recall that this is just a formal computation of bare amplitudes):

$$\int d^d p_l e^{-a_l p_l^2 - i \sum_v y_v \varepsilon_{l,v} p_l} = \frac{1}{(\prod a_l)^{d/2}} e^{-\left(\sum_v y_v \varepsilon_{l,v}\right)^2 / 4a_l}. \tag{I.4.23}$$

Therefore:

$$A_G(p_1, \ldots, p_N) = \int \prod_v \frac{dy_v}{(2\pi)^d} e^{-iy_v \cdot P_v} \prod_l \frac{da_l}{(\pi a_l)^{d/2}} e^{-a_l m^2 - \left(\sum_v y_v \varepsilon_{l,v}\right)^2 / 4a_l}. \tag{I.4.24}$$

To integrate over y's variables, we shift by y_n, the last variable, defining $z_v = y_v + y_n$ for $v < n$, and $z_n = y_n$. The Jacobian is one. Since $\sum_v \varepsilon_{v,l} = 0$, the integration over z_n simply yields the overall momentum conservation $\delta\left(\sum_v P_v\right)$ and we are left with $n-1$ quadratic integrations. We define the $n-1$ by $n-1$ square matrix $[d_G]_{i,j}$ for $i \leq n-1, j \leq n-1$ as

$$d_G(a)_{i,j} = \sum_l \varepsilon_{i,l} \frac{1}{a_l} \varepsilon_{j,l}. \tag{I.4.25}$$

The Gaussian integration in (I.4.24) corresponds to

$$e^{-\left(\frac{1}{4}\right)(y + 2iPd_G^{-1}) d_G (y + 2iPd_G^{-1})^t - Pd_G^{-1} P}, \tag{I.4.26}$$

with obvious matrix notations, and the result is $\det(d_G^{-d/2} e^{-Pd_G^{-1}P})$. Combining this with the factor $\prod_l a_l^{-d/2}$ in (I.4.24) achieves the proof of (I.4.19)–(I.4.21), provided one can show that d_G is non singular and that

$$\det d_G(a) = \sum_T \prod_{l \in T} \frac{1}{a_l} = \frac{U_G(a)}{\prod_l a_l}, \tag{I.4.27}$$

$$\sum_{i,j=1}^{n-1} P_i [d_G^{-1}]_{ij} P_j = \frac{V_G(p,a)}{U_G(a)}. \tag{I.4.28}$$

These results are called topological formulas. A detailed study of such formulas is in [Nak]. We will limit ourselves to a quick proof of (I.4.27), using the Binet–Cauchy formula. This formula computes the determinant of a p by p matrix C which is the product of two rectangular matrices A and B, the first p by q and the second q by p. We may assume $p < q$ for non triviality and the formula is simply:

$$\det C = \sum_M \det A_M \det B_M, \tag{I.4.29}$$

where M runs over all subsets of p indices among q, and the minors A_M and B_M are obtained by deleting in A (respectively in B) the columns (respectively the lines) with indices not in M. Now let us call ε_n the reduced incidence matrix obtained from ε by deleting the n-th line (corresponding to vertex n). We have $d_G = \varepsilon_n a \varepsilon_n^t$, where a is the diagonal matrix $a_l \delta_{ll}$. From (I.4.29), (I.4.27) obviously follows if we can check that ε_n has rank $n-1$ and that its $n-1$ by $n-1$ minors $\varepsilon_{n,M}$ are ± 1 or 0 depending on whether the lines kept in M form a tree or not. If in ε_n we keep a subset of lines M which is not a tree, it has to contain a closed circuit, and the sum of incidence numbers along such a circuit is 0; this gives a linear relation which proves that the corresponding minor is 0. Finally when M is a tree, the fact that the minor is ± 1 can be checked by induction on the number of vertices of G; we consider a line l of M hooked to the vertex n deleted in ε_n; then the corresponding column in ε_n has only one nonzero element $[\varepsilon_n]_{il}$. Expanding the corresponding minor with respect to that column we obtain up to a sign the same problem for a smaller graph G/l in which line l has been deleted and vertices n and i are collapsed. Indeed $M - l$ is still a spanning tree of G/l and we may assume the collapsed vertex to be the distinguished one in G/l so that the induction hypothesis applies.

The parametric representation is perhaps the most elegant of all and is well suited for the study of particular amplitudes, and also for general results like analyticity properties, existence and nature of asymptotic expansions in various regimes, and also for writing renormalization operators which act directly on the integrand [BL][BZ]. It is in this representation that large order bounds for the renormalized φ^4 theory were first obtained [dCR1]. Nevertheless we will not use it in these notes, because constructive theory relies heavily on x space which is lost in the parametric representation.

In the next chapter we will use a mixed representation based on (I.3.6) but with a further discrete slicing of the α integration. This representation which we call the multiscale representation, was introduced and used for perturbative and constructive studies in the series of papers [FMRS1-5] which is at the origin of this book. Many variations over this theme are worth being studied. For instance with the pure x and α propagator (without the discrete slicing), perhaps the most elegant and dense study of the perturbative renormalization of φ^4 has been given recently in [Hu]. Nevertheless throughout this book we stick as much as possible to the original representation of [FMRS1], because at least up to now a discrete slicing seems necessary for constructive purposes.

Chapter I.5

Borel Summability

Borel summability is one of many possible substitutes for ordinary summability. It has proved very useful for the analysis of many divergent series met in physics, and provides in particular a natural framework for the study of the perturbative series met in this book. Therefore we include this brief section to recall what it means.

An analytic function inside its domain of analyticity is the sum of its Taylor series, so that all information about this function is embedded in the list of Taylor coefficients at an interior point. Ordinary summation then provides a one to one correspondence (at least inside a convergence disk) between convergent power series and analytic functions. Borel summability is a method to extend this one to one correspondence to the case of an analytic function expanded in Taylor series at a point on the *border* of the domain of analyticity. Such a Taylor series is no longer summable in the ordinary sense, but under precise conditions it may still be associated in a unique way to a function with sufficiently large analyticity domain and sufficiently strong asymptoticity to this Taylor series. In this sense, again all the information about the function (the "Borel sum") is still embedded in the series (the "Borel series").

In [Ha] there is a version of Borel summability which was popular among physicists, until A. Sokal found a somewhat simpler and more

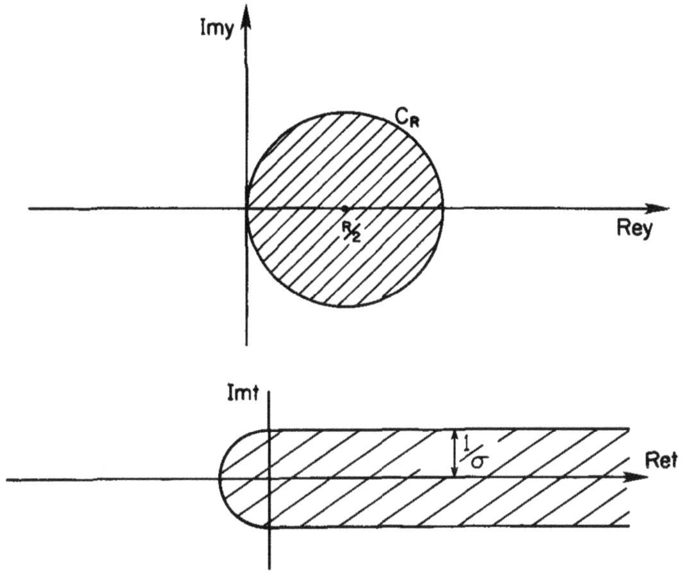

Figure I.5.1 (Top) The direct plane and the disk C_R. (Bottom) *Nevanlinna's theorem*: The Borel plan and the strip S_σ.

natural version [Sok1]. This version was in fact a rediscovery of a theorem of Nevanlinna [Ne], published in 1919:

Theorem I.5.1 (Nevanlinna–Sokal)

Let f be analytic in the disk $C_R = \{y \mid \mathrm{Re}\, y^{-1} > 1/R\}$ of Fig. I.5.1. Suppose f admits an asymptotic power series $\sum a_k y^k$ (its Taylor series at the origin) hence:

$$f(y) = \sum_{k=0}^{r-1} a_k y^k + R_r(y) \tag{I.5.1}$$

such that the bound

$$|R_r(y)| \leq C\sigma^r r! |y|^r \tag{I.5.2}$$

holds uniformly in r and $y \in C_R$, for some constants σ and C. Then f is Borel summable, which means that the power series $\sum_k a_k t^k/k!$ converges for $|t| < 1/\sigma$, that it defines a function $B(t)$ which has an analytic continuation in the strip $S_\sigma = \{t \mid \mathrm{dist}(t, \mathbb{R}^+) < 1/\sigma\}$ of Fig. I.5.1, and that this function satisfies the bound

$$|B(t)| \leq \mathrm{const.} e^{t/R} \qquad \text{for} \quad t \in \mathbb{R}^+ \tag{I.5.3}$$

Finally f is represented by the following absolutely convergent integral:

$$f(y) = \frac{1}{y} \int_0^\infty e^{-t/y} B(t)dt \qquad \text{for} \quad y \in C_R \tag{I.5.4}$$

Under these conditions, f is said to be Borel summable, and B is called its Borel transform. The complex t plane is called the Borel plane. There is a reciprocal to this theorem which states that starting with a given power series $\sum a_k y^k$ if the power series $\sum a_k t^k/k!$ converges in a disk $|t| < 1/\sigma$, admits an analytic continuation $B(t)$ in the strip S_σ and satisfies the bound (I.5.3) in this strip, then the function f defined by the integral representation (I.5.4) is analytic in C_R, has $\sum a_k y^k$ as Taylor series at the origin and satisfies the uniform remainder estimates (I.5.1–2) (with σ in (I.5.2) replaced by any $\sigma' < \sigma$). In this case we say that the series $\sum a_k y^k$ is Borel summable, that the series $\sum a_k t^k/k!$ is its Borel transform, and that the function f is its Borel sum. In conclusion Borel summable series and Borel summable functions are in correspondence just as are ordinary series and germs of analytic functions. However the analytic continuation in the Borel strip involved in the construction of the function from its series is usually intractable, so that it is the direct theorem which is used in practice.

This theorem is by no means optimal, in the sense that for many typical power series it does not reconstruct the maximal analyticity domain of the Borel sum; this can be checked even in the simplest non-trivial case $a_k = (-1)^k k!$. One can use modified versions to recover in this kind of situations a larger domain of analyticity; also for power series with different large order behavior, such that $(n!)^a$ for instance, the Nevanlinna–Sokal theorem should be modified (for instance by applying some conformal maps of the disk of Fig. I.5.1 onto other domains). We do not attempt a review on the extensive mathematical literature on this subject. There are also many results of a general nature on Borel summable functions, which state for instance that products, derivatives, inverse, composition etc.... of Borel summable functions remain Borel summable basically provided these operations make sense (see e.g., [AM][BCS]). Regular Borel summability is adequate for the problems discussed in this book; in particular the Borel plane is the natural setting to study the large order behavior of φ^4 (section II.6) and the relationship between asymptotically free models and their perturbation expansion (sections II.5, III.3–4).

PART II

PERTURBATIVE RENORMALIZATION

Chapter II.1

The Multiscale Representation and a Bound on Convergent Graphs

Suppose we had a strictly finite theory,
with bounded propagators, bounded integrals and all that.
Individual diagrams in such a theory are then bounded
by a pure power law as a function of their order n.

—G. 't Hooft, "Can we make sense out of QCD?"

In this section we prove in detail a bound which we think is the best pedagogical introduction to multiscale expansions. It is a kind of "uniform" Weinberg theorem. The original Weinberg theorem [We] proves that any graph G for which each connected subgraph S has a positive superficial degree of convergence $\omega(S)$ has indeed a finite amplitude (hence does not require renormalization). From (I.4.7) this means in the case of φ_4^4 that any graph G such that each connected subgraph S has at least 6 external lines (5 being forbidden) is finite. Such graphs will be called completely convergent; some examples of them are shown in Fig. II.1.1.

The best proof of the Weinberg theorem is in the parametric representation, by use of Hepp's sectors [He1-2]. Although the result agrees with intuition based on "power counting," it is not trivial and

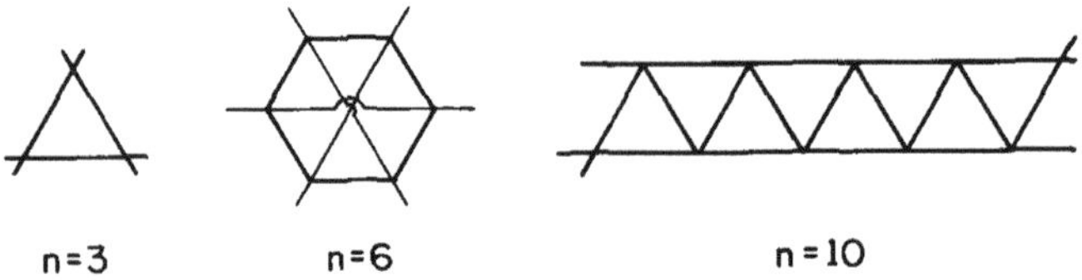

Figure II.1.1 Some completely convergent φ_4^4 graphs.

rests on the particular structure of the Feynman integrand, more precisely on the structure of Symanzik's polynomials. It is of course not true that any multidimensional integral converges when each subset of parameters has convergent scaling properties; for instance the integral

$$\int_0^\infty \int_0^\infty \int_0^\infty \frac{da_1 da_2 da_3 \sqrt{a_2} a_3 e^{-a_1-a_2-a_3}}{(a_1^2 + a_2 a_3)^2} \tag{II.1.1}$$

is not absolutely convergent, although it is superficially convergent under scaling of any given subset of a_1, a_2, a_3 (one could find more complicated examples, superficially more resembling to actual Feynman amplitudes). But the Symanzik polynomial U_G of the last section cannot be of this type. The key property of U_G, not shared by the example above, was called the *fine* (Factorized IN Each sector) property in [dCM]. For σ any permutation of $[1, \ldots, l]$, the associated Hepp's sector H_σ is the region $a_{\sigma(1)} \leq \cdots \leq a_{\sigma(l)}$. By definition, a fine polynomial P of l variables $a_1, \ldots, a_l$ is such that in each Hepp's sector H_σ after the change of variables $a_{\sigma(i)} = \prod_{j=i}^{l} \beta_j$ is performed, P factorizes as $\prod_i \beta_i^{\omega(\sigma,i)}(a_\sigma + Q_\sigma(\beta))$ where a_σ is a non-zero constant, and Q_σ is a polynomial in β's. It is not too hard to check from (I.4.20) that U_G is fine; one identifies a particular spanning tree T_σ in G whose lines have maximal numbering with respect to the permutation σ; the corresponding monomial gives the leading term of the fine property, hence in this case $a_\sigma = 1$; then one checks that all other trees in G give non-trivial monomials in β, hence their sum is the polynomial Q_σ.

With these indications the reader can easily recover Weinberg's theorem. But it is harder to give a bound on the value of the corresponding Feynman amplitudes which is of the type K^n where K is a constant and n the order of the graph (a good measure of its "size"). Such a bound, which we call the "uniform" Weinberg theorem, was in fact first proved in the α representation [dCR] but this first proof is admittedly not very transparent. The goal of this section is to give in detail a more transparent proof, based on what we call a multiscale representation. Multiscale decomposition inspired early work of K. Wilson on the renormalization group; introduced in constructive field theory by Glimm and Jaffe [GJ1], it was in particular developed and applied in [MS1-3][FO][Fed1]; the version we use now evolved from these works and follows [FMRS1] with some simplifications and improvements.

We consider a sequence of "momentum slices" which follow a geometric progression of ratio M, where $M > 1$ is fixed (later it might be often convenient to choose M integer). If we use an ultraviolet cutoff in α space of the type (I.3.7–8), it is natural to cut these slices in α parametric space. So we start from (I.3.7), with $\kappa = M^{-2\rho}$. The propagator C_κ, now written C_ρ with a slight abuse of notation is cut into a discrete sum of propagators C^i by the rule:

$$C_\rho = \sum_{i=0}^{\rho} C^i, \tag{II.1.2}$$

$$C^i = \int_{M^{-2i}}^{M^{-2(i-1)}} e^{-m^2\alpha-(|x-y|^2/4\alpha)} \frac{d\alpha}{\alpha^{d/2}}, \tag{II.1.3}$$

$$C^0 = \int_1^\infty e^{-m^2\alpha-(|x-y|^2/4\alpha)} \frac{d\alpha}{\alpha^{d/2}}. \tag{II.1.4}$$

α being dual to p^2, one should consider each propagator C^i as corresponding to a theory with both an ultraviolet and an infrared cutoff, which differ by the fixed multiplicative constant M, the momentum slice "thickness."

The decomposition (II.1.2–4) is the fundamental technical tool used in this book. By the general structure of Gaussian measures, there are associated decompositions of the Gaussian measure $d\mu_\rho$ corresponding to the propagator C_ρ into a product of independent Gaussian measures $d\mu^i$ and of the field φ_ρ, as a random variable distributed according to $d\mu_\rho$, into a sum of independent random variables φ^i, each φ^i distributed according to $d\mu^i$:

$$\varphi_\rho = \sum_{i=0}^{\rho} \varphi^i; \qquad d\mu_\rho(\varphi_\rho) = \otimes_{i=0}^{\rho} d\mu^i(\varphi^i). \tag{II.1.5}$$

There are of course other technical ways to introduce momentum slices; in statistical mechanics or for lattice regularized models, the technique of block spin transformations [Ka] is very general and powerful; in constructive theory it has been used for instance in [BCGNOPS][GK1-4][Ba2-9][Fed2-7] etc ...; it consists in writing, in a sequence of scaled lattices, each field variable as an averaged field in some cube of the next scale, plus a fluctuation field. Another elegant method close to (II.1.2–5) consists in defining the sliced fields as the coefficients of the full field on an orthonormal ondelettes basis [Bat2-3]. Each method has some advantages and drawbacks; the decomposition (II.1.2–5) is not the most general since it is only adapted to expansions around a Gaussian measure, a frequent but not universal situation. Nevertheless it is simple, and has over the more general block spin methods the advantage of independence of the Gaussian measures in various slices. We summarize this property by saying that the covariance is diagonal (in the space of values i); this leads to a clear picture of the respective rôle of the Gaussian measure and of the interaction, as will be stressed below.

In this part II devoted to perturbation theory, the language of fields and Gaussian measures may be most of the time avoided; in Feynman diagrams the functional integral over the fields is already performed and only propagators do appear, so that we will use mostly (II.1.2–4). However it is a good thing to keep also in mind the related decomposition (II.1.5), since the field point of view becomes more and more important as harder problems are considered, and is absolutely necessary for constructive theory (part III).

It is an easy exercise to derive the following bound:

Lemma II.1.1.

There are positive constants † $K > 1$ and $\delta < 1$ such that:

$$C^i(x,y) \leq KM^{(d-2)i} e^{-\delta M^i |x-y|}. \qquad \text{(II.1.6)}$$

---

† **Important convention:** We use δ, δ', ε, ζ, ... as generic names for small constants and K, K_1, c, ... for large ones. We may also write *const.* for a constant in some equations. We try to avoid any ambiguity, but these "constants" are often constant only in the context of the statement, and may depend in fact of the value of some parameters in a larger context.

This bound captures the significant aspect of both cutoffs; an overall factor which shows that the singularity of C at coinciding points has been smoothed by the ultraviolet cutoff at a certain scale, and a scaled spatial decrease which comes from the infrared cutoff. (In the case $i = 0$, δ can be taken as any number less than m, the mass appearing in C). This is optimal from the point of view of Fourier analysis; better "spatial resolution" costs a worse overall power counting factor.

Of course changing by a fixed factor M or M^{-1} the values of K and δ, we could as well rewrite (II.1.6) as:

$$C^i(x,y) \leq KM^{(d-2)(i+1)}e^{-\delta M^{i+1}|x-y|} \qquad \text{(II.1.7)}$$

which is later convenient because $i+1$ is exactly the number of slices between 0 and i.

Using the slice decomposition we rewrite the bare amplitude for a Feynman graph as:

$$A_G = \sum_{\mu \in \mathbb{N}^{l(G)}} A_{G,\mu},$$

$$A_{G,\mu} = \int \prod_v dx_v \prod_{l \text{ internal line of } G} C^{i_l}(x_l, y_l) \prod_{l \text{ external line of } G} C_p(x_l, y_l), \qquad \text{(II.1.8)}$$

where μ is called a "momentum assignment" (or "index assignment") and is a list of integers, which gives for each internal line l of G the index $i_l(\mu)$ (or simply i_l) of the corresponding slice. $A_{G,\mu}$ is the amplitude associated to the pair (G,μ), and (II.1.8) is called the multiscale representation of Feynman amplitudes. For the moment it is convenient to consider that external lines have an associated fictitious index -1, but that their propagator C_p is not decomposed into slices. Decomposition of external lines into slices may become useful for detailed results, e.g., on asymptotic behavior as some set of external momenta are scaled, but we do not use it in this book since we focus on basic existence problems.

The bare amplitude in (II.1.8) appears as a sum over the position of vertices and over momentum assignments. Together, these two sums make a kind of sum over "phase space;" but this traditional name is misleading, because *this* phase space is not the $2d$-dimensional cotangent bundle associated to d-dimensional space! It is rather a sort of $d+1$-dimensional space with d continuous spatial dimensions and one discrete dimension, called the index space (or momentum space, with

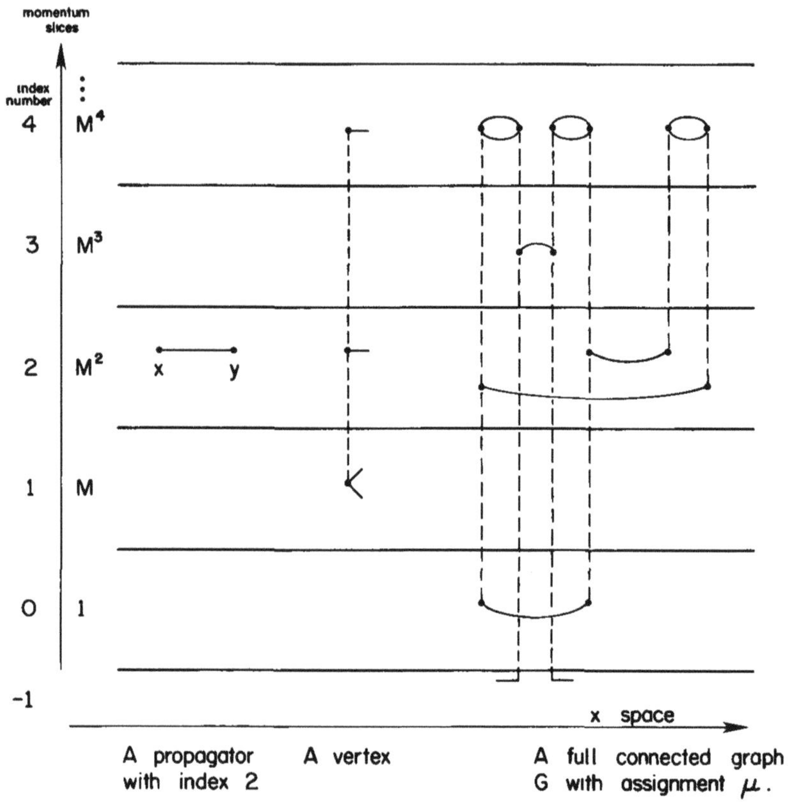

Figure II.1.2 The multiscale or phase space decomposition.

some abuse of language). This discrete space is moreover in most problems only a half-space, because we are usually interested either in an ultraviolet or in an infrared limit, but not in both together.

A key fact to realize is that in this phase space lines and vertices play a dual rôle. To support this intuition we draw a picture, Fig. II.1.2, of this phase space, which is our fundamental way of thinking to the decomposition (II.1.8); so we urge the reader to get some familiarity with it before going on, since we use the underlying intuition all the time in the rest of this book.

In the two dimensional plane of the figure, we use the horizontal direction to picture the d dimensions of space-time, and the vertical one to picture the discrete momentum slices, with the highest ones at the top. Then a propagator belongs to the slice of its index and appears as an horizontal line joining two vertices. Internal vertices sit at a particular point in space and join 4 half-lines which may be located in different slices. Hence it is convenient to picture them as vertical lines connecting the 4 horizontal half-lines hooked to them. These lines are dotted to distinguish them from the first ones. Finally the external lines are pictured in the fictitious "-1" slice, hence at the bottom of the picture.

Our goal is now to obtain the following theorem:

Theorem II.1.1: Uniform Weinberg theorem for φ_4^4

There exists a constant K such that for any connected completely convergent φ_4^4 graph (i.e., $\omega(S) > 0$, $\forall S \subset G$) the Feynman amplitude (I.4.8) of G is absolutely convergent and the following bound holds:

$$|A_G| \leq K^{n(G)}.\,\mathrm{Ext}, \tag{II.1.9}$$

where the function Ext depends on the way external arguments are treated.

This theorem extends to a very general class of completely convergent graphs which have power counting properties of the just renormalizable kind, but are not necessarily φ^4 graphs. In this more general version, in which an unbounded number of lines may be hooked to a single vertex, it is $l(G)$, not $n(G)$ which is the natural measure of the size of the graph, so that the corresponding generalized bound is similar to (II.1.9) but with a factor $K^{l(G)}$ instead of $K^{n(G)}$; by formula (I.4.3), both bounds are equivalent for φ^4. For this more general result we refer to [FMRS1].

By factorization, bounds for non-connected graphs can also be derived easily from (II.1.9). The point to emphasize in (II.1.9) is the uniform exponential character of the bound at large order expressed by the factor $K^{n(G)}$; the particular form of Ext is secondary. In [FMRS1] the set of external vertices is called V_E, and three possible treatments for external cases are considered (the reader is invited to try its own favorite):

- H1 Each external vertex is integrated over a standard unit cube of $\mathbb{R}^4$,

- H2 Each external vertex v is integrated against a test function f_v on $\mathbb{R}^4$,
- H3 Each external vertex has a fixed external momentum k_v entering it.

In these three cases a possible choice for Ext would be:

- H1 $\text{Ext} = \sup_{x_v, v\in V_E} \prod_{l\in G} e^{-m(1-\zeta)|x_l - y_l|}$,
- H2 $\text{Ext} = \inf_{v\in V_E} |f_v|_{L^1} \prod_{w\neq v\in V_E} |f_w|_{L^\infty}$,
- H3 $\text{Ext} = \delta\left(\sum_{v\in V_E} k_v\right)$.

To underline the essential part of the argument, let us prove the theorem in the slightly simpler case of an amplitude for which external propagators have been amputated, and exactly one internal vertex v_0 is fixed to the origin (not integrated over $\mathbb{R}^4$). This is in essence the case H3, because each external propagator $(p^2+m^2)^{-1}$ is bounded by m^{-2} and $N(G) \leq 4n(G)$, so that amputation does not change the nature of the bound at large order, and fixing v_0 at the origin is the same as taking into account the overall δ function of Ext, which simply reflects translation invariance.

We consider (II.1.8) and perform first the integration over the positions of the vertices in $\mathbb{R}^4$. To do this in the best possible way according to the momentum assignment, one should use as much as possible the decay of the lines with highest possible index (see (II.1.6–7)). This leads naturally to consider, for a given μ and $i \in \mathbb{N}$, the connected components of G^i, the subgraph of G made of all lines with index $j \geq i$. Let us call these connected components G^i_k, $k = 1, \ldots, l(i)$. There is a systematic way to know whether a given connected subgraph $g \subseteq G$ is a G^i_k for some i and k. We define the internal and external index for g in the assignment μ as:

$$i_g(\mu) = \inf_{l\in g} i_l(\mu), \tag{II.1.10}$$

$$e_g(\mu) = \sup_{l \text{ external line of } g} i_l(\mu) \tag{II.1.11}$$

(with the μ dependence sometimes omitted for shortness). The condition that in the assignment μ the subgraph g is a G^i_k for some values of i and k is simply:

$$i_g(\mu) > e_g(\mu) \qquad \text{(almost local condition).} \tag{II.1.12}$$

Subgraphs verifying condition (II.1.12) are called *almost local* (with respect to the assignment μ). For such a g and each value of i such that $e_g(\mu) < i \leq i_g(\mu)$ there exists a value of k such that g is equal to G^i_k.

The G_k^i's are partially ordered by inclusion and form in fact a forest, i.e., a set of connected graphs such that any two of them are either disjoint or included one in the other (disjoint here means not only no line but also no vertex in common). This remark is essential both here and for the next section, which deals with renormalization. Since $G = G^0$ is itself connected, the forest is in fact a tree, whose root is the full graph G. This tree as been pictured in Fig. II.1.3 for the case corresponding to G and μ as in Fig. II.1.2.

Each G_k^i is a node on this picture and a line between G_k^i and $G_{k'}^{i-1}$ simply means that $G_k^i \subseteq G_{k'}^{i-1}$. For g almost local there is a sequence of vertical lines in this tree with $i_g - e_g$ nodes corresponding to g. This sequence may be collapsed to a single line for simplicity and we obtain the tree of almost local subgraphs pictured in Fig. II.1.3.

The tree of Fig. II.1.3 is not exactly the same but closely related to the Gallavotti–Nicoló trees which are the key tools in their analysis of renormalization theory [GaNi][Gal].

To integrate over the positions of internal vertices (save one, v_0) requires at least the decay of a spanning tree, which is a minimal set of lines connecting together all the vertices For such a spanning tree T, defining the x_0 vertex to be the root induces a particular partial ordering on the vertices of G, as explained in Sect. I.4. Such a particular tree for the situation of Fig. II.1.2–3 is pictured in Fig. II.1.4.

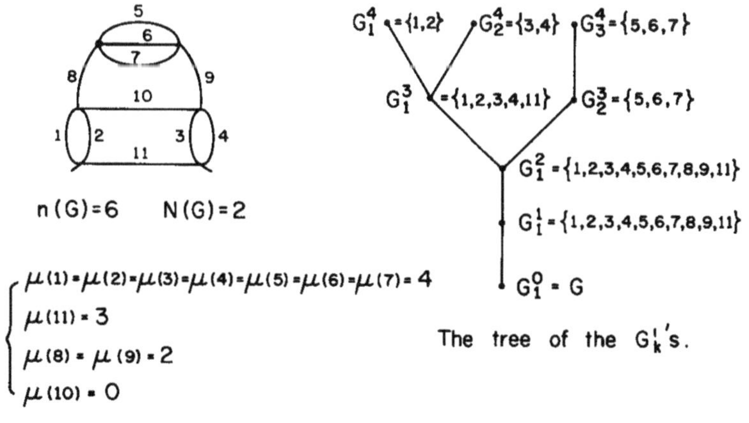

The graph G and assignment μ.

The tree of the G_k^i's.

Figure II.1.3

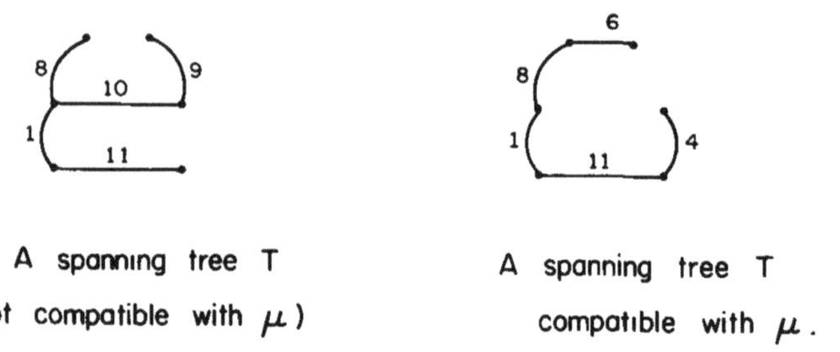

Figure II.1.4 (Left) A spanning tree T (not compatible with μ). (Right) A spanning tree T compatible with μ.

Remark that the two trees of Fig. II.1.3 and Fig. II.1.4. have very different meaning; the one in Fig. II.1.3 is an abstract picture of the inclusion relations derived from assignment μ, but the tree of Fig. II.1.4 is a concrete set of lines of G. Let us call $i_v(T)$ the index in μ of the line directly under vertex v in the partial ordering shown in Fig. II.1.4. If we use the decay of the lines of T to integrate over the positions of the vertices, starting from the "leaves" at the top of the tree towards the root which is fixed, we have to forget the decay of the other lines of G (the ones not in T) in order to have a simple expression. This is possible for an upper bound (but would be difficult for a lower bound). This means that for any T we can write:

$$\int \prod_{v\neq v_0} d^4x_v \prod_{l\in G} e^{-\delta\sum M^{i_l+1}|x_l-y_l|}$$

$$\leq \int \prod_{v\neq v_0} d^4x_v \prod_{l\in T} e^{-\delta\sum M^{i_l+1}|x_l-y_l|}$$

$$= \prod_{v\neq v_0} KM^{-4(i_v(T)+1)} = \prod_{v\neq v_0} K \prod_{j=0}^{i_v(T)} M^{-4}, \qquad \text{(II.1.13)}$$

where K is a constant (here $\delta^{-4}.\Omega_3$ where Ω_3 is the volume of the unit three dimensional sphere). Now we should choose T so as to optimize (II.1.13), i.e., to make the $i_v(T)$'s as large as possible. The best possible choice would be to require that T restricted to every almost local subgraph is still a spanning tree of this subgraph. In this case there is a kind of maximal compatibility between the trees of Fig. II.1.3 and II.1.4. Such an optimal choice is always possible (usually in a non-unique way). Indeed we can first choose spanning trees T_k^ℓ

in each "leaf" G^{ρ}_{k} where ρ is the maximal index appearing in μ; then by induction, when several $G^{i}_{k'}$ are included in a single G^{i-1}_{k}, we can always add one line in G^{i-1}_{k} to the union of the corresponding $T^{i}_{k'}$'s without creating loops in G^{i-1}_{k}; we can repeat this until it is impossible to add lines without creating a closed loop; but then we must have built a spanning tree T^{i-1}_{k} of G^{i-1}_{k}. In this way a global T is built inductively, which has the required property of spanning each G^{i}_{k}. Remark that the tree structure of the G^{i}_{k} themselves is crucial for this optimization to be possible; obviously for the bubble graph of Fig. I.1.1 there is no way to choose a tree whose restriction to the two "overlapping" subsets made of one line of G is a spanning tree of both.

Once this (non-unique) choice of T is completed, for any G^{i}_{k} every vertex v save one has $i_v(T) \geq i$. (This is an easy consequence of the connectedness of $T \cap G^{i}_{k}$). We use (II.1.8) and (II.1.13) with this choice of T. Apart from overall constants of type $K^{n(G)}$ or $K^{l(G)}$ (which are allowed in the bound since $l(G) \leq 2n(G)$), the contribution $A_{G,\mu}$ is therefore bounded by:

$$\prod_{i\geq 0}\prod_{k=1}^{l(i)}\left\{\left[\prod_{l\in G^{i}_{k}} M^{2}\right] M^{-4[n(G^{i}_{k})-1]}\right\} = \prod_{i\geq 0}\prod_{k=1}^{l(i)} M^{-\omega(G^{i}_{k})}. \tag{II.1.14}$$

To get this bound one should first rewrite for each line l:

$$M^{2(i_l+1)} = \prod_{(i,k)\text{ such that } l\in G^{i}_{k}} M^{2}, \tag{II.1.14a}$$

because for every $i \leq i_l$, the line l belongs to exactly one G^{i}_{k}, and to none for $i > i_l$. Then one should combine this factor with the right hand side of (II.1.13), noticing that for every G^{i}_{k} since T is a tree in G and in G^{i}_{k} there is a single vertex $v_0(G^{i}_{k})$ which is connected to v_0 by a path of lines of T which do not contain any line of $T^{i}_{k} = T \cap G^{i}_{k}$. This vertex is therefore such that $i_{v_0(G^{i}_{k})} < i$ and is the only one in G^{i}_{k} with this property; every $v \in G^{i}_{k}$, $v \neq v_0(G^{i}_{k})$ is such that $i_v(T) \geq i$. Therefore $\prod_{v\neq v_0} prod_{j=0}^{i_v(T)} M^{-4}$ is equal to $\prod_{i,k} M^{-4[n(G^{i}_{k})-1]}$ which proves (II.1.14).

Now recall that since G is completely convergent we have:

$$\omega(G^{i}_{k}) = N(G^{i}_{k}) - 4 \geq \frac{N(G^{i}_{k})}{3}. \tag{II.1.15}$$

Let us define

$$e_v(\mu) = \sup_{l \text{ hooked to } v} i_l(\mu), \tag{II.1.16}$$

$$i_v(\mu) = \inf_{l \text{ hooked to } v} i_l(\mu) \tag{II.1.17}$$

where the inf in (II.1.17) is over *every* line hooked to v, including external lines, which by convention have index -1.

We remark that for any i, a given vertex v belongs to exactly one G^i_k for $i \le e_v(\mu)$ and to none otherwise. Furthermore some external lines of this G^i_k are hooked precisely at v if and only if $i_v(\mu) < i \le e_v(\mu)$. Hence, using (II.1.15):

$$\prod_{i\geq 0}\prod_{k=1}^{l(i)} \{M^{-\omega(G^i_k)}\} \le \prod_{i\geq 0}\prod_{k=1}^{l(i)} \{M^{-N(G^i_k)/3}\} \le \prod_v M^{-|e_v(\mu)-i_v(\mu)|/3}. \quad \text{(II.1.18)}$$

The meaning of this bound is that for a completely convergent graph, i.e., one which has favorable power counting, after spatial integration the vertices pictured as dotted lines in Fig. II.1.2 acquire an exponential decay in their length $|e_v(\mu) - i_v(\mu)|$ in the vertical direction. This is in a sense simply power counting viewed on a logarithmic momentum scale (the space of discrete indices of the slices), but a power counting ready to be exploited. Therefore the main advantage of the multiscale decomposition emerges: it may be viewed as a machine which at each scale effectuates some spatial integration affected to this scale and, if power counting is favorable, gives as a reward a small factor for each dotted line (vertex) which crosses this scale. These small factors build up a vertical exponential decay which is very much the dual of the horizontal exponential decay of the ordinary lines in each slice. It is intuitively obvious that this decay should make the sum over momentum assignments easy, the external lines with their index -1 breaking vertical translation invariance, and playing therefore a rôle dual to the one of the fixed vertex x_0.

There are several ways to make this intuition more precise; let us describe one of them, with no efforts to find optimal constants. Using the fact that there are at most 4 half-lines, hence at most 6 pairs of half-lines hooked to a given vertex, and that for such a pair obviously $|e_v(\mu) - i_v(\mu)| \ge |i_l - i_{l'}|$, we can convert the decay in vertical length of (II.1.18) into a decay associated to each pair of half-lines hooked to the same vertex:

$$\prod_v M^{-|e_v(\mu)-i_v(\mu)|/3} \le \prod_v \left\{ \prod_{(l,l')\text{hooked to } v} M^{-|i_l - i_{l'}|/18} \right\}, \quad \text{(II.1.19)}$$

where again lines l, l' in (II.1.19) are either internal or external, and the factor 18 is not optimal (it can be improved to 12 with negligible effort ...). The analog of picking a tree to perform the spatial integration is to pick a total ordering of the internal lines of G as $l_1, \ldots, l_{l(G)}$ such that l_1 is hooked to an external line of G and such that each subset $\{l_1, \ldots, l_m\}$, $m \leq l(G)$ is connected, which is clearly possible. Using only a fraction of the decay in (II.1.19) we have:

$$\prod_v \left\{ \prod_{(l,l')\text{hooked to } v} M^{-|i_l - i_{l'}|/18} \right\} \leq \prod_{j=1}^{l(G)} M^{-|i_{l_j} - i_{l_{j'(j)}}|/18}, \qquad \text{(II.1.20)}$$

where $j'(j)$ is the index of a line hooked at j but of lower index: $1 \leq j'(j) < j$ if $2 \leq j \leq l(G)$ and by convention $i_{l_{j'(1)}} = -1$.

Finally in (II.1.20) performing the sum over μ we obtain:

$$A_G \leq K_1^{n(G)} \sum_{\mu=\{i_1,\ldots,i_{l(G)}\}} M^{-|i_{l_j} - i_{l_{j'(j)}}|/18} \leq K_1^{n(G)} K_2^{l(G)} \leq K^{n(G)} \qquad \text{(II.1.21)}$$

which achieves the proof of the theorem.

This proof is elementary and quite compact, but a subtle point remains, which is the optimization over the trees T. It introduces some sort of global choice which is not very transparent. Therefore without giving all details we describe now another slightly different way of getting the same result, which although in a sense more complicated, has the advantage of being a single induction which works from higher frequencies towards the lower ones. One can at each scale i establish an inductive bound in which spatial integration is completed for positions of all vertices save one in each G^i_k, *without knowing the structure at lower scales.* The general outline is as above: In each G^i_k all vertices save one should be integrated with lines of scales higher or equal to i, and this gives a volume effect naively in $M^{-4i\sum_k (n(G^i_k)-1)}$, but when powers of M are distributed over all scales as above, it is rather $M^{-4\sum_k (n(G^i_k)-1)}$ which should be attributed to scale i. Similarly out of the M^{2i_l} associated to a line l should come for scale i a factor $M^{+2\sum_k l(G^i_k)}$; together they reconstruct the desired factor $M^{-\sum_k \omega(G^i_k)}$, from which vertical exponential decrease for dotted vertical lines and a uniform bound on the sum over momentum assignments follows again. However the difficulty in this point of view in which no global choice of T is made first, is that one should preserve a piece of the internal decay between the vertices of G^i_k for later use, otherwise spatial integrations over vertices in some $G^{i'}_{k'}$ with $i' < i$ may be blocked later (if these vertices were joined in $G^{i'}_{k'}$ only by paths passing through lines of G^i_k, which is perfectly possible). But how to remember a decay when points are supposed to have been already integrated out? It is of course possible but there is no canonical way. In the case of the exponential decay (II.1.6) or (II.1.7) a possible technical solution is to duplicate the decrease of each line l into a product over all scales smaller or equal to i_l, since a geometric sum is of same order as its leading term. Hence changing δ one writes:

$$C^i(x,y) \leq KM^{(d-2)(i+1)} e^{-\delta \sum_{0 \leq j \leq i} M^{j+1}|x-y|}. \tag{II.1.22}$$

Then as integration is made with the help of the higher scale decay, a simple supremum may be taken over the lower scales decay. This supremum becomes available later to reconstruct tree decay of the right scale between the remaining vertices of the $G^{i'}_{k'}$ even after the ones of G^i_k (save one) have been integrated [FMRS1]. In spite of this technical complication, the inductive point of view is important because it adapts better to constructive field theory and to the general philosophy of the renormalization group, in which high frequencies should be integrated out without information about lower ones.

One should emphasize that the exponential spatial decay (II.1.6) is not crucial to the phase space machinery. In particular a scaled power decay of the type:

$$C^i(x-y) \leq KM^{2i}[1+M^i|x-y|]^{-p} \tag{II.1.23}$$

is sufficient provided $p > 4$ (summability of the propagator is what really matters), and it becomes almost as convenient as exponential decay if p is very large. A different type of decay may occur if different rules for the momentum slicing are used; in particular one may prefer to slice with a C_0^∞ cutoff in momentum space, because one can then bound with probability one norms on the field which contain derivatives, like Sobolev norms, by ordinary norms like L^2 norms. For such cutoffs exponential decay in x-space is no longer possible, but power decay like (II.1.23) can still hold. An interesting exercise suggested to the reader is to rewrite in this case the inductive proof sketched above; the replica trick in the simple form of (II.1.22) is no longer allowed!

Finally it is interesting to compare this multiscale proof with earlier proofs of similar results by constructivists for φ^4_3 [Gl]. Here as in any superrenormalizable theory one can exploit the fact that after some minimum size is attained a graph has to be superficially convergent. But amplitudes for subgraphs can be viewed as functions of the positions of their "border vertices" or as kernels for integral operators. In superrenormalizable theories one can devise a way to break a big graph into a lot of pieces of finite size, large enough so that the kernel for each piece has a finite Hilbert–Schmidt norm. Using convex analysis for instance in the form of some Schwarz inequalities a uniform bound similar to (II.1.9) can be reached, the constant K being related to a supremum over the *finite* set of these typical pieces. This method and its extension to functional integrals was widely used in the first period of constructive theory.

But no refinement of this method is going to work for a theory like φ_4^4, because Schwarz or other convex inequalities cannot be applied without losing a fraction of the power counting, and this loss is fatal to a just renormalizable theory. Multiscale chopping becomes the right tool, and once adopted it seems simpler to apply it to the superrenormalizable case as well. We leave as an exercise to the reader to check that in this superrenormalizable case the vertical decrease occurs for every vertex from $e_v(\mu)$ all the way down to $i = 0$; hence, as far as their decay is concerned, dotted lines of decay may be drawn in this case from the highest line at a vertex down to the bottom (the slice with $i = 0$). This result will be used in section III.1D to construct the ultraviolet limit of the φ_2^4 theory in a unit square.

Comparing the dotted lines to some sort of springs, the kind of intuition to be gained from this exercise and the rest of this section is that for completely convergent graphs, in the superrenormalizable case any subgraph is tied to the "ground" of low energies (the $i = 0$ slice) by springs attached to all its vertices, whereas in the just renormalizable case it is tied only by springs at its "border vertices," and not directly to the ground; chains of such springs will ultimately reach the ground at some true external lines of the full graph but the itinerary for that may be arbitrarily long and complex, involving many intermediary other subgraphs. Nevertheless in both cases the same uniform bound ultimately holds. This intuitive picture is also very useful for the next section in which multiscale slicing will lead us to an effective analysis of renormalization.

Chapter II.2

Renormalization Theory for Φ_4^4

We start with an overview of the situation and some examples before to introduce the full formalism of renormalization, which unfortunately even in its most recent and transparent versions still involves some heavy notations.

Our first remark is that the proof we gave of the "uniform Weinberg" theorem also gives uniform bounds for many momentum assignments corresponding to graphs which are no longer completely convergent. Indeed the only convergence degrees which appear in a given μ are the $\omega(G_k^i)$'s. This leads us to call a momentum assignment μ for a graph G *convergent* if $\omega(g) > 0$ for any g almost local with respect to μ. Then in the preceding section we have proved in fact:

$$|A_G^c| = \left| \sum_{\mu \text{ convergent}} A_{G,\mu} \right| \leq K^{n(G)}. \tag{II.2.1}$$

Remark that this result, strictly speaking, is non-trivial only for superficially convergent G's, otherwise there are no convergent assignments (since G itself is always almost local, by our convention that the external lines of G are considered to have index -1).

When some G_k^i have $N(G_k^i) = 4$ or 2 (0 is excluded if G is connected and $N(G) > 0$) the proof of the last section breaks down. In the intuitive language we used, the phase space machinery, instead of rewarding by a small factor for each corresponding pair i and k becomes neutral (if $N(G_k^i) = 4$) or even costs a large factor (if $N(G_k^i) = 2$). In the language of the renormalization group these behaviors are called respectively marginal and relevant; they are the source of the famous ultraviolet divergences.

What does the phase space point of view teach us about these divergences? There are two important facts to remark, which pave the way for the full solution of the problem of perturbative renormalization, and both facts appear particularly clearly in the multiscale representation.

First the divergences arise from the appearance of superficially divergent (or "divergent," for short) almost local subgraphs G_k^i's. Their name has been chosen to suggest that these objects are almost point-like from the "external world" point of view. By this we mean that their internal lines, because of the decay (II.1.6) may be thought as horizontal springs which constrain them to extend only over a region in space typically of diameter of the order of M^{-i}. In contrast the external lines which in an intuitive sense carry out information about these subgraphs to the rest of the graph, are in a lower horizontal slice, hence they distinguish only larger scales. So for them the G_k^i's appear *almost* point-like, and this effect becomes stronger as the gap grows between the internal and external scales which is precisely the source of ultraviolet divergences. It is therefore not surprising that these divergences can be cancelled by comparing these contributions to the ones of a purely local (*exactly* point-like) counterterm, which is precisely what renormalization does.

The second and more subtle remark is that the divergent almost local subgraphs have obviously a forest structure inherited from the tree structure of the inclusion relations of all G_k^i's. This simple remark is the key to the solution of one of the historically most subtle and debated points in the theory of perturbative renormalization, the one of "overlapping divergences." It was early understood and later formalized by Bogoliubov and others [BP][BS][He1][He2][Zim] that every renormalization implemented by counterterms in the Lagrangian gives rise to contributions in the amplitudes indexed by forests of connected divergent subgraphs. This is not really difficult to grasp. Indeed local counterterms in the Lagrangian are like vertices; branching propagators on them will create contributions which can match any structure of divergent disjoint subgraphs (see Fig. II.2.1).

Figure II.2.1

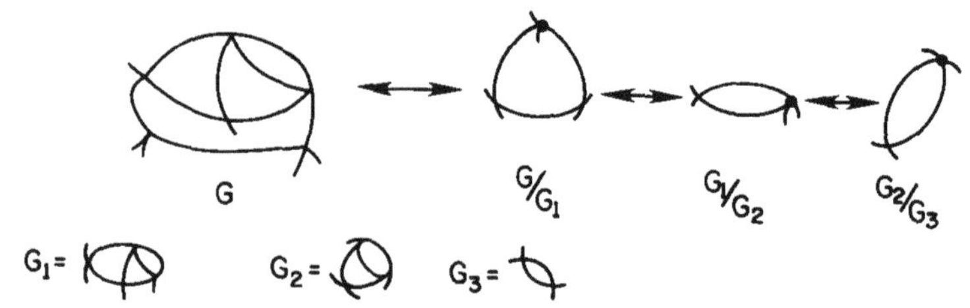

Figure II.2.2

Since the definition of counterterms may be inductive one can imagine that their definition includes earlier subtractions inside them, and in this way contributions may be associated with stacks of subgraphs included one into the other as in Fig. II.2.2.

Combining both ideas one gets contributions for each forest. But there is certainly no way to associate a specific contribution which in the perturbative expansion matches with two divergent overlapping subgraphs (i.e., with nontrivial intersection, see Fig. II.2.3.).

For many years experts worried about these overlapping divergences, since there was no way to build a specific contribution for them from local counterterms. The fact that the almost local subgraphs have forest structure is precisely the solution to this point; "overlapping divergences" simply never occur simultaneously in phase space, hence there is no need to associate any specific counterterm to them. This statement will be fully substantiated below.

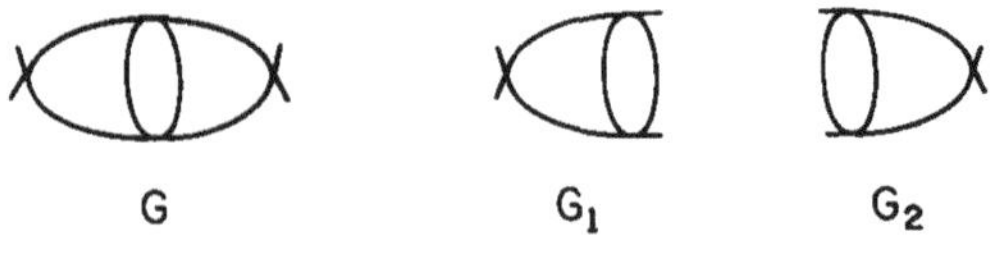

Figure II.2.3

After this general overview, we start by showing how the divergent "bubble" graph of Fig. I.1.1 is renormalized by a local counterterm in the multiscale representation. A subgraph $g \subset G$, after integration of its inside vertices (it may have none) becomes a function $g(x_1, \ldots, x_{v_e})$ of its v_e border vertices (see Sect. I.4 for these notions). By translation invariance its Fourier transform is of the form $\delta(k_1 + \cdots + k_{v_e})\hat{g}(k_1 + \cdots + k_{v_e-1})$. The Zimmermann prescription for renormalizing it at 0 external momenta is to subtract from it a counterterm $\tau_g g$ which is $\delta(k_1 + \cdots + k_{v_e})$ times a Taylor expansion of $\hat{g}$ around 0 momenta:

$$\sum_{j=0}^{D(g)} \frac{1}{j!} \frac{d^j}{dt^j} \hat{g}(tk_1, \ldots tk_{v_e-1})|_{t=0} \tag{II.2.2}$$

where $D(g)$ is the largest integer less than or equal to $-\omega(g)$ the superficial degree of *divergence* of g. Thanks to translation invariance, the formula does not really depend on v_e. Remark that the 0-momentum subtraction scheme is valid only for a massive theory; otherwise it would be ill defined, due to infrared singularities, and some subtraction scale should be introduced by hand to represent the scale of physical phenomena one is interested in. This is for instance the case for non-Abelian gauge theories, in which the gauge boson is massless. (A subtraction scheme for massless φ_4^4 and corresponding bounds may be found in [dCPR]).

In the multiscale representation we want to compute in x space; therefore we should apply a subtraction operator τ_g^* to the test function smearing g rather than to g itself. This is possible through the formula:

$$\int \tau_g g(x_1, \ldots, x_{v_e}) a(x_1, \ldots, x_{v_e}) dx_1 \ldots dx_{v_e}$$
$$= \int g(x_1, \ldots, x_{v_e}) \tau_g^*(v_e) a(x_1, \ldots, x_{v_e}) dx_1 \ldots dx_{v_e}, \tag{II.2.3}$$

where

$$\tau_g^*(v_e) a(x_1, \ldots, x_{v_e}) = \sum_{j=0}^{D(g)} \frac{1}{j!} \frac{d^j}{dt^j} a(x_1(t), \ldots, x_{v_e}(t))|_{t=0} \tag{II.2.4}$$

with $x_i(t) = x_{v_e} + t(x_i - x_{v_e})$. Here v_e really plays a distinguished rôle, so that there are several noncanonical ways of defining the "adjoint" $\tau_g^*(v_e)$, one for each border vertex v_e; they are equally good.

Let us consider a bubble subgraph g in a bigger graph G. In this case $\omega(g) = 0$ and there are two border vertices at x_1 and x_2. Let us suppose to simplify that the two internal lines are propagators of the same scale i and that the four external half-legs are four different propagators of the same scale j. The test function a is therefore the product $C^j(x_1,y_1)C^j(x_1,y_2)C^j(x_2,y_3)C^j(x_2,y_4)$ and the subgraph is almost local (divergent case) if and only if $j < i$ (see Fig. II.2.4).

As we saw in the last section there is no decay in i-j in this case, which would allow to sum over the i index, holding j fixed, hence there is a "logarithmic" divergence in energy. To renormalize, we apply $(1 - \tau_g^*)$ to a, which gives the bare amplitude minus its counterterm. The result may be written as the following sum:

$$[C^j(x_1,y_1) - C^j(x_2,y_1)]C^j(x_1,y_2)C^j(x_2,y_3)C^j(x_2,y_4)$$
$$+ C^j(x_2,y_1)[C^j(x_1,y_2) - C^j(x_2,y_2)]C^j(x_2,y_3)C^j(x_2,y_4). \quad \text{(II.2.5)}$$

Each term contains exactly one difference which may be written as a Taylor remainder. For instance:

$$C^j(x_2,y_1) - C^j(x_1,y_1) = \int_0^1 dt \frac{d}{dt} C^j(x_1 + t(x_2 - x_1), y_1). \quad \text{(II.2.6)}$$

Returning to the definition (II.1.3–4) of C^j it is easy to derive the bound:

$$\frac{d}{dt} C^j(x_1 + t(x_2 - x_1), y_1) \le K|x_2 - x_1| M^{3j} e^{-\delta M^j |x_1 + t(x_2 - x_1) - y_1|} \quad \text{(II.2.7)}$$

where K and δ are new constants, which may be taken slightly worse than those of (II.1.6). Hence the net effect of the derivation is to add to

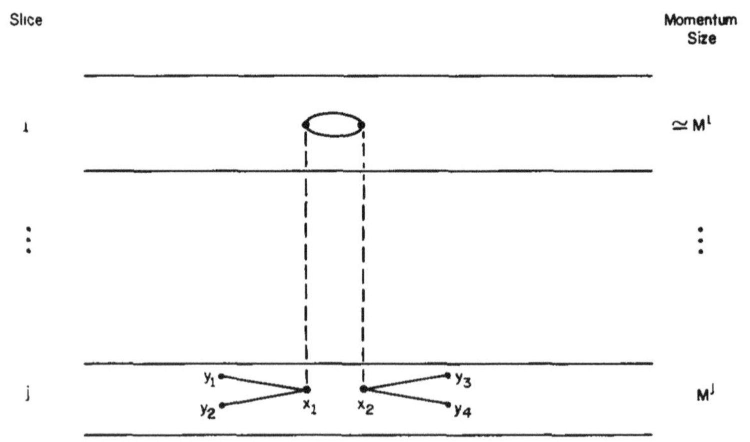

Figure II.2.4 An almost local bubble.

the usual bound a factor M^j times the distance $|x_2 - x_1|$. This distance, between the two border vertices of the bubble may be then estimated by using a piece of the decay of the two internal lines of the bubble. Remember that these lines have decay of scale $i > j$. Hence this gives a bound:

$$|x_2 - x_1| e^{-2\delta M^i |x_2 - x_1|} \leq \frac{2}{\delta} M^{-i} e^{-(3/2)\delta M^i |x_2 - x_1|}. \tag{II.2.8}$$

Similarly this internal decay can be used to replace $|x_1 + t(x_2 - x_1) - y_1|$ in (II.2.7) by $|x_1 - y_1|$; then the integral over t in (II.2.6) is just bounded by 1. Altogether renormalization has delivered an extra factor $M^{-(i-j)}$ at the cost of making constants like K or δ slightly worse. This does not change the nature of the estimates, but the extra factor $M^{-(i-j)}$ has restored a vertical exponential decrease previously missing; more concretely it allows now to sum over i with respect to j.

In the generic case recall that the condition for a divergent subgraph g to be almost local with respect to μ hence to be a source of ultraviolet divergence, is $i_g(\mu) > e_g(\mu)$. For instance in the case of a bubble subgraph, the condition of almost locality is simply that i_g, the smallest of the two indices i_1 and i_2 in μ of its internal lines, has to be larger than e_g, the largest of the 4 indices e_1, e_2, e_3 and e_4 in μ of its external lines. For g a general divergent subgraph, assignments μ satisfying the condition $i_g(\mu) > e_g(\mu)$ are called *dangerous* (with respect to g) and the ultraviolet divergence for g solely comes from such assignments. In the case of a single bubble subgraph, the reader can easily extend the previous argument to the sum over such dangerous assignments and restore the missing decay in $i_g - e_g$ by subtracting again the counterterm τ_g. In conclusion and anticipating a bit, the divergence of the bubble is cancelled successfully by a counterterm in the Lagrangian of the form:

$$\int dx_1 \left[\int dx_2 \sum_{i_g > e_g} C^{i_1}(x_1 - x_2) C^{i_2}(x_1 - x_2) \varphi_{e_1}(x_1) \varphi_{e_2}(x_1) \varphi_{e_3}(x_1) \varphi_{e_4}(x_1) \right] \tag{II.2.9}$$

where we recall that the random field φ^i of frequency i (respectively φ_i) is distributed according to the gaussian measure of covariance C^i (respectively $C_i = \sum_{j=0}^{i} C^j$) (see (II.1.5)). We refer to pieces of counterterms like the one of (II.2.9) loosely as to "useful" counterterms, because they really kill a divergence. But objects like (II.2.9) cannot be considered local because of the restriction in the sum, which makes the value of the counterterm dependent of the energy scale of its external lines. If we want to preserve the formal locality of the Lagrangian, we have to introduce the counterterm for g also in the case of assignments which are not dangerous for g, hence satisfy the opposite condition in the sum (II.2.9); we call this piece of the counterterm the "useless" counterterm (it is "useless" from the point of view of cancellation of divergences, not locality).

To combine useless counterterms with the bare amplitude in assignments which are not dangerous does not make sense from the point of view of estimates, essentially because subgraphs which are not G^i_k's are not almost local from the external point of view, and there is no interesting gain to compare them with a completely local counterterm.

An interesting "naive" question is: for these assignments, where they do not match, which one is biggest, the bare amplitude or the counterterm? A good short answer is: "the counterterm." To motivate this answer we remark that the condition to be "useless," $i_g \leq e_g$, puts an ultraviolet cutoff at scale e_g on the divergent sum over internal scales in the counterterm. This cutoff makes this sum finite, but there is a subtle point: apart from the small breaking of scale invariance due to the mass m in the propagator C, any counterterm for a marginal operator (like φ^4 here) is invariant by vertical translation. Hence when e_g is large, the useless counterterm has to behave asymptotically as e_g. But there is no such effect for the corresponding piece of the bare amplitude! This may seem a bit mysterious, but the reader may convince himself of this important fact by checking again the proof of the last section and its extension to (II.2.1) above: no linear divergence in the summation over scales can occur for subgraphs which are not G^i_k's, because such subgraphs simply never appear in the argument. We conclude that at least for e_g large the useless counterterm must be much larger than the corresponding bare contribution.

It is a bit uncomfortable that something which is useless from the point of view of convergence dominates. It is not hard to guess more precisely to which problem this phenomenon will lead us. Useless counterterms will not make amplitudes divergent; they are big, but finite. But what will happen if a large number of such big objects are inserted into a convergent bare amplitude or an amplitude with a useful renormalization? The convergence we found for such amplitudes is exponential in the space of momentum indices, hence it cannot be destroyed by multiplying by any number of linear factors like those associated to the useless counterterms. This is in essence why renormalization works and renormalized amplitudes are finite. However since:

$$\sum_{e=0}^{\infty} M^{-\delta e} e^n \simeq K^n n! \tag{II.2.10}$$

we should expect the production of factorials in the number of useless counterterms, and this leads to a violation of the uniform bound (II.2.1) for renormalized amplitudes. This violation, called the "renormalon" phenomenon, leads to new difficulties if one tries to sum up renormalized perturbation theory. We have identified the useless counterterms as the source of this trouble, and this diagnosis already suggests that the renormalized series are not the ones that one should try to sum up, a point of view which will be developed in Sect. II.4–5.

Having completed this brief sketch of the situation, based on the sole example of the bubble, we are ready to enter into the heart of renormalization to substantiate the corresponding ideas with some proofs. We will focus on a proof of finiteness of renormalized amplitudes with reasonable estimates which imply the existence of a finite disk of analyticity for the Borel transform of the renormalized series. This result, the "uniform BPH theorem" below, was first obtained in [dCR1] using the a space representation. Later it was rederived in several different ways [GaNi][FMRS2][FHRW][Hu]. As remarked already, a version like [Hu] may be the most compact. But the one given here, in the style of [FMRS2] is well suited to the extension to constructive theory. Small details have been improved so that the final estimates are slightly better than the ones of [dCR1] or [FMRS2] The general statement that φ_4^4 is perturbatively renormalizable of course does not mean simply that it is possible to make amplitudes finite by some subtraction process, since this would be obvious for any theory in any dimension. It means that the subtraction process must correspond solely to counterterms in the action of the form φ^4, φ^2 and $\partial_\mu\varphi\partial^\mu\varphi$, as the original ones in (I.3.1).

More precisely it means that in (I.3.1) we can replace the constants g, m^2 and a of the theory by three formal power series in a renormalized coupling constant g_r, respectively $g_r + \sum_{n=2}^{\infty} c_n g_r^n$, $m^2 + \sum_{n=1}^{\infty} d_n g_r^n$ and $a + \sum_{n=2}^{\infty} e_n g_r^n$ such that the perturbative expansion in g_r of any Schwinger function is finite order by order. The three corresponding renormalizations are called respectively coupling constant, mass and wave function renormalization. (For connected amplitudes there is no need to discuss the vacuum energy renormalization.) But counterterms c_n, d_n or e_n should diverge in the continuum limit, so that this is not yet a rigorous definition. What is finally required is that for the well defined theory with cutoffs like (Λ, κ) (I.3.9) (or (Λ, δ) (I.3.12)) there exist three such formal power series with coefficients $c_n(\Lambda, \kappa)$ $d_n(\Lambda, \kappa)$ and $e_n(\Lambda, \kappa)$ such that the perturbative expansion for this theory with cutoffs has a finite limit, order by order in g_r, when both cutoffs are removed.

In the standard literature, renormalizability consists therefore of two steps: the first one is the definition of the coefficients c_n, d_n and e_n, at the heart of which lies the Bogoliubov recursion; the second step is to expand the full theory in powers of g_r and to match the contributions obtained from the counterterms in the action coming from this recursion with the bare contributions so that every renormalized amplitude remains finite when cutoffs are removed. At the heart of this second step is a technical tool, Zimmermann's forest formula, which gives for a given graph the complete list of counterterms which should be combined with it; the forest formula appears therefore as a kind of inverse solution to the Bogoliubov recursion, and it has the advantage of being global, not inductive. Both tools unfortunately were invented before the multiscale representation and do not take advantage of it in their design.

This defect is alleviated in the recent versions of the renormalizability theorem [GaNi][Gal][FHRW][Hu]; they neither focus on the Bogoliubov formula and Zimmermann's forest, nor on the individual renormalized graphs. To the decomposition of the propagator into slices corresponds the decomposition of the field as a sum of random variables (II.1.5); then integrating over the field in each slice, an effective potential for the sum of slices lower than a given one is obtained and is renormalized in an inductive way. This is a beautiful formalism, the most natural from the point of view of the multiscale representation, and it avoids some of "the combinatoric mess to relate the counterterms to the individual graphs" [Ros]. The main technical combinatoric tool, the Gallavotti–Nicolò tree, plays however the same organizing rôle as Zimmermann forests or the tree structure of almost local subgraphs; hence everybody may agree that the key technical aspect of renormalization lies in such a structure (recall that a rooted tree minus its root is a forest, so that the difference is extremely tiny; in particular it is simply a mistake induced by their names to believe that forests are more complicated than trees ...). Here we refer the reader to the above literature but choose to still explain the basis of the Bogoliubov recursion and of the forest formula in the old fashioned way. We do not want however to focus too much attention on them, first because it is by now rather standard material, but also because in the broader context of constructive theory, the renormalized series are not the relevant objects anyway; they are rather a natural dead end to explore before passing to the effective series and their constructive generalizations.

The Bogoliubov induction is on the size of the graphs; it defines the counterterms to be associated to each connected superficially divergent graph at order $n+1$ when the same thing has been done up to order n. Most of the "combinatoric mess" is simply due to the use of symmetry factors $S(G)$ in formulas for graphs and disappears if one returns instead to the notion of contraction schemes, which we abbreviate in this section as CS. Also the presence of an ultraviolet cutoff is necessary for well defined formulas and we assume it.

The best thing is before the general rule to gain some practice with simple examples derived from the perturbation theory of the 4 point function, in which we do not consider graphs which contain subgraphs with $N = 2$ (called "bipeds" in [dCR1]). The lowest-order divergent graph is then the bubble. It corresponds to many CS, several for each of the three traditional s, t and u channels of scattering theory. We introduce a counterterm in the Lagrangian $c_{\text{bubble}}\varphi^4$ with c_{bubble} defined by the subtraction prescription; in the BPHZ scheme it is minus the value of the bare amplitude at 0 momenta, but the principle of the Bogoliubov recursion would be the same for other subtraction schemes. With this new term in the Lagrangian, the four point function becomes finite up to second order. What is less obvious is that for each graph containing a single bubble subgraph, there is exactly one associated subgraph containing a single vertex of type c_{bubble}; the combinatoric to check, at the level of CS, is simply the multinomial formula which allows to choose a pair of vertices among n and to build the bubble with them. Similarly for a graph with several disjoint bubbles one new contribution is generated for each non-empty subset of these bubbles, where each bubble in the subset is replaced by c_{bubble}. Again the combinatoric is checked through the multinomial formula for the (unordered) choice of several pairs of vertices among n. For instance at third order we get the contributions of Fig. II.2.5.

Grouping together the contributions in the natural way we obtain partially renormalized amplitudes A^{PR}. They are pictured in a symbolic, but hopefully transparent way in Fig. II.2.6.

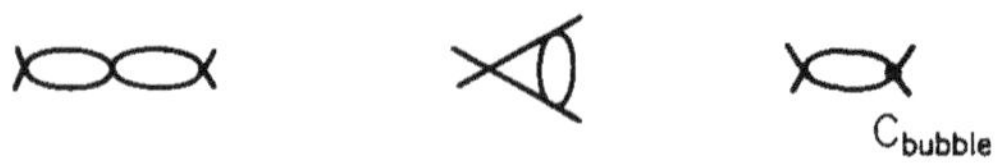

Figure II.2.5 Third order 4-point contributions.

These partially renormalized amplitudes are not convergent when the cutoff is removed, but partial subintegrations previously divergent are now convergent. We introduce a new third order counterterm c_G for each third order graph G which is minus the value of A_G^{PR} at 0 external momenta. With this new Lagrangian, the amplitudes associated to the graphs of Fig. II.2.5–6 become the renormalized ones $A_G^R = A_G^{PR} + c_G$. Again after some use of multinomial coefficients new structures appear at higher order which contain insertions of third order counterterms for each possible set of reductions of a third order divergent subgraph to a single vertex. This notion will be made soon more precise by the use of forests, but for the moment we invite the reader to try other particular examples and to check the multinomial coefficients. We can now state the general principle of the Bogoliubov induction:

$$c_G = - \sum_{g_1,\dots,g_k} \tau_G A_{G/\{g\}} \prod_{i=1}^{k} c_{g_i}|_{p=0} \tag{II.2.11}$$

which means:

- For each possible family $\{g\} = \{g_1, \dots, g_k\}$ of *disjoint* connected superficially divergent subgraphs g_i of G, multiply the corresponding counterterms, which by induction have been defined at an earlier stage,
- multiply for each such family by the bare amplitude for the "reduced graph" $G/\{g\}$ obtained by reducing each g_i to a single vertex (of the correct type, i.e., with $N(g_i)$ lines hooked to it),
- take minus the beginning of a Taylor expansion in the external momenta (at the value 0 in this subtraction scheme) as indi-

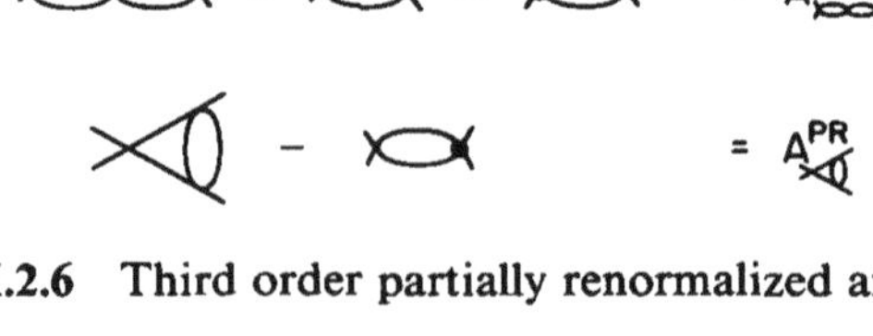

Figure II.2.6 Third order partially renormalized amplitudes.

cated by the operator τ_G and sum over all possible such families, including the empty one.

We can remark that we may use other subtraction prescriptions, for we may oversubtract; for instance in the φ_3^4 theory we could use the same scheme as for φ_4^4 and introduce counterterms for the coupling constant, although they would be in this case finite as the cutoff is removed, and therefore not necessary from the point of view of power counting. In this way the Bogoliubov recursion treats finite or infinite renormalizations on the same footing.

The renormalized functional integral is then formally given by:

$$\frac{1}{Z} e^{-(1/4!)\left(g_r+\sum_n g_r^n c_n\right) \int \varphi^4 - (1/2)\left(m^2+\sum_n g_r^n d_n\right) \int \varphi^2}$$
$$e^{-(1/2)\left(a+\sum_n g_r^n e_n\right) \int \partial_\mu\varphi\partial^\mu\varphi} D\varphi \qquad \text{(II.2.12)}$$

where c_n is given by a sum over connected graphs G with n vertices and 4 external legs (called "quadrupeds") of counterterms c_G defined by (II.2.11); and d_n and e_n correspond respectively to the first and the second subtraction of the Taylor operator τ_G for connected graphs G with n vertices and two external legs (called "bipeds").

We insist on the purely formal character of the functional integral (II.2.12). It is worse than the bare functional integral (I.3.1), which becomes the well defined functional integrals (I.3.9) or (I.3.12) when cutoffs are applied. Even with cutoffs, (II.2.12) remains ill defined because the infinite series c_n, d_n and e_n are formal and not expected to be convergent.

If one accepts this enormous drawback, we are now in a position to check Zimmermann's forest formula, which derives the renormalized perturbation series from (II.2.12). The result is given in terms of renormalized amplitudes in which a certain set of subtractions is performed directly on the *integrand*, so that the resulting integrals are absolutely convergent. The advantage is that this derivation, although formal, works as well for the theory with and without ultraviolet cutoff. The renormalized amplitudes A_G^R in [Zim] are then similar to the bare amplitudes in momentum space (I.4.18) but with the insertion of a renormalization operator:

$$A_G^R(p_1,\ldots,p_N) = \int \mathbf{R} \prod_{l\in G} \frac{d^4p_l}{p_l^2+m^2} \prod_{v\in G} \delta\left(\sum_l \varepsilon_{v,l} p_l\right), \qquad \text{(II.2.13)}$$

$$\mathbf{R} = \sum_{\mathbf{F}} \prod_{g\in\mathbf{F}} (-\tau_g). \qquad \text{(II.2.14)}$$

The sum runs over all possible forests **F** of subgraphs which are connected and superficially divergent, hence all possible forests of quadrupeds and bipeds, *including the empty one*, which corresponds to the bare amplitude. We recall that a forest is a set of subgraphs so that any two elements are either disjoint or included one in the other. The reader is invited to look for some examples, for instance to the twelve forests which contribute to the sum for the graph G in Fig. II.2.7.

The BPHZ scheme (II.2.13–14) is characterized by the following normalization conditions on the connected functions in momentum space:

$$C^4(0,0,0,0) = -g_r, \tag{II.2.15}$$

$$C^2(p^2 = 0) = \frac{1}{m^2}, \tag{II.2.16}$$

$$\frac{d}{dp^2} C^2|_{p^2=0} = -\frac{a}{m^4}. \tag{II.2.17}$$

These conditions are often stated in terms of the vertex functions (one particle irreducible amputated functions).

We sketch now the derivation of (II.2.13–14) from (II.2.12). The fact that forests are the solution to Bogoliubov's recursion is quite obvious, because considering families of disjoint subgraphs among the members of a family of disjoint subgraphs, etc ... produces obviously forests. We have to check that each forest is produced exactly once for each graph, at the level of Wick contractions. The corresponding combinatoric to check reduces to the multinomial formula for choosing the sequence of n_i vertices $i = 1, \ldots, k$ among n in each maximal subgraph g_i in the forest (the "root" of each tree, from the point of view

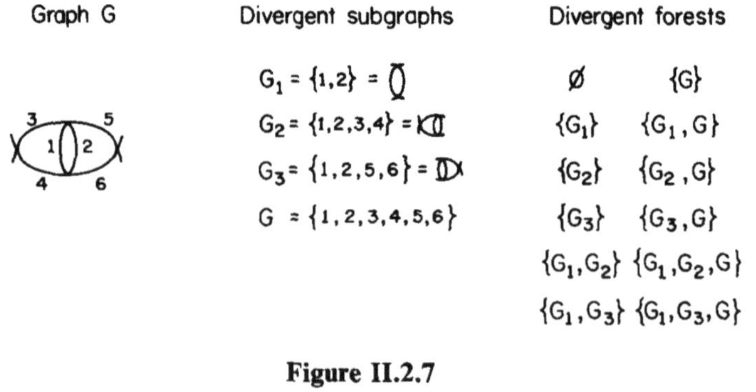

Figure II.2.7

of the inclusion relation). The combinatoric inside each maximal such subgraph has not to be checked since it is treated automatically by the induction.

Therefore the last subtle point is to check that in formula (II.2.13) the **R** operator may be taken to act really on the integrand. This was accomplished in [Zim]. In (II.2.13–14) the elementary operators τ_g for instance for forests made of a single subgraph g are really given by (II.2.2), namely for logarithmically divergent subgraphs the operator τ_g simply puts to zero their external momenta; for more divergent subgraphs a Taylor expansion around zero momenta is taken as in (II.2.2). However when we want to combine these elementary operations into products over subgraphs of a forest some technical ambiguities have to be fixed, and this makes the true definition of the **R** operator in (II.2.13–14) rather complicated. The product of the Taylor operators (II.2.2) has to be applied in the natural order of the forest, starting from the smallest graphs (the "leaves"). But one has also in fact to eliminate the δ functions in (II.2.13) by choosing a momentum routing rule which must be "admissible;" then one can define the action of the **R** operator on the reduced integrand and check that the result is independent of the admissible routing chosen. The insertion of the **R** operator leads in the end to well defined absolutely convergent integrals [Zim].

These complications were bypassed in the parametric representation, where Bergère, Lam and Zuber [BL][BZ] proved that there is an equivalent but completely canonical definition of the **R** operator, which also acts by direct subtractions on the integrand, and makes the Feynman integrals absolutely convergent, provided one simply works in the α representation (I.4.19) rather than in the momentum representation (I.4.18). Moreover the Taylor operators defined in α space may be shown to commute precisely when the graphs belong to a common forest, so that there is no ordering ambiguity in formula (II.2.14) in this case. This formalism, which was used in [dCR1][Ri1], is certainly the most elegant and compact one for a purely perturbative definition of renormalization.

However we want neither to work in the momentum, nor in the parametric representation, but in phase space. Hence we will define and use below a third equivalent version in which the **R** operator acts in x-space. In this formalism, which was developed in [FMRS2], the momentum space Taylor operations τ_G must be replaced by x-space "adjoints" τ_G^*, which generalize (II.2.3–4) and whose exact definition is given in the next section. This definition, like (II.2.3–4), implies the non-canonical choice of one particular vertex v_e among the border vertices of each subgraph of a given forest; there is also a condition that this non-canonical choice must be done coherently throughout the forest, which can be considered a dual of the admissibility condition on Zimmermann's routings of momenta. This sounds like a backward-step with respect to the a space formalism of [BL][BZ]; but the advantage is that this formalism is fully compatible with our multiscale representation; therefore it can be extended naturally to constructive field theory.

Starting from such a well defined formula for renormalized amplitudes, our goal is to show not only that these amplitudes are indeed finite (the "BPH theorem") but again to find also good uniform bounds at large order. The simplest such bound is obtained for amplitudes with fixed external momenta, but of course it is possible to derive analog results for amplitudes with external points smoothed against given test functions (see H1–H3 in the preceding section). Let us define $f(G)$ as the supremum over all forests appearing in (II.2.14) of $|\mathbf{F}|$, the number of subgraphs in $\mathbf{F}$. It is easy to check that $f(G) \leq n(G)$. As remarked already, a bound similar to (II.1.9) is hopeless because of the renormalon factorials. Instead we want to prove:

Theorem II.2.1: The uniform BPH theorem

There exists some constant K such that:

$$A_G^R(p_1, \ldots, p_N) \leq K^n f(G)! \left(1 + \sup_j |p_j|\right)^{\hat{N}} \tag{II.2.18}$$

where $\hat{N}$ is a function of $N(G)$ which may be taken to be $\hat{N} = N(G)/2$ if $N \geq 6$, $\hat{N}(G) = 1$ if $N(G) = 4$ and $\hat{N}(G) = 3$ if $N(G) = 2$

This bound is more accurate than those of [dCR1] and [FMRS2] as far as the external momenta dependence is concerned. (II.2.18) should not be considered to reflect the true large momentum behavior of I_G^R. If necessary one can prove with the multiscale expansion a more precise bound for the right hand side of (II.2.18). For instance in the case of quadrupeds ($N(G) = 4$), we leave as an exercise to check the bound:

$$K^n \sum_{k=1}^{f(G)} (f(G) - k)! [\text{Log}(1 + \sup |p_i|)]^k \tag{II.2.19}$$

for the right hand side of (II.2.18).

Nevertheless, the bound (II.2.18) for the first time gives a radius of convergence in the Borel plane which is uniform in the external momenta. This is an improvement on [FMRS2] due to our use of the replica trick (II.1.22). Even without this trick it should be possible to get a Borel radius uniform in the external momenta, but the analysis is more tedious and we do not try it here.

When later combined with an analysis of the number of graphs with fixed n and f, (II.2.18) leads to a finite "Borel radius" of the renormalized series, and one which does not shrink at large momenta like those of [dCR1][FMRS2]. An improved analysis sketched in Sect. II.6 even leads to bounds which give the optimal expected radius of convergence in the Borel plane. For the moment however we do not try to find optimal constants K in (II.2.18).

The next section is devoted to a full proof of the theorem. For pedagogical reasons it is divided in two steps; the proof is given first for graphs without bipeds (connected 2 point subgraphs), then in the general case. Indeed bipeds of φ_4^4 are the source of technical complications. To avoid redundant subtractions which result in worse bounds, it is natural to push for them the analysis beyond the level of connected subgraphs, to the level of one-particle irreducible subgraphs. The natural forests associated with this notion are the forests of "closed graphs" [dCR1]. However the corresponding technicalities divert the attention from the core of the proof, which appears more clearly in the biped-free case. Hence the hope is that with this presentation, the reader will *not* skip at least the first part of the next section.

Chapter II.3

Proof of the Uniform BPH Theorem

A The biped-free case

For the proof to work in the multiscale representation, the key point is to extract additional "index space" decay when some almost local subgraphs are divergent; in the biped free case they can only be quadrupeds. This decay should be extracted as in the example of the bubble subgraph treated in the preceding section. We need to give first the precise definition of the subtraction operators which are the equivalents in x-space of Zimmermann's operators in momentum space; in other words we must give the rule for "adjoints" τ_g^* which generalize (II.2.3–4). Let us fix some forest of quadrupeds **F**. We want to define the equivalent of the product $\prod_{g \in \mathbf{F}} \tau_g$ acting on the momentum space integrand $\prod_l (p_l^2 + m^2)^{-1} \prod_v \delta(\sum \varepsilon_{v,l} p_l)$ as a product $\prod_{g \in \mathbf{F}} \tau_g^*(v_e(g))$, acting on the x-space integrand, so that the renormalized amplitudes computed with the x-space τ^* operators agree with the ones computed with Zimmermann's τ operators. (Since $C(x, y)$ is ill defined at coinciding points, we define in fact the τ^* operators as acting always on a regularized integrand $\prod_l C_\kappa(x_l, y_l)$, and the preceding statement will be true only in the limit $\kappa \to \infty$.) The definition of the τ^* operators simplifies in our case because we meet only quadrupeds, for which $\omega = 0$. Therefore the sum (II.2.2) reduces to a single term, the 0 momentum value. We show now that in x-space, the equivalent of taking external momenta to 0 is to integrate over the position of vertices, save one (which corresponds to global translation invariance, hence to the overall δ function of momentum conservation for the subgraph in Zimmermann's scheme). Hence we must do a consistent choice for all the subgraphs g of a given forest of quadrupeds $\mathbf{F}$ of a preferred or "fixed" border-vertex $v_e(g,\mathbf{F})$ and define the corresponding $\tau_g^*(v_e(g,\mathbf{F}))$ operators. The following rule is a correct one (not unique).

Choose a border vertex arbitrarily for any of the maximal subgraphs of $\mathbf{F}$ (the trunks), but one which, if possible, is also a border vertex for G itself. Then choose inductively the other border vertices according to the natural rule: if g' is the immediate ancestor of g in the forest, which we note $g' = B_{\mathbf{F}}(g)$ and $v_e(g',\mathbf{F})$ is also a border vertex of g, choose $v_e(g,\mathbf{F}) = v_e(g',\mathbf{F})$. If it is not the case but there are some border vertices of g which are also border vertices of g', choose $v_e(g,\mathbf{F})$ among them; otherwise choose $v_e(g,\mathbf{F})$ arbitrarily.

With this simple rule we may picture in a graphic way our definition of the action of the product $\prod_{g\in\mathbf{F}} \tau_g^*(v_e(g,\mathbf{F}))$ on the integrand $\prod_l C_\kappa(x,y) \prod_{j=1}^N e^{i\sum p_j x_j}$. Each τ_g^* operation simply changes every external line $C(x,z)$ of g into $C(x_{v_e(g)},z)$ which means that it moves each external line of g to attach it to the single border vertex $v_e(g,\mathbf{F})$. These operations are consistent and commute because whenever $g \subseteq g'$ our rule ensures that an external line common to g and g' is never moved by the τ_g^* operator to an inside vertex of g' (which would be bad because in the definition of the τ_g' operator the inside vertices of g' have to be integrated out and one could not apply both the τ_g^* and $\tau_{g'^*}$ operators). With our rule the product $\prod_{g\in\mathbf{F}} \tau_g^*(v_e(g,\mathbf{F}))$ results in a well defined set of "moves" for the lines of g. Taken literally as in Fig. II.3.1 these moves would lead to unpleasant vertices with more than 4 lines attached to them so it is better to view the τ^* operators as creating thick "reduction" vertices which are then expanded separately to show their "inside" (the subgraph to which they correspond) as in Fig. II.3.2 where arrows show which "inside" is associated with which vertex.

Returning to phase space, we apply the decomposition (II.1.8) into momentum assignments and obtain:

$$A_G^R(p_1,\ldots,p_N) = \sum_\mu \int \prod_v dx_v \mathbf{R} \prod_l C^{i_l(\mu)}(x_l,y_l) \prod_{j=1}^N e^{ip_j x_j}, \quad \text{(II.3.1)}$$

$$\mathbf{R} \equiv \sum_{\mathbf{F}} \prod_{g\in\mathbf{F}} [-\tau_g^*(v_e(g,\mathbf{F}))] \quad \text{(II.3.2)}$$

(for simplicity we may often forget to write the dependence on $v_e(g,\mathbf{F})$).

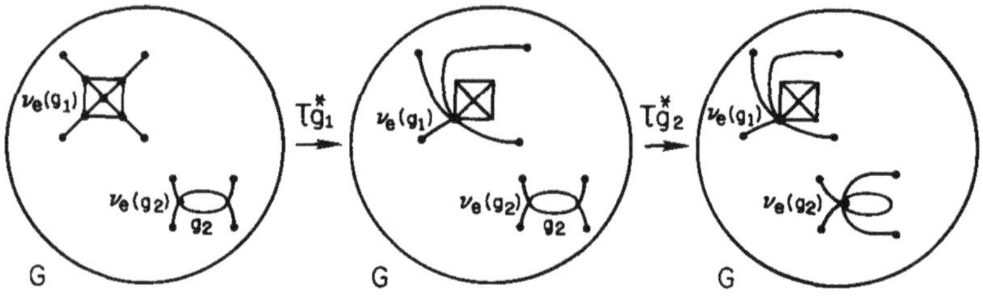

Figure II.3.1 Some "moves." Only some parts of the full graph G are shown.

Suppose now that we have a fixed assignment μ. The basic problem is to organize the forests appearing in the definitions (II.2.17) or (II.3.2) of **R** according to the fundamental tree structure induced by μ, namely the tree of the almost local G^i_k's. The almost local divergent subgraphs form a particularly obvious subforest of this tree, which we call $\mathbf{D}_\mu$, the "dangerous" forest for μ (since we remarked already at length that it is responsible for all the ultraviolet divergences in the assignment μ). Recalling (II.1.12), this forest may be characterized as the set of quadrupeds g with $i_g(\mu) > e_g(\mu)$ (see (II.1.10–11) for the definition of these indices). In (II.3.2) the sum over all forests which are subforests of $\mathbf{D}_\mu$, reconstructs exactly the operator $\prod_{g\in\mathbf{D}_\mu}(1-\tau^*_g)$. Intuitively we may guess that this is exactly what we need to restore vertical (index space) decay and to cure the ultraviolet divergences. But why are there other forests in (II.3.2), and what should we do with them?

We saw that graphically the τ^*_g operators extract the subgraph g from G and replace it by a reduction vertex. Hence intuitively these operators should prevent any contact across the boundary of g. Therefore when an operator τ^*_g, or more precisely a forest of such operators

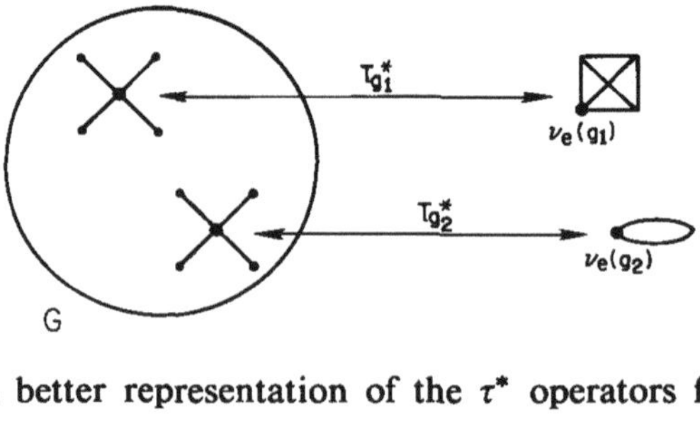

Figure II.3.2 A better representation of the τ^* operators for the situation above.

is applied we should try to define a new version of $\mathbf{D}_\mu$ relative to it, which respects this constraint of no information passing through any boundary of any g. This natural idea is formalized as follows.

Given any quadruped forest $\mathbf{F}$ and any g compatible with it (such that $\mathbf{F} \cup g$ is still a forest) we define $B_\mathbf{F}(g)$ as the ancestor of g in $\mathbf{F} \cup G$ and $A_\mathbf{F}(g)$ as $\cup_{h:g \supset h \in \mathbf{F}} h$.

Then for μ and $\mathbf{F}$ given, we define two subforests of $\mathbf{F}$, called respectively the safe and dangerous parts of $\mathbf{F}$. The safe part $\mathbf{T}_\mu(\mathbf{F})$ of $\mathbf{F}$ is the complement in $\mathbf{F}$ of the dangerous part $\mathbf{D}_\mu(\mathbf{F})$, which is defined by:

$$g \in \mathbf{D}_\mu(\mathbf{F}) \Leftrightarrow g \in \mathbf{F}, \qquad i_g(\mathbf{F}) > e_g(\mathbf{F}) \tag{II.3.3}$$

where

$$i_g(\mathbf{F}) = \min\{i_l(\mu) \mid l \in g - A_\mathbf{F}(g)\}, \tag{II.3.4}$$

$$e_g(\mathbf{F}) = \max\{i_l(\mu) \mid l \in E(g) \cap B_\mathbf{F}(g)\}, \tag{II.3.5}$$

$E(g)$ being the set of external lines of g internal in G. $E(g) \cap B_\mathbf{F}(g)$ is empty only when $g = G$, in which case we set $e_g(\mathbf{F}) = -1$. Remark that these definitions indeed generalize (II.1.10–11) since $i_g = i_g(\emptyset)$ and $e_g = e_g(\emptyset)$. Remark also that the dangerous graphs in $\mathbf{D}_\mu(\mathbf{F})$ are exactly those g in $\mathbf{F}$ such that $g/A_\mathbf{F}(g)$ is a connected component of $[B_\mathbf{F}(g)/A_\mathbf{F}(g)]^i(\mu)$ for some i, hence they generalize the notion of almost locality to the case of the reduced graph $B_\mathbf{F}(g)/A_\mathbf{F}(g)$. We recall that the notation g/h means that in g every connected component of h has been reduced to a single vertex, and that $g^i(\mu) = \{l \in g \mid i_l(\mu) \geq i\}$.

Obviously if $\mathbf{F}_1 \subseteq \mathbf{F}_2$ we have:

$$i_g(\mathbf{F}_1) \leq i_g(\mathbf{F}_2); \qquad e_g(\mathbf{F}_1) \geq e_g(\mathbf{F}_2) \quad \forall g \in \mathbf{F}_1. \tag{II.3.6}$$

We have also for any $g \in \mathbf{F}$:

Lemma II.3.1

$$i_g(\mathbf{F}) = i_g(\mathbf{T}_\mu(\mathbf{F}) \cup \{g\}), \tag{II.3.7}$$

$$e_g(\mathbf{F}) = e_g(\mathbf{T}_\mu(\mathbf{F}) \cup \{g\}). \tag{II.3.8}$$

Proof [FMRS2] Suppose $A_{\mathbf{T}_\mu(\mathbf{F})}(g) \subset A_\mathbf{F}(g)$ and let l_0 be any line in $A_\mathbf{F}(g)/A_{\mathbf{T}_\mu(\mathbf{F})}(g)$. Let $d_0 \subset d_1 \cdots \subset d_n$ be the set of all dangerous elements of $\mathbf{F}$ containing l_0 and contained in $g \equiv d_{n+1}$. For each $k = 0, \ldots, n$ there must exist a line l_{k+1} which is an external line of d_k and is contained in d_{k+1}. Since each d_k is dangerous, $i_{l_k} > i_{l_{k+1}}$, hence $i_{l_0} > i_{l_{n+1}}$, with

$l_{n+1} \in g/A_{\mathbf{F}}$. Hence l_0 cannot bear the minimal index in $g/A_{\mathbf{T}_\mu(\mathbf{F})}(g)$, which proves (II.3.7).

Similarly if $E(g) \cap B_{\mathbf{F}}(g) \subset E(g) \cap B_{\mathbf{T}_\mu(\mathbf{F})}(g)$, let l_0 be any line in the difference, and l_1 be any line in $E(g) \cap B_{\mathbf{F}}(g)$. Let d be the largest element of $\mathbf{F}$ such that $l_0 \in E(d)$ and $g \subset d \subset B_{\mathbf{T}_\mu(\mathbf{F})}(g)$. d must be dangerous for $\mathbf{F}$ so using (II.3.8):

$$i_{l_0} \leq e_d(\mathbf{F}) < i_d(\mathbf{F}) = i_d(\mathbf{T}_\mu(\mathbf{F}) \cup \{d\}) \leq i_{l_1}, \qquad \text{(II.3.9)}$$

hence l_0 cannot provide the maximum in $E(g) \cap B_{\mathbf{T}_\mu(\mathbf{F})}(g)$.

As a consequence of the Lemma, $\mathbf{T}_\mu(\mathbf{T}_\mu(\mathbf{F})) = \mathbf{T}_\mu(\mathbf{F})$ and the set $F^D(G)$ of all quadruped forests decomposes according to classes under the action of the $\mathbf{T}_\mu$ projector:

$$F^D(G) = \cup_{\mathbf{F}|\mathbf{T}_\mu(\mathbf{F})=\mathbf{F}} \{\mathbf{F}' \mid \mathbf{T}_\mu(\mathbf{F}') = \mathbf{F}\}. \qquad \text{(II.3.10)}$$

The forests $\mathbf{F}$ satisfying $\mathbf{T}_\mu(\mathbf{F}) = \mathbf{F}$ form the set $\mathbf{Safe}(\mu)$ of the so called safe forests (with respect to μ).

For any such safe forest $\mathbf{F}$, the next lemma characterizes completely the equivalence class $\{\mathbf{F}' \in F^D(G) \mid \mathbf{T}_\mu(\mathbf{F}') = \mathbf{F}\}$ in terms of the forest:

$$\mathbf{H}_\mu(\mathbf{F}) = \{g \subseteq G \mid g \text{ quadruped compatible with } \mathbf{F} \text{ and } g \in \mathbf{D}_\mu(\mathbf{F} \cup \{g\})\}. \qquad \text{(II.3.11)}$$

Lemma II.3.2 (Classification of Forests)
For any $\mathbf{F} \in \mathbf{Safe}(\mu)$ one has:

$$\mathbf{F} \cup \mathbf{H}_\mu(\mathbf{F}) \in F^D(G); \qquad \text{(II.3.12a)}$$

$$\forall \mathbf{F}' \in F^D(G), \mathbf{T}_\mu(\mathbf{F}') = \mathbf{F} \Leftrightarrow \mathbf{F} \subseteq \mathbf{F}' \subseteq \mathbf{F} \cup \mathbf{H}_\mu(\mathbf{F}). \qquad \text{(II.3.12b)}$$

Proof [FMRS2] (a) We have to show that two subgraphs g, g' in $\mathbf{H}_\mu(\mathbf{F})$ cannot overlap, i.e., have a nontrivial intersection. If they do, since g is connected, $g - g'$ must contain a line l which is internal for g and external for g'; this line l cannot be in $A_{\mathbf{F}}(g)$ (otherwise g' would not be compatible with $\mathbf{F}$) and must be in $B_{\mathbf{F}}(g')$ (by compatibility of g with $\mathbf{F}$). Similarly g' must contain a line l' in $g'/A_{\mathbf{F}}(g') \cap E(g) \cap B_{\mathbf{F}}(g)$. This enforces a contradiction:

$$i_l(\mu) \geq i_g(\mathbf{F} \cup \{g\}) > e_g(\mathbf{F} \cup \{g\}) \geq i_{l'}(\mu); \qquad \text{(II.3.13a)}$$

$$i_{l'}(\mu) \geq i_{g'}(\mathbf{F} \cup \{g'\}) > e_{g'}(\mathbf{F} \cup \{g'\}) \geq i_l(\mu). \qquad \text{(II.3.13b)}$$

(b) $\Leftarrow$ Note first that if $\mathbf{F}_1, \mathbf{F}_2 \in F^D(G)$ and $\mathbf{F}_1 \subseteq \mathbf{F}_2$, then by (II.3.6):

$$g \in \mathbf{D}_\mu(\mathbf{F}_1) \Rightarrow g \in \mathbf{D}_\mu(\mathbf{F}_2). \qquad \text{(II.3.14)}$$

Hence if $g \in \mathbf{F}' - \mathbf{F} = \mathbf{F}' \cap \mathbf{H}_\mu(\mathbf{F})$, then $g \in \mathbf{D}_\mu(\mathbf{F} \cup \{g\}) \subseteq \mathbf{D}_\mu(\mathbf{F}')$. So $\mathbf{T}_\mu(\mathbf{F}') \subseteq \mathbf{F}$. If this inclusion was a strict one, it would mean that there is a $g \in \mathbf{F}$ with $g \in \mathbf{D}_\mu(\mathbf{F}')$, hence using Lemma II.3.1:

$$i_g(\mathbf{F}') > e_g(\mathbf{F}') = e_g(\mathbf{T}_\mu(\mathbf{F}' \cup g)) \tag{II.3.15}$$

and using (II.3.6) and the fact that $\mathbf{F} = \mathbf{T}_\mu(\mathbf{F})$:

$$e_g(\mathbf{T}_\mu(\mathbf{F}' \cup g)) \geq e_g(\mathbf{F}) \geq i_g(\mathbf{F}) = i_{l_0}(\mu). \tag{II.3.16}$$

Combining (II.3.15) and (II.3.16), we see that the line l_0 must have collapsed under reduction of $A_{\mathbf{F}'}(g)$ but not under reduction of $A_{\mathbf{F}}(g)$. Therefore l_0 must belong to some h in $\mathbf{H}_\mu(\mathbf{F})$ with $h \subset g$, but cannot belong to $A_{\mathbf{F}}(g)$. Since $h \cap A_{\mathbf{F}}(g) = A_{\mathbf{F}}(h)$ we have:

$$i_{l_0}(\mu) = \min\{i_l \mid l \in h/A_{\mathbf{F}}(h)\} > \max\{i_l \mid l \in E(h) \cap B_{\mathbf{F}}(h)\}. \tag{II.3.17}$$

This contradicts the last part of (II.3.16) because $E(h)$ must contain a line that is internal to $B_{\mathbf{F}\cup\{h\}}(h) \subseteq g$ (or else $B_{\mathbf{F}\cup\{h\}}(h)$ would not be connected) and this line cannot be in $A_{\mathbf{F}}(g)$, since h is compatible with $\mathbf{F}$ and is not contained in $A_{\mathbf{F}}(g)$.

$\Rightarrow$ If $\mathbf{T}_\mu(\mathbf{F}') = \mathbf{F}$ then $\mathbf{F} \subseteq \mathbf{F}'$ and using Lemma II.3.1:

$$g \in \mathbf{F}' - \mathbf{F} \Rightarrow g \in \mathbf{D}_\mu(\mathbf{F}') \Rightarrow g \in \mathbf{D}_\mu(\mathbf{T}_\mu(\mathbf{F}') \cup \{g\}) \Rightarrow g \in \mathbf{H}_\mu(\mathbf{F}). \tag{II.3.18}$$

Lemma II.3.2 allows us to reorganize the $\mathbf{R}$ operator in the assignment μ as:

$$\mathbf{R} = \sum_{\mathbf{F} \in \mathrm{Safe}(\mu)} \prod_{g \in \mathbf{F}} (-\tau_g^*) \prod_{h \in \mathbf{H}_\mu(\mathbf{F})} (1 - \tau_h^*). \tag{II.3.19}$$

This is a beautiful rearrangement (different for each μ) because the product $\prod_{g \in \mathbf{H}_\mu(\mathbf{F})}(1 - \tau_g^*)$ will provide the desired cancellations for dangerous subgraphs within each particular counterterm of $\mathbf{F}$ and within $G/\mathbf{F}$ itself, and the counterterms of $\mathbf{F}$ themselves, corresponding to safe subgraphs, will have some external scale acting as a cutoff.

To establish the uniform BPH theorem, we write (II.3.1) in the form:

$$A_G^R = \sum_{\mathbf{F} \in F^D(G)} A_{G,\mathbf{F}}^R, \tag{II.3.20}$$

$$A_{G,\mathbf{F}}^R \equiv \sum_{\mu \mid \mathbf{F} \in \mathrm{Safe}(\mu)} \int \prod_v dx_v \prod_{g \in \mathbf{F}} (-\tau_g^*)$$

$$\prod_{h \in \mathbf{H}_\mu(\mathbf{F})} (1 - \tau_h^*) \prod_l C^{i_l(\mu)}(x_l, y_l) \prod_{j=1}^N e^{ip_j x_j}, \tag{II.3.21}$$

and we give a bound of the form (II.2.18) on $A^R_{G,\mathbf{F}}$ for every fixed $\mathbf{F} \in F^D(G)$. Then we can conclude because it has been shown ([dCR1], Lemma A.2) that the number of connected divergent forests in a biped-free φ^4_4 graph is at most $8^{n(G)}$. (Hint: bound first the number of maximal such forests. The only way quadrupeds may overlap in a biped-free graph is shown in the left of Fig. II.3.3 and by writing the 2-particle reducibility structure of G as in the right of Fig. II.3.3, one derives an inductive bound:

$$d_n \leq \sum_{p=1}^{n-1} d_p d_{n-p} \tag{II.3.22}$$

for the maximal number d_n of such maximal forests over graphs G with $n(G) = n$ (with initial condition $d_1 = 1$). Hence $d_n \leq 4^n$ and one can conclude).

We proceed to evaluate the action of the τ^* and $(1-\tau^*)$ operators in (II.3.21). The τ^*_g operators are applied first, for every $g \in \mathbf{F}$. They attach all external lines of g internal in $B_\mathbf{F}(g)$ to a single reduction vertex whose position is $v_e(g, \mathbf{F} \cup \mathbf{H}_\mu(\mathbf{F}))$, and which we abbreviate as $v_e(g)$ in what follows (see Fig II.3.1–2). Then we apply the $(1 - \tau^*_h)$ operators for each $h \in \mathbf{H}_\mu(\mathbf{F})$ and like in (II.2.5) we decompose the corresponding difference of products of external propagators into a sum of at most three terms, each of which contains exactly one difference concerning a single external propagator taken at two different end arguments. The generic case is:

$$\prod_{i=1}^{4} C(x_i, y_i) - \prod_{i=1}^{4} C(x_1, y_i) = C(x_1, y_1)$$
$$\cdot \sum_{i=2}^{4} \prod_{2 \leq j < i} C(x_j, y_j)[C(x_i, y_i) - C(x_1, y_i)] \prod_{i<j\leq 4} C(x_1, y_j). \tag{II.3.23}$$

Figure II.3.3 Overlap of quadrupeds.

Remark that the line bearing the difference must be in $E(h) \cap B_{\mathrm{F}}(h)$, because common external lines for h and $B_{\mathrm{F}}(h)$ are all attached to the same vertex after the $\tau^*_{B_{\mathrm{F}}(h)}$ operation, hence the corresponding difference in (II.3.23) gives 0 for this case and can be discarded.

Therefore the product of the $(1-\tau^*_h)$ operators factorizes as

$$\prod_{g \in \mathrm{F} \cup \{G\}} \left[\prod_{h \in \mathrm{H}_\mu(\mathrm{F}) | B_{\mathrm{F}}(h) = g} (1-\tau^*_h) \right],$$

each internal product acting only on the integrand corresponding to $g/A_{\mathrm{F}}(g)$. Finally when we compute such a product, any difference $C(x_1,y) - C(x_2,y)$ created at one end of a line by a $(1-\tau^*_h)$ operator cannot be modified by an other $(1-\tau^*_{h'})$. This is because with our rule for the choice of external vertices, the $\tau^*_{h'}$ operator would either leave both $C(x_1,y)$ and $C(x_2,y)$ untouched or move them both to $C(x_{v_e(h')},y)$, and the difference would give 0 and can be discarded. Of course this concerns only one given end of a line, and at the other end another difference may appear (this would be the case for Fig. II.3.1–2). Hence after all operators τ^*_g and $(1-\tau^*_h)$ have been applied we get an integrand similar to the initial one, with ordinary lines (some ends of which may have been moved to some $v_e(g)$ vertex), and lines bearing one difference at one end or two differences, one at both ends. The lines with one difference are then replaced by the interpolating formula (II.2.6) and the lines with two differences by a similar formula:

$$\begin{aligned} &C^i(x,y) - C^i(u,y) - C^i(x,z) + C^i(u,z) \\ &= \int_0^1\int_0^1 dt_1 dt_2 \frac{d}{dt_1}\frac{d}{dt_2} C^i(x + t_1(u-x), y + t_2(y-z)). \end{aligned} \tag{II.3.24}$$

Some of the lines bearing a (single!) difference may be true external lines of G; there is one such line if $N(G) = 4$ and at most $N(G)/2$ if $N(G) \geq 6$ (a pair of external lines of G could be external to some quadruped h, and one line in the pair could bear a difference; this is not true for single lines, because the rule in this case chooses the corresponding border vertex of h to be the fixed one $v_e(h)$ and this external line does not bear any difference).

We evaluate interpolating lines of the type (II.2.6) by (II.2.7), and lines of the type (II.3.24), by a similar estimate, based on representation (II.1.3–5). A line of index i bearing 0, 1 or 2 differences will therefore have power counting M^{2i}, M^{3i} or M^{4i} and 0, 1 or 2 corresponding multiplicative distances (like $|x-u||y-z|$ for (II.3.24)); it will have also interpolating decay (see (II.2.7). Before integrating over the positions of vertices we have to bound these distance factors and replace the interpolating decays by ordinary ones, as in (II.2.8). This must be done by using the decay of the internal lines connecting x to u (or y to z), which must be of higher scales for harvesting some net gain.

If the difference $|x-u|$ to bound has been created by a $(1-\tau_h^*)$ operator with $B_{\mathbf{F}}(h)=g$, the line bearing the difference has index at most $e_h(\mathbf{F})$. The internal decay to use takes place entirely within $g/A_{\mathbf{F}}(g)$. But h belongs to $\mathbf{H}_\mu(\mathbf{F})$, which means that $h/A_{\mathbf{F}}(h)$ is a connected component of $(g/A_{\mathbf{F}}(g))^i$ for $i=i_h(\mathbf{F})$. Therefore using the triangular inequality and the decay of the internal lines of $h/A_{\mathbf{F}}(h)$, we can bound the difference in external arguments by $KM^{-i_h(\mathbf{F})}$ as in (II.2.8); combining this with the extra power counting factor of the line bearing the difference results in a net gain of at least:

$$M^{-|i_h(\mathbf{F})-e_h(\mathbf{F})|}. \tag{II.3.25}$$

We can also replace the interpolating decay in $|x+t(u-x)-y|$ by the regular one $|x-y|$ and still keep a fraction of the internal decay of $h/A_{\mathbf{F}}(h)$ for later use. But a subtlety arises: what if several distance factors $|x_1-u_1|$, $|x_2-u_2|$ etc .... made use of the decay of the same internal lines? In this case they would correspond to graphs h_1, h_2, ..., which must be ordered by inclusion, hence have $i_{h_1}(\mathbf{F})>i_{h_2}(\mathbf{F})>\cdots$. But we can duplicate the decay of any internal line by formula (II.1.22) before using it. With this trick, different distance factors will always use different copies of any given internal decay, which solves the problem.

There is no dependence on interpolating parameters t any more and we can therefore bound the corresponding integrals by 1. This achieves the preparation of the integrand, and the rest will be very similar to Sect. II.1. We are ready to perform the integral over the positions of vertices by choosing a spanning tree T whose restriction to $g/A_{\mathbf{F}}(g)$ is still a spanning tree T_g of $g/A_{\mathbf{F}}(g)$ for each $g\in\mathbf{F}\cup\{G\}$; this is possible because of the tree structure of $\mathbf{F}\cup\{G\}$. Moreover we require that for each i,k the restriction of T_g to $(g/A_{\mathbf{F}}(g))_k^i$ is again a spanning tree of $(g/A_{\mathbf{F}}(g))_k^i$. This is the natural generalization of the choice of T in Sect. II.1, and is possible because of the tree structure of the $(g/A_{\mathbf{F}}(g))_k^i$ for each g. Then the spatial integration inside each $g/A_{\mathbf{F}}(g)$

goes exactly as for the full G in Sect. II.1, and delivers altogether the factor

$$\prod_{g\in \mathbf{F}\cup\{G\}} \prod_{(i,k)} M^{-\omega((g/A_{\mathbf{F}}(g))^i_k)}. \tag{II.3.26}$$

Moreover the combined effect of the extra factors (II.3.25) adds to this estimate a factor which is at worst:

$$\prod_{h\in \mathbf{H}_\mu(\mathbf{F})} M^{-|i_h(\mathbf{F})-e_h(\mathbf{F})|}. \tag{II.3.27}$$

More precisely we must take into account the case where external lines of G bear differences, where we write $|e^{ip_jx}-e^{ip_ju}| \leq (1+\sup_j |p_j|)|x-u|$ instead of the regular estimate. $|x-u|$ is a distance factor similar to the previous one, and the bound after integration of internal vertices is therefore:

$$|A^R_{G,\mathbf{F}}(p_1,\ldots,p_N)| \leq (1+\sup_j |p_j|)^{\hat{N}} \cdot \sum_{\mu|\mathbf{F}\in \mathbf{Safe}(\mu)} \prod_{g\in \mathbf{F}\cup\{G\}} \prod_{(i,k)} M^{-\omega'((g/A_{\mathbf{F}}(g))^i_k)}, \tag{II.3.28}$$

where

$$\omega'((g/A_{\mathbf{F}}(g))^i_k) \equiv \sup\{1, \omega((g/A_{\mathbf{F}}(g))^i_k)\} \tag{II.3.29}$$

except when $g\in \mathbf{F}$ and $(g/A_{\mathbf{F}}(g))^i_k = g/A_{\mathbf{F}}(g)$, in which case:

$$\omega'((g/A_{\mathbf{F}}(g))^i_k) \equiv \omega((g/A_{\mathbf{F}}(g))^i_k) = 0. \tag{II.3.30}$$

Let $i_{\max}(\mu)$ be the largest index appearing in the assignment μ. From half of the index space decay in (II.3.28) we can extract a factor $M^{-\delta i_{\max}(\mu)}$ patching together the internal vertical decays inside each $g/A_{\mathbf{F}}(g)$, $g\in \mathbf{F}\cup\{G\}$ with the condition $i_g(\mathbf{F})\leq e_g(\mathbf{F})$ for each $g\in\mathbf{F}$. Then with the rest of the decay we can sum over every assignment of each $g/A_{\mathbf{F}}(g)$ as in Sect II.1, provided one index of $g/A_{\mathbf{F}}(g)$ for each $g\in\mathbf{F}$ is kept fixed; indeed (II.3.30) implies overall vertical translation invariance for each $g/A_{\mathbf{F}}(g)$, $g\in\mathbf{F}$. The result is bounded by:

$$\prod_{g\in \mathbf{F}\cup\{G\}} K^{n(g/A_{\mathbf{F}}(g))} \leq K^{2n(G)}. \tag{II.3.31}$$

Each "translation invariant" sum over the fixed index of a given $g/A_{\mathbf{F}}(g)$ is nevertheless obviously bounded by $i_{\max}(\mu)$, hence the corresponding sums combined are bounded by:

$$\sum_{i_{\max}(\mu)} (i_{\max}(\mu))^{|\mathbf{F}|} M^{-\delta i_{\max}(\mu)} \leq |\mathbf{F}|! K^{|\mathbf{F}|}. \tag{II.3.32}$$

Combining (II.3.28) (II.3.31) and (II.3.32) achieves the proof of Theorem II.2.1 in the biped free case.

In fact this proof again achieves more than the theorem it is designed for. It is indeed worth to consider as a separate theorem the particular case of an empty **F**, because the empty forest is the only safe forest common to all assignments, and by (II.3.32) there is no factorial in the estimates for it. Since $\mathbf{H}_\mu(\emptyset) \equiv \mathbf{D}_\mu$, we have proved that in the biped free case:

Theorem II.3.1: Uniform bound for usefully renormalized amplitudes

$$\begin{aligned} &|A_G^{UR}(p_1,\ldots,p_N)| \\ &\quad = \left| \sum_\mu \int \prod_v dx_v \prod_{h\in\mathbf{D}_\mu} (1-\tau_h^*) \prod_l C^{i_l(\mu)}(x_l,y_l) \prod_{j=1}^N e^{ip_jx_j} \right| \\ &\quad \le (1+\sup_j |p_j|)^{\hat N} K^{n(G)}. \end{aligned} \tag{II.3.33}$$

A_G^{UR} is a piece of the renormalized amplitude, called the "usefully" renormalized amplitude because it has no insertion of "useless" counterterms which correspond to the elements of safe forests. Theorem II.3.1 shows that it does not contain any renormalon effect. Hence we can conclude, as announced in the previous section, that the renormalon effects are solely due to the useless counterterms.

We give now the proof of the BPH uniform theorem (Theorem II.2.1) in the general case where bipeds are present. We will also obtain the generalization of Theorem II.3.1, proving that it also holds with bipeds. We will not rephrase what is similar to the above analysis but concentrate on the new technicalities created by the bipeds.

B The general case, with bipeds

To get a reasonably accurate bound when bipeds are present, we must trim some redundant subtractions in the definition (II.2.14) of the **R** operator due to one particle reducible divergent graphs. Indeed for a graph with only n vertices like the one of Fig. II.3.4 there are forests of connected divergent subgraphs with about $3n/2$ elements, and the

Figure II.3.4

corresponding estimate (II.2.18) would not even allow a finite disk of analyticity in the Borel plane (see Sect. II.6).

We recall that a subgraph was called proper or 1PI, if it cannot be broken into two disconnected pieces by cutting a single line. A quadruped (connected subgraph q with $N(q) = 4$) is called *open* if it is proper and there exists a proper biped $b \supset q$ such that both border vertices of b are border vertices of q. b is then called the closure q^* of q and is obtained by adding to q a line or a chain of proper bipeds between these two vertices (see Fig. II.3.5). More generally a subgraph g is called closed if by definition:

$$\forall q \text{ open quadruped, } q \subseteq g \Rightarrow q^* \subseteq g$$

and the closure g^* of a proper subgraph g is the smallest closed subgraph containing it [dCR1]. This definition is consistent with the first one when g is a quadruped.

Then we can restrict, in the definition (II.2.14) of the **R** operator the sum to run over forests of closed divergent subgraphs (in short

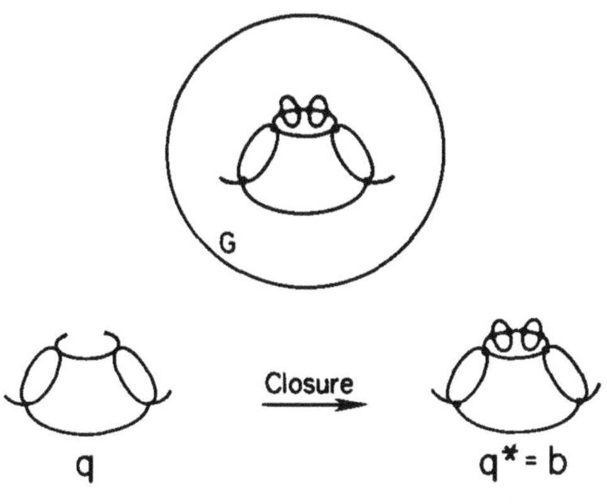

Figure II.3.5

"closed forests"). The fact that subtractions for one particle reducible subgraphs are redundant is rather standard (see for example [BL]); it follows intuitively from the observation that putting to 0 the external momenta of a graph also puts to 0 the external momenta of its proper parts by momentum conservation. The fact that subtractions for open quadrupeds are redundant is less well known but can be grasped as follows. Let q be an open quadruped and $b = q^*$ be its closure. Then $(1 - \tau_b)\tau_q = 0$ since the τ_q operator hooks both external lines of b to the same reduction vertex, and it is therefore not necessary to introduce counterterms for both q and $b = q^*$. From this idea and some induction, one can restrict the sum over forests in (II.2.14) to the closed forests without actually changing the action of $\mathbf{R}$ on a graph ([dCR1], Lemma II.3). As a general definition of the factor $f(G)$ appearing in Theorem II.2 we take the supremum over all closed forests $\mathbf{F}$ of $|\mathbf{F}|$, the number of elements in $\mathbf{F}$†.

For any given closed forest $\mathbf{F}$ and assignment μ, we introduce relative indices which generalize (II.3.4)–(II.3.5):

$$i_g^c(\mathbf{F}) = \max_h \min \{i_l(\mu) \mid l \in h/A_{\mathbf{F}}(g)\} \tag{II.3.34}$$

where the max is taken over all h compatible with $\mathbf{F}$ which obey $h^* = g$ and such that $h/A_{\mathbf{F}}(g)$ is a proper component of $(g/A_{\mathbf{F}}(g))^i$ for some i. The h which realizes this maximum is called $I_g^c(\mathbf{F})$. We define e_g^c as:

$$\begin{aligned} e_g^c(\mathbf{F}) &= \max \{i_l(\mu) \mid l \in E(g) \cap B_{\mathbf{T}_\mu^c(\mathbf{F})}(g)\} \qquad \text{if} \quad N(g) = 4, \\ e_g^c(\mathbf{F}) &= \min \{i_l(\mu) \mid l \in E(g) \cap B_{\mathbf{T}_\mu^c(\mathbf{F})}(g)\} \qquad \text{if} \quad N(g) = 2, \end{aligned} \tag{II.3.35}$$

where $\mathbf{T}_\mu^c(\mathbf{F})$ and its complement $\mathbf{D}_\mu^c(\mathbf{F})$ are defined inductively, starting from the largest graphs in $\mathbf{F}$ and proceeding towards the smallest ones, using the conditions:

$$g \in \mathbf{T}_\mu^c(\mathbf{F}) \Leftrightarrow e_g^c(\mathbf{F}) \geq i_g^c(\mathbf{F}), \tag{II.3.36}$$

$$g \in \mathbf{D}_\mu^c(\mathbf{F}) \Leftrightarrow e_g^c(\mathbf{F}) < i_g^c(\mathbf{F}). \tag{II.3.37}$$

Since the definition of $\mathbf{T}_\mu^c(\mathbf{F})$ is inductive there is no logical loophole between (II.3.35) and (II.3.36).

The lemmas necessary to classify the forests are the following generalizations of Lemmas II.3.1–2:

---

†This is *not* the definition of $f(G)$ in [dCR1], and the factorial bound of [dCR1] is therefore not optimal in this respect.

Lemma II.3.3

For $\mathbf{F}$ a closed forest of G and $g \in \mathbf{F}$:

$$i_g^c(\mathbf{T}_\mu^c(\mathbf{F}) \cup \{g\}) = i_g^c(\mathbf{F}), \qquad \text{(II.3.38)}$$

$$e_g^c(\mathbf{T}_\mu^c(\mathbf{F}) \cup \{g\}) = e_g^c(\mathbf{F}), \qquad \text{(II.3.39)}$$

$$\mathbf{T}_\mu^c(\mathbf{T}_\mu^c(\mathbf{F})) = \mathbf{T}_\mu^c(\mathbf{F}). \qquad \text{(II.3.40)}$$

Furthermore for $\mathbf{F}$ a safe forest, i.e., such that $\mathbf{T}_\mu^c(\mathbf{F}) = \mathbf{F}$, we define $\mathbf{H}_\mu^c(\mathbf{F}) = \{g \subseteq G,\ g$ proper closed divergent subgraph compatible with $\mathbf{F}$, and $g \in \mathbf{D}_\mu^c(\mathbf{F} \cup \{g\})\}$. Then:

Lemma II.3.4

Lemma II.3.2 still holds with the generalized definitions $\mathbf{T}_\mu^c(\mathbf{F})$ and $\mathbf{H}_\mu^c(\mathbf{F})$ replacing the former $\mathbf{T}_\mu(\mathbf{F})$ and $\mathbf{H}_\mu(\mathbf{F})$.

Sketch of proof (for a detailed proof of these two Lemmas we refer to [dCR1]). (II.3.39) is obvious from definition (II.3.35), and with (II.3.38) it implies (II.3.40). (II.3.38) is non-trivial. By induction, it is enough to show that for any $g' \neq g, g' \in \mathbf{D}_\mu^c(\mathbf{F})$, we have $i_g^c(\mathbf{F}) = i_g^c(\mathbf{F} - \{g'\})$. This is obvious except when $g = B_\mathbf{F}(g')$. In this last case we remark first that $h_0 = I_g^c(\mathbf{F} - \{g'\})$ satisfies all the conditions appearing on the list over which the max is taken in (II.3.34) for $i_g^c(\mathbf{F})$. Since $h_0/A_\mathbf{F}(g) \subseteq h_0/A_{\mathbf{F}-\{g'\}}$, we have the inequality $i_g^c(\mathbf{F}) \geq i_g^c(\mathbf{F} - \{g'\})$. The opposite inequality requires some care. An easy case is when $I_g^c(\mathbf{F})$ is disjoint from g'; then it appears also in the list for $\mathbf{F} - \{g'\}$ and one concludes easily. The last possibility is $g' \subset I_g^c(\mathbf{F})$ (since $I_g^c(\mathbf{F})$ is compatible with $\mathbf{F}$). In this case $I_g^c(\mathbf{F})$ has to contain at least two external legs of g', hence

$$i_g^c(\mathbf{F}) \leq e_{g'}^c(\mathbf{F}) < i_{g'}^c(\mathbf{F}) \qquad \text{(II.3.41)}$$

and $h_1 = I_g^c(\mathbf{F}) - [g' - I_{g'}^c(\mathbf{F})]$ is in the list over which a maximum is taken in the definition (II.3.34) of $i_g^c(\mathbf{F} - \{g'\})$. Hence we conclude also that $i_g^c(\mathbf{F}) \leq i_g^c(\mathbf{F} - \{g'\})$, which achieves the proof of (II.3.38).

Remark that Lemma II.3.3 remains also true if we modify slightly the definition of $i_g^c(\mathbf{F})$ when $g/A_\mathbf{F}(g)$ is the bubble graph of Fig I.1.1, defining it in this special case by:

$$i_g^c(\mathbf{F}) = \max\{i_1, i_2\} \qquad \text{(II.3.42)}$$

where i_1 and i_2 are the two indices in μ of the two lines of $g/A_\mathbf{F}(g)$. This small change is not fundamental at all but is necessary to have a nice rule for the "bubble resummations" of [DFR] which will be briefly considered in Sect. II.6.[†] The reason for which renormalization still works with this definition is that a single line in the bubble is enough to ensure spatial decay between the two border vertices of the bubble.

For more general graphs the real constraint on i_g^c to ensure such a decay is that there is at least a spanning tree of $g/A_{\mathbf{F}}(g)$ made of lines with indices higher or equal to $i_g^c(\mathbf{F})$; both (II.3.34) and (II.3.42) satisfy this rule, as does the construction in [GaNi]; clearly there remains some flexibility in the choice of i_g^c, and some details are a matter of convenience. This remark applies also to the expansions of constructive theory, in which the mechanism of convergence is fundamental but some particular technical details are not.

Lemmas II.3.3–4 allow us to reorganize the operator **R** exactly as in (II.3.19). Also the number of closed divergent forests in a graph is still bounded by $8^{n(G)}$ ([dCR1], Lemma A.2). Hence we can again restrict our attention to a single $A_{G,\mathbf{F}}^R$ in (II.3.20).

We must make precise the choice of $v_e(g)$ when g is a biped. Recall that in this case, renormalization is performed (i.e., a factor $(1 - \tau_g^*)$ appears in (II.3.19)) even when one of the two external legs of the biped has an index bigger than the internal index i_g^c (see (II.3.35)). Hence we have to ensure that the τ_g^* operator in this case applies to the external leg of the biped with lowest index e_g^c, otherwise there would be no net gain. This can be done by adding to our previous rule for choosing the $v_e(g)$ vertices the prescription that when g is a biped the fixed vertex $v_e(g)$ is the one to which the leg of highest index in μ is hooked. This prescription does not interfere with the former rules because a border-vertex of a proper biped cannot be also border-vertex of any other proper closed divergent subgraph containing it, hence the choice of the fixed vertex for it was arbitrary in the former rules.

We have also to supply a new formula for the action of τ_g^* when g is a biped; since $\omega(g) = -2$ in this case, the Taylor expansion at 0 momenta is pushed to second order. When the corresponding adjoints are applied to the external propagators $C(x, u)C(y, z)$ of the biped, with

---

[†]It has also some advantages over the "2nd max" definition for e_g^c that the curious reader will find in [Ri1][DFR]; this definition was introduced also solely to allow explicit bubble resummations, but is worse than the simple rule (II.3.42).

$x = x_{v_e(g)}$ and y the position of the other border vertex of g, one obtains:

$$\tau_g^*[C(x,u)C(y,z)] = C(x,u)\left\{C(x,z) + (y-x)^\mu \frac{\partial C}{\partial x^\mu}(x,z) + \frac{1}{2}(y-x)^\mu(y-x)^\nu \frac{\partial^2 C}{\partial x^\mu \partial x^\nu}(x,z)\right\}, \qquad \text{(II.3.43)}$$

where partial derivatives apply to the first argument x of $C(x,z)$. By a parity argument the second term in the sum (II.3.43) vanishes when integrated over y, so that we may forget it and write $\tau_g = \tau_g^0 + \tau_g^1$. τ_g^0 is the mass counterterm, whose adjoint is defined by:

$$\tau_g^{0*}[C(y,z)C(x,u)] = C(x,z)C(x,u), \qquad \text{(II.3.44)}$$

and τ_g^1 is the wave function counterterm, whose adjoint is defined by:

$$\tau_g^{1*}[C(y,z)C(x,u)] = \frac{1}{2}(y-x)^\mu(y-x)^\nu \frac{\partial^2 C}{\partial x^\mu \partial x^\nu}(x,z)C(x,u). \qquad \text{(II.3.45)}$$

These counterterms are pictured in Fig. II.3.6. Again by parity, only the terms with $\mu = \nu$ survive after the internal integration over $y-x$ (holding x fixed) is performed. Finally we can remark that $y-x$ and $z-x$ are independent in (II.3.45), and we can use Euclidean invariance to replace each integral containing a particular $[(y-x)^\mu]^2 \partial^2 C(x,z)/\partial x^\mu \partial x^\mu$

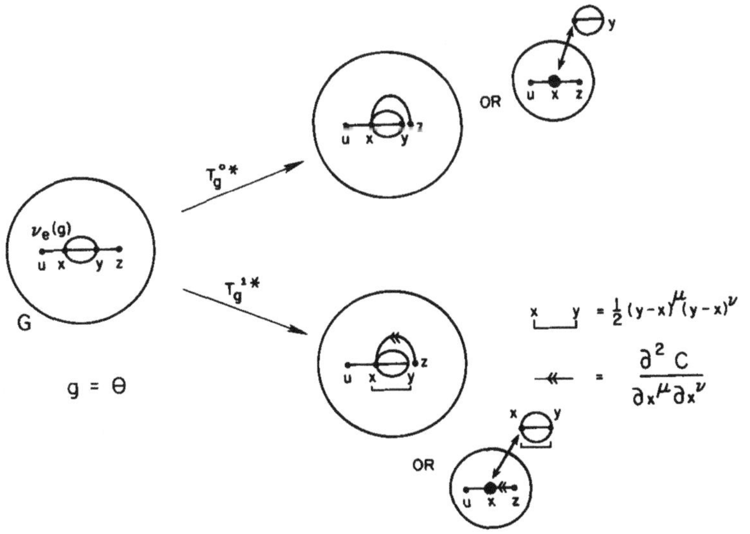

Figure II.3.6

by $(1/4)|y-x|^2\partial^2 C(x,z)/\partial x^\mu \partial x^\mu$. Writing $\Delta = \sum_\mu \partial^2/\partial x^\mu \partial x^\mu$ we can therefore rewrite (II.3.45) as:

$$\tau_g^{1*}[C(y,z)C(x,u)] = \frac{1}{8}|y-x|^2[\Delta C(x,z)]C(x,u), \tag{II.3.46}$$

which is simpler and, we hope, suggests more clearly that the wave function counterterm really corresponds to the renormalization of a, the parameter in (I.3.1) which corresponds in the action to $\int \partial_\mu \varphi \partial^\mu \varphi$ hence, integrating by parts, to $-\int \Delta\varphi.\varphi$.

It is also useful to write the Taylor remainder formula:

$$(1-\tau_g^*)[C(x,u)C(y,z)] = C(x,u)\int_0^1 \frac{(1-t)^2}{2}\frac{d^3}{dt^3}C(x+t(y-x),z). \tag{II.3.47}$$

As before, our rules on the choice of fixed vertices prevents any end of line to bear the action of several $(1-\tau_g^*)$ operators. To the preceding argument for biped free graphs, one has to add the observation that an external line of a biped $g \in \mathbf{F} \cup \mathbf{H}_\mu^c(\mathbf{F})$ cannot be the external line of other elements of $\mathbf{F} \cup \mathbf{H}_\mu^c(\mathbf{F})$, except maybe for quadrupeds $h \subset g$, and for these h our rule for choosing $v_e(h)$ ensures that this external line common to g and h is hooked to $v_e(h)$, hence cannot bear the action of the $(1-\tau_h^*)$ operator.

We can evaluate the net effect of operations (II.3.43–47). In (II.3.47) we earn a factor at least $M^{-3[i_g^c(\mathbf{F})-e_g^c(\mathbf{F}]}$, again using representation (II.1.3–4), and the internal decay of a spanning tree of g with lines of indices greater or equal to $i_g^c(\mathbf{F})$. Duplicating internal lines decay like in (II.1.22) again avoids using too many times the same decay of the same line.

Similarly evaluating (II.3.44) is neutral, and (II.3.45–46) results in a *loss* of at most $M^{2[e_g^c(\mathbf{F})-i_g^c(\mathbf{F}]}$. (Remember that g is in a safe forest if (II.3.44–46) apply).

Let g be a biped of a safe forest $\mathbf{F}$, and let us admit for a moment that inside $g/A_\mathbf{F}(f)$ everything proceeds as in the preceding case. In the multiscale representation, this means that after internal spatial integrations are performed in $g/A_\mathbf{F}(f)$ and internal line indices are summed with respect to the internal scale $i_g^c(\mathbf{F})$ we obtain a quadratically divergent overall factor $M^{2i_g^c(\mathbf{F})}$, since $\omega(g) = -2$ (apart from an overall $K^{n(g)}$ and renormalon factors which will be considered later). The indices of the two external lines of g are $e_g^c(\mathbf{F})$ and j, with $j \geq e_g^c(\mathbf{F})$ by (II.3.35). We want to compare the two lines $C^j(y-x)C^{e_g^c(\mathbf{F})}(x,z)$ with insertion of

Figure II.3.7

the biped counterterm for g at x, to a single ordinary line $C^{e_g^c(\mathbf{F})}(y-z)$ (see Fig. II.3.7).

We have to add to the quadratically divergent factor $M^{2i_g^c(\mathbf{F})}$ the effect of spatial integration over the position x of the φ^2 reduction vertex corresponding to g. This integration may be performed with the line of index j and brings a factor M^{-4j}. We combine this factor with the M^{2j} power counting factor of the line of index j and sum over $j > e_g^c(\mathbf{F})$; this results in a factor $M^{-2e_g^c(\mathbf{F})}$. Using the triangular inequality we can also reconstruct the regular spatial decay of scale $e_g^c(\mathbf{F})$ between y and z. Combining all these factors with formulae (II.3.44–46) we reconstruct the ordinary estimate (II.1.6) for $C^{e_g^c(\mathbf{F})}(y-z)$, plus a total power counting factor of $M^{-2[e_g^c(\mathbf{F})-i_g^c(\mathbf{F})]}$ for the mass counterterm and of 1 for the wave function counterterm. When we sum over $i_g^c(\mathbf{F})$, respecting the constraint of safeness for g which is $i_g^c(\mathbf{F}) \le e_g^c(\mathbf{F})$, we get a constant in the first case and a factor $e_g^c(\mathbf{F})$ in the second case (which may be bounded by $i_{\max}(\mu)$).

The conclusion is that useless wave function counterterms (which correspond to marginal operators) generate the same kind of renormalon effects as useless coupling constant counterterms, and that useless mass counterterms do not. This is consistent with the bound (II.2.18) and the general definition of $f(G)$.

It remains to explain what we assumed, namely that inside each separate $g/A_{\mathbf{F}}(g)$, $g \in \mathbf{F} \cup \{G\}$ an analysis similar to the biped-free case takes place. The key point is to check that after spatial integration, vertical index space decay is correctly generated. This means in particular that each connected component $[g/A_{\mathbf{F}}(g)]_k^i$ with four or two external legs is properly renormalized, i.e., provides at each scale a reward at least M^{-1} instead of costing respectively a constant or M^2. Consider a proper closed subgraph $h \in \mathbf{H}_\mu^c(\mathbf{F})$ such that $B_{\mathbf{F}}(h) = g$. The main remark is that the less stringent definitions (II.3.34–35) now allow renormalization of h on a wider range of indices than the minimal range in which $h/A_F(h)$ appears as a $[g/A_F(g)]^i_k$ component. Hence the bonuses M^{-1} or M^{-3} (respectively for h quadruped or biped) occur on this wider range, and this results in renormalized power counting not only for the corresponding proper closed graphs, but also for one-particle reducible bipeds and quadrupeds and for 1PI open quadrupeds when they occur as $[g/A_F(g)]^i_k$'s. As expected, renormalization of proper closed subgraphs takes care of all other divergences. We will check this fact on typical examples.

Let us for simplicity forget about reduction by **F**, i.e., write simply g for $g/A_F(g)$ etc .... We consider first the case of a one particle reducible divergent subgraph h of g. There must exist a proper biped $h' \subset h$ as in Fig. II.3.8, i.e., h' is a proper component of h and there is a common border vertex for h and h'; otherwise h could not be divergent.

In this case consider the set I of indices i for which h is a connected component of g^i. We claim that if I is not empty, $h' \in \mathbf{H}^c_\mu(\mathbf{F})$ and I is a piece of the extra range in which renormalization of h' provides an M^{-3} bonus. Indeed if $i \in I$ one must have $j_1 \geq i > j_2$ where j_1 and j_2 are the external indices of h' as shown in Fig. II.3.8. Therefore $e^c_{h'}(\mathbf{F}) = j_2$. Since $i_{h'}(\mathbf{F}) \geq i$ (here we do mean $i_{h'}(\mathbf{F})$ as defined in (II.3.4), *not* $i^c_{h'}(\mathbf{F})$), we have $i^c_{h'}(\mathbf{F}) \geq i_{h'}(\mathbf{F}) > j_2 = e^c_{h'}(\mathbf{F})$, hence $h' \in \mathbf{H}^c_\mu(\mathbf{F})$. The factors earned in the renormalization of h' which cover the range $]j_2, \min\{j_1, i_{h'}(\mathbf{F})\}]$, in which obviously h' itself is not a connected component of g^i, can then be safely attributed to the renormalization of h; they cover the desired set I of indices.

Finally we examine the case of an open quadruped h, for which the renormalization of its closure h^* should save the day (see Fig. II.3.5). Indeed by definition (II.3.34), if the set I of indices where h is a connected component of g^i is non-empty we can conclude that h^* belongs to $\mathbf{H}^c_\mu(\mathbf{F})$; and I is covered again by the index range $i^c_{h^*}(\mathbf{F}) \geq i > i_{h^*}(\mathbf{F})$ in which we get bonus factors M^{-3} from the renormalization of h^*.

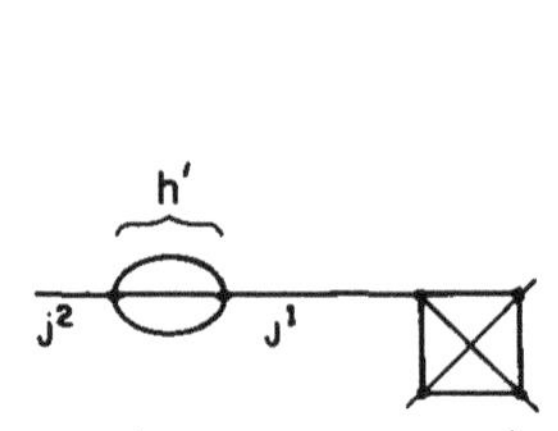

Figure II.3.8

To complete the argument, let us check that there is no overlap, for a given biped h, between the factors used in the different cases considered in Fig. II.3.8 and II.3.5. Indeed the range $i_h < i \leq i_h^c$ takes care of open quadrupeds inside h; the range $e_h < i \leq i_h$ takes care of h itself, and the range $e_h^c < i \leq \min\{i_h, e_h\}$ takes care of one particle reducible graphs with h at one of their "ends," and there is no overlap between these three ranges.

As a last remark, consider that when G itself is a biped, the derivatives corresponding to the $(1-\tau_G)$ operator apply to a true external leg of G; in the formula analogous to (II.3.47) factors of degree 3 appear, which can be bounded by $|p|^3$. This explains the external momentum dependence of (II.2.18) and the definition of $\hat{N}$ in this case. This completes our sketch of the general proof of Theorem II.2.1, and we refer to [dCR1][FMRS2] for a somewhat more formalized but less pedagogical proof.

As a technical comment, we note that the proof given above relies on the particular properties of φ_4^4 amplitudes and is therefore in some respects less general and systematic than the one of [FMRS2]. But in [FMRS2] a more inductive approach is used, in which a large number of Taylor operators may apply to the same line. This makes the inductive bound to be proven more complicated, and also results in a bound which is not uniform in n as far as the external momentum dependence is concerned, which for applications may be considered a significant drawback.

Again a byproduct of the analysis is that Theorem II.3.1 holds in the general case as well: usefully renormalized amplitudes I_G^{UR}, which are defined by the same formula (II.3.33) than in the biped free case, but with $\mathbf{D}_\mu$ replaced by the generalized definition $\mathbf{D}_\mu^c \equiv \mathbf{H}_\mu^c(\emptyset)$ still do not develop any factorial and satisfy the bound (II.3.33). Hence it is a general phenomenon that "useless counterterms" are solely responsible for renormalon effects.

In fact more precise statements can be derived from the analysis above. We remarked that mass counterterms do not contribute to any factorial effect; so there should be a way to write amplitudes with full mass renormalization and useful coupling constant and wave function renormalization which does not display any factorial behavior. It is easy to check (by trying some examples) that a proper biped can never overlap with any proper closed divergent subgraph, and this was proved rigorously in [dCR1]. In particular the set of all proper bipeds of G is itself a closed forest called $\mathbf{B}(G)$, and the product $\prod_{b\in\mathbf{B}(G)}(1-\tau_b)$ factorizes in the $\mathbf{R}$ operator. We are interested in factorizing only the mass renormalizations, so we write:

$$\mathbf{R} = \prod_{b\in\mathbf{B}(G)} (1-\tau_b^{0*}) \sum_{\mathbf{F}} \prod_{b\in\mathbf{F}} (-\tau_b^{1*}) \prod_{q\in\mathbf{F}} (-\tau_q^{*}), \tag{II.3.48}$$

where the first product runs over all proper bipeds b of G, the second over all bipeds of $\mathbf{F}$ and the third over the quadrupeds of $\mathbf{F}$, and τ_b^{0*} and τ_b^{1*} are defined in (II.3.44–46). We may now apply the classification of forests only to the sum over forests in (II.3.48) and obtain amplitudes $A_{G,\mathbf{F}}^{MR}$ in which the superscript MR means that all the mass renormalizations are fully performed. The case $A_{G,\emptyset}^{MR} \equiv A_G^{MR,UR}$ is what we are looking for, and we obtain:

Theorem II.3.2

$$\begin{aligned} |A_G^{MR,UR}&(p_1,\ldots,p_N)| \\ &= \Bigg| \sum_{\mu} \int \prod dx_v \prod_{b\in\mathbf{B}(G)} (1-\tau_b^{0*}) \prod_{b\in\mathbf{D}_\mu^c} (1-\tau_b^{1*}) \cdots \\ &\quad \cdots \prod_{q\in\mathbf{D}_\mu^c} (1-\tau_q^{*}) \prod_l C^{i_l(\mu)}(x_l,y_l) \prod_{j=1}^{N} e^{ip_jx_j} \Bigg| \\ &\le (1+\sup_j |p_j|^{\hat{N}}) K^{n(G)}. \end{aligned} \tag{II.3.49}$$

Theorems II.3.1 and II.3.2 are powerful motivations to find the expansions corresponding to these usefully renormalized amplitudes which have the big advantage to be free of renormalon effects, hence to behave at large order like the amplitudes of a superrenormalizable theory. The next section is devoted to this problem.

From now on we drop the superscript c most of the time; in the biped free case the definitions of the first part of this section are to be used, and in the general case, the general definitions with superscript c.

Chapter II.4

The Effective Expansion

Tout ce qui est simple est inexact
mais tout ce qui est compliqué est inutilisable.
—P. Valéry

Let us summarize the conclusions of the preceding section. The bare expansion does not have a limit when the ultraviolet cutoff is removed. The renormalized expansion has cured this defect, but the price to pay for that is definitely too heavy. The "useless counterterms" make the study of renormalization quite painful (because one has to use forestry and to work always in reduced subgraphs $g/A_F(g)$); they also generate renormalon behavior which puts in danger the constructive program. Intuitively a factor K^n in large order estimates may be compensated by requiring the coupling constant to be very small, but this is not true for factorials (renormalon effects).

So we search for an expansion "in between" the bare and the renormalized expansion, as shown in Fig. II.4.1, one which is just a reshuffling of both, but with amplitudes which are the usefully renormalized ones. Hence it would have the advantage of ultraviolet finiteness without the renormalon effects. The idea of renormalization is that counterterms are hidden in the bare coupling constants; so

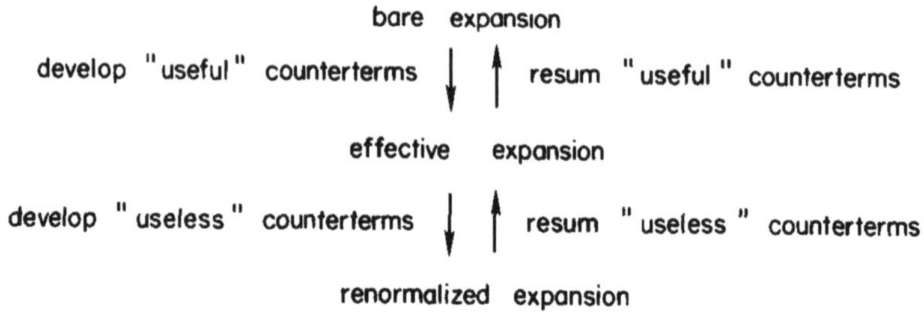

Figure II.4.1

we should get rid of the useless counterterms in this way. But since the useless counterterms depend on an index, or momentum scale, the corresponding coupling constants must also depend on it. Therefore the corresponding expansion cannot be a power series in a single bare or renormalized constant, but must be a series in a whole sequence of index dependent constants called the effective constants. This is particularly clear on equations (II.4.1–2) below: the separation of counterterms into useful and useless pieces being index dependent, there should be a compensating index dependence in the effective constants in order for the left hand side of (II.4.1–2) to be scale independent.

We pause briefly to remark that we have reached the typical renormalization group concept of effective or running constants from the unusual point of view of simply organizing counterterms so as to get the best large order bounds. The historical and more traditional road to these concepts has been to write renormalization group equations which investigate the dependence of the theory upon the subtraction scale, or the somewhat equivalent problem of finite changes of the counterterms [SP][GL]. This standard road paved the way for an enormous amount of work, from the discovery of the particularly interesting Callan–Symanzik equation [Ca1][Sy1], to Wilson's generalization of the renormalization group program [Wil][KW], the investigation of renormalization group fixed points and their stability for various models, and led even by a back-reaction to new proofs of renormalizability [Ca2]. We do not try any review of this vast subject. In particular we do not enter into the definition of normal ordered products, the Zimmermann identities, the renormalization group equations which investigate how the theory changes with a change of cutoff or a change of the subtraction scale. This is because these topics and the Callan–Symanzik equation are explained in great detail in most books devoted to quantum field theory or renormalization theory: a dedicated reference is [Co]. For a rather complete overview of the renormalization group and its application to critical phenomena, we refer to [Am]. The reader may be disappointed not to find this standard material here; but the traditional presentation involves defining the corresponding concepts and equations as bare or renormalized power series, and we are about to argue that the proper framework is neither the bare nor the renormalized expansions but an effective expansion in between. We are also going to define the discrete version of these equations which is naturally adapted to the multiscale representation. In particular we will find natural discrete analogue of the famous renormalization group functions such as the β function, and we prefer to proceed directly toward this goal. Discrete equations are less elegant than differential equations, but again they seem to be required by the constructive point of view to be developed in part III.

However let us warn the reader that the effective expansion derived below is more limited in scope than the general philosophy of the renormalization group, and also stress again that the multiscale slicing of the propagator is a somewhat simpler but less general method than Kadanoff's and Wilson's block-spin methods; as remarked already, it requires indeed a Gaussian measure in the problem. We hope however that the presentation below may have an advantage over the more standard one, at least for the beginner: it applies to the internal lines "inside" the Feynman amplitudes rather than to the external legs. Hence it develops the correct intuition that renormalization group behavior is an essential piece for the *construction* of the theory, not just a device to analyze its behavior in various regimes.

Let us make precise the paths by which the three expansions of Fig. II.4.1 communicate. The bare expansion leads to the effective one by developing selectively some of the bare constants into effective constants plus the useful piece of the counterterms. This is the path used in chapter III, where the bare theory is always the constructive starting point. Conversely the renormalized expansion leads to the effective one by resumming the useless counterterms, or equivalently by absorbing them into effective constants. Hence in a schematic way:

$$\text{bare constant} = \text{effective constants} + \text{useful counterterms}, \qquad \text{(II.4.1)}$$

$$\text{renormalized constant} = \text{effective constants} - \text{useless counterterms}, \qquad \text{(II.4.2)}$$

these equations being consistent with the standard rule:

$$\text{bare constant} = \text{renormalized constant} + \text{full counterterms}. \tag{II.4.3}$$

The two paths are not exactly symmetrical, however, and one can consistently argue that the first one is shorter and simpler than the second one. This is because (forgetting for the moment the subtleties due to bipeds) the definitions (II.1.10–11)–(II.3.4–5) of indices i_g and e_g make the condition of almost locality $i_g > e_g$ rather stringent: *every* internal line of g has to have higher index than *every* external line of g. In other words there is not much renormalization performed in the usefully renormalized expansion and in this way it is closer to the bare one than to the renormalized one. Perhaps this argument may not seem serious at first sight. Nevertheless it is also the reason for which the first path is simpler; we do not expect any sum over forests to appear along it, since these sums neither appear in the formulas for the starting point (the bare amplitudes) nor for the end point (the usefully renormalized amplitudes, see (II.3.33), and also (II.3.49)). Sums over various forests in fact only arise when counterterms are developed for intermediate situations in which internal and external indices are quite mixed.

It is therefore natural to start with a theorem relating the bare and effective expansion. As in the preceding section and for the same reasons we state it first for the biped-free piece of the perturbative expansion, then extend it to the general case.

We fix a cutoff index ρ and the bare coupling g_ρ. For each vertex v of a graph G it is useful to define;

$$e_v(\mu) = \max\{i_l(\mu) \mid l \text{ hooked to } v\}. \tag{II.4.4}$$

Recall that by convention the index of external lines is -1, so $e_v(\mu) = -1$ is possible but only for the unique vertex of the trivial graph with a single vertex, $N = 4$, and no internal lines.

The bare expansion for a connected Schwinger function with cutoff ρ is written in analogy with (I.4.15) as:

$$C^{\rho}_{N,bf} = \sum_{G,\mu|\mu\leq\rho} \frac{(-g_\rho)^{n(G)}}{S(G)} A_{G,\mu} \tag{II.4.5}$$

where the sum is over assignments $\mu \in [0,\rho]^{l(G)}$ (in short $\mu \leq \rho$) and over connected biped free graphs with $N(G) = N$, as indicated by the index bf (*b*iped-*f*ree). (II.4.5) defines $C^{\rho}_{N,bf}$ as a formal power series in g_ρ.

Theorem II.4.1: Existence of the effective expansion
There exist $\rho+1$ formal power series in $g_\rho \equiv g_\rho^\rho$, called $g_{\rho-1}^\rho, g_{\rho-2}^\rho, \dots, g_0^\rho$ and g_{-1}^ρ (the upper index is to remind the reader that the entire theory has ultraviolet cutoff ρ) such that the formal power series (II.4.5) is the same as:

$$C_{N,bf}^\rho = \sum_{G,\mu\leq\rho} \left[\prod_{v\in G}(-g_{e_v(\mu)}^\rho)\right] \frac{1}{S(G)} A_{G,\mu}^{UR} \tag{II.4.6}$$

where we recall (see (II.3.33)):

$$A_{G,\mu}^{UR} \equiv \int \prod_v dx_v \prod_{h\in \mathbf{D}_\mu} (1-\tau_h^*) Z_{G,\mu}. \tag{II.4.7}$$

$Z_{G,\mu}$ is by definition the integrand for the graph G and the assignment μ; for instance if we take the external arguments to be fixed momenta we have:

$$Z_{G,\mu} \equiv \prod_l C^{i_l(\mu)}(x_l, y_l) \prod_{j=1}^N e^{ip_j x_j} \tag{II.4.8}$$

and the effective constants g_i^ρ obey the following inductive definition:

$$g_i^\rho = g_{i+1}^\rho - \sum_{\substack{H \text{ quadruped, } \mu\leq\rho \\ i_H(\emptyset)=i+1}} \frac{1}{S(H)} \prod_{v\in H}(-g_{e_v(\mu)}^\rho) \int \prod_{v\in H} dx_v \cdot \left[\prod_{h\in \mathbf{D}_\mu(H), h\neq H} (1-\tau_h^*)\right] \tau_H Z_{H,\mu}, \tag{II.4.9}$$

where $\tau_H Z_{H,\mu} = Z_{H,\mu}|_{p=0} = \prod_{l\in H} C^{i_l(\mu)}(x_l, y_l)$ is the integrand for H taken at 0 external momenta, and $\mathbf{D}_\mu(H)$ is simply the forest $\mathbf{D}_\mu = \mathbf{H}_\mu(\emptyset)$ of the preceding section, but relative to the graph H. In (II.4.9), the summation over quadrupeds does not include the trivial case of the graph reduced to a single vertex, which corresponds in fact to the first factor g_{i+1}^ρ in the right hand side of (II.4.9).

The minus sign in (II.4.9) is consistent with the minus sign in (II.4.2), because the counterterm is $\int [\prod_{h\in \mathbf{D}_\mu(H), h\neq H}(1-\tau_h^*)](-\tau_H) Z_H$, but the vertex has really a value $-g$, so the reader should think of (II.4.9) as a more convenient form of the equation $-g_i = -g_{i+1} - \sum_H \text{counterterm}(H)$.

(II.4.9) defines each g_i^ρ (by inductive substitution) as a formal power series in g_ρ of the form $g_\rho + \sum_{n\geq 2} \gamma_n^i (g_\rho)^n$. The induction stops at g^{ρ}_{-1} which is the last one for which the sum in (II.4.9) is not empty. Let us apply the result (II.4.6) to $N = 4$ and put to 0 the four external momenta. When G is a non-trivial quadruped, G itself always belongs to $\mathbf{D}_\mu(G)$, and the $(1 - \tau_G)$ operator makes $A^{UR}_{G,\mu}$ vanish at 0 external momenta. For the trivial graph with a single vertex v we remarked that $e_v(\mu) = -1$. Hence the formal power series in g_ρ (II.4.6) for $C^{\rho}_4(0,0,0,0)$ reduces exactly to $-g_{-1}$. This means that in the sense of formal power series in g_ρ we must identify g^{ρ}_{-1} with the renormalized coupling g_r, which by definition of our subtraction scheme is precisely minus the connected four point function at 0 external momenta $C^{\rho}_4(0,0,0,0)$.†

The proof of Theorem II.4.1 is a simple combinatoric exercise; no analysis is involved, since all integrals involved have cutoffs and are therefore obviously absolutely convergent. Again the combinatoric has to be checked at the level of contraction schemes. We go from (II.4.5) to (II.4.6) by pulling out inductively the useful counterterms hidden in g_ρ, one slice after the other. At slice i an intermediate version of Theorem II.4.1 is obtained:

$$C^{\rho}_{N,bf} = \sum_{G,\mu \leq \rho} [\prod_{v \in G} (-g^{\rho}_{\sup(i,e_v(\mu))})] \frac{1}{S(G)} A^{UR,i}_{G,\mu} \qquad \text{(II.4.10)}$$

where

$$A^{UR,i}_{G,\mu} \equiv \int \prod_v dx_v \prod_{h \in \mathbf{D}^i_\mu} (1 - \tau^*_h) Z_{G,\mu} \qquad \text{(II.4.11)}$$

and

$$\mathbf{D}^i_\mu \equiv \{h \in \mathbf{D}_\mu \mid i_h > i\}. \qquad \text{(II.4.12)}$$

(II.4.10) is obviously nothing but (II.4.5) if $i = \rho$. Assuming it at scale $i+1$, we prove it at scale i by simply adding and subtracting the counterterms which change $A^{UR,i+1}_{G,\mu}$ into $A^{UR,i}_{G,\mu}$. These are the counterterms corresponding to the quadrupeds $\{H_1, \ldots, H_k\} = \{H \in \mathbf{D}_\mu \mid i_H = i+1\}$.

---

†The renormalization condition of BPHZ define in fact g_r as the 0 momentum value of $-[\partial\Gamma_2/\partial p^2]^{-2}(0)\cdot\Gamma_4(0)$, hence include a field strength renormalization factor which disappears only if the renormalized parameter a_r is 1. However we do not need to discuss this subtlety for the moment, since in the biped-free theory there is obviously no wave function renormalization.

Hence we add and subtract to each $A^{UR,i+1}_{G,\mu}$ the quantity:

$$\sum_{\substack{S\subseteq\{H_1,\ldots,H_k\}\\ S\neq\emptyset}} \prod_{H_j\in S}(-\tau_{H_j}) \prod_{h\in \mathbf{D}^{i+1}_\mu}(1-\tau^*_h)Z_{G,\mu}. \tag{II.4.13}$$

The piece added changes $\prod_{h\in\mathbf{D}^{i+1}_\mu}(1-\tau^*_h)$ into $\prod_{h\in\mathbf{D}^{i}_\mu}(1-\tau^*_h)$ in each amplitude, hence it changes $A^{UR,i+1}_{G,\mu}$ into $A^{UR,i}_{G,\mu}$ The piece "subtracted" should be developed as a sum over S, so as to get:

$$C^\rho_{N,bf} = \sum_{\substack{(G,\mu,S)\\ \mu\leq\rho, S\subseteq\mathbf{D}^i_\mu-\mathbf{D}^{i+1}_\mu}} [\prod_{v\in G}(-g^\rho_{\sup(i+1,e_v(\mu))})] \frac{1}{S(G)} A^{UR,i}_{G,\mu,S} \tag{II.4.14}$$

with

$$A^{UR,i}_{G,\mu,S} \equiv A^{UR,i}_{G,\mu} \tag{II.4.15}$$

if $S=\emptyset$ and

$$A^{UR,i}_{G,\mu,S} \equiv \int \prod_v dx_v \prod_{H_j\in S}(-\tau_{H_j}) \prod_{h\in\mathbf{D}^{i+1}_\mu}(1-\tau^*_h)Z_{G,\mu} \tag{II.4.16}$$

otherwise.

Remark that this induction is really nothing but the Bogoliubov induction, but with the additional element that the multiscale decomposition provides at each scale a well defined natural family $H_1, \ldots, H_k$ of disjoint subgraphs, which are the ones to which the Bogoliubov induction should be applied, instead of being performed blindly with respect to scales.

We can now define, since the elements of S are disjoint, the collapse φ_i as an operation which is defined on triplets (G,μ,S), $S\subseteq \mathbf{D}^i_\mu-\mathbf{D}^{i+1}_\mu$, and which sends (G,μ,S) to $(G',\mu',\emptyset)$, G' being obtained from G by reducing each $H_j\in S$ to a single vertex, and μ' being the assignment derived from μ by simple restriction to the lines of G'. Remark that every vertex of G' corresponding to such a reduction must have $e_v(\mu)=e_v(\mu')\leq i$. We reorder now (II.4.14) as:

$$C^\rho_{N,bf} = \sum_{(G',\mu')} \left\{ \sum_{\substack{(G,\mu,S),\mu\leq\rho\\ \varphi_i(G,\mu,S)=(G',\mu',\emptyset)}} \left[\prod_{v\in G}(-g^\rho_{\sup(i+1,e_v(\mu))})\right] \frac{1}{S(G)} A^{UR,i}_{G,\mu,S} \right\}. \tag{II.4.17}$$

For each (G',μ') the corresponding sum in (II.4.17) is an infinite power series which in fact replaces exactly, at each vertex v of G' satisfying

$e_v(\mu') \leq i$, the coupling g^ρ_{i+1} by the right hand side of (II.4.9), hence by g^ρ_i; the sum over H in (II.4.9) indeed corresponds exactly to the sum over all possible insertions of an H_j which is collapsed by the φ_i operation to the vertex v, in the above notation. To check that combinatoric factors agree, one has again to perform this analysis at the level of contraction schemes, in which case it becomes the same problem as for the Bogoliubov induction considered above.

This achieves the proof of (II.4.10) at scale i, hence by induction, the proof of Theorem II.4.1.

It remains to show that the effective expansion has the advantages but not the drawbacks of the renormalized expansion, and so we should derive a version in which the ultraviolet limit has been taken. This is not straightforward, since we can no longer use the bare constant. But since $g_r = g^\rho_{-1}$ is a formal power series in g_ρ starting with g_ρ, it is possible to invert it at the level of formal power series. This is also true for each effective constant g^ρ_i which by substitution becomes a formal power series in g_r starting with g_r. (Remark that the power series g^ρ_i obtained in this way still depend on ρ, the global cutoff).

Now the series (II.4.6) when considered as formal power series in g_r through such substitutions is exactly the ordinary (biped-free) fully renormalized power series in g_r with cutoff ρ (i.e., with full renormalization operator $\mathbf{R}$, but propagators C^ρ instead of C).

An ultraviolet limit of the effective expansion may then be obtained in the following sense:

Theorem II.4.2 *Ultraviolet limit of the effective expansion*

The effective constants g^ρ_i have a limit as $\rho \to \infty$ order by order as formal series in g_r. This limit is called g^∞_i, or simply g_i. Furthermore, in the sense of formal power series in g_r, the biped free part of the BPHZ renormalized expansion for C_N satisfies:

$$C_{N,bf} = \sum_{G,\mu} \left[\prod_{v \in G} (-g_{e_v(\mu)}) \right] \frac{1}{S(G)} A^{UR}_{G,\mu}. \qquad \text{(II.4.18)}$$

The subtle point is to show the convergence (order by order in g_r) of g^ρ_i to g_i. This may be done by rewriting $g^{\rho+1}_i - g^\rho_i$ as a sum over renormalized graphs which have at least one line at scale ρ, and then use the vertical exponential decay of such graphs in $\rho - i$. Formula (II.4.18) then follows from the similar statement for the theory with cutoff and the absolute convergence of the renormalized amplitudes with cutoff to the ones without cutoff.

Theorem II.4.2 achieves our goal of an effective expansion which is ultraviolet finite, like the renormalized one, but free of renormalons and of sums over forests.

When bipeds are added, we may choose between several generalizations of Theorems II.4.1–II.4.2. The most natural generalization is to derive an effective expansion with three types of effective parameters, the effective coupling constant, the effective mass and the effective wave function constant. This requires to reexpress that theory as a sum over generalized φ^4 graphs with two point mass and wave function insertions, as in Fig. I.4.2 (recall that the initial mass and wave function parameters are not expressed as coupling constants, but used to build the propagator of the theory).

To implement this idea it is convenient to use the general definitions of Sect. II.3B for internal and external indices. We have now generalized graphs $\hat{G}$ with regular vertices $v \in V(\hat{G})$ with 4 legs, for which definition (II.4.4) is adequate, and two other sets of vertices $W^0(\hat{G})$ and $W^1(\hat{G})$ with 2 legs, corresponding respectively to mass and wave function insertions. For $w \in W^0 \cup W^1$, in view of (II.3.35) one should define

$$e_w(\mu) = \min\{i_l(\mu) \mid l \text{ hooked to } w\}. \tag{II.4.19}$$

Then we have the generalization of Theorem II.4.1:

Theorem II.4.3

There exist $3(\rho+1)$ formal power series in g_ρ, called g_i, δm_i^2 and δa_i, $i = \rho - 1, \ldots, 0, -1$ (they depend on ρ, like those of Theorem II.4.1, but we drop this dependence to avoid too heavy notations), such that the formal power series in g_ρ for C_N^ρ can be rewritten as:

$$C_N^\rho = \sum_{\hat{G},\mu} \left[\prod_{v \in V(\hat{G})} (-g_{e_v(\mu)})\right] \left[\prod_{w \in W^0(\hat{G})} (-\delta m^2_{e_w(\mu)})\right] \cdot \left[\prod_{w \in W^1(\hat{G})} (-\delta a_{e_w(\mu)})\right] \frac{1}{S(\hat{G})} A^{UR}_{\hat{G},\mu}, \tag{II.4.20}$$

where the formula for $A^{UR}_{\hat{G},\mu}$ is:

$$A^{UR}_{\hat{G},\mu} \equiv \int \prod_{v \in V \cup W^0 \cup W^1} dx_v \prod_{h \in \mathbf{D}_\mu} (1 - \tau_h^*) \prod_{w \in W^1(\hat{G})} (-\Delta) Z_{\hat{G},\mu}. \tag{II.4.21}$$

$Z_{\hat{G},\mu}$ is the integrand for $\hat{G}$, and μ, namely at fixed external momenta:

$$Z_{\hat{G},\mu} \equiv \prod_{l} C^{i_l(\mu)}(x_l, y_l) \prod_{j=1}^{N} e^{ip_j x_j}, \tag{II.4.22}$$

and the operator $\Delta \equiv \partial_\nu \partial_\nu$ acts, for each $w \in W^1(\hat{G})$, on one of the two propagators hooked to w, in agreement with formula (II.3.46) for the action of the operator τ^{1*}.

Furthermore the effective constants g_i, δm_i^2, δa_i obey inductive relations generalizing (II.4.9). One starts with g_ρ being the bare coupling, $\delta m_\rho^2 \equiv 0$ and $\delta a_\rho \equiv 0$ and one defines by recursion:

$$g_i = g_{i+1} - \sum_{\substack{\hat{H} \text{ quadruped }, \mu \leq \rho \\ i_{\hat{H}}(\emptyset) = i+1}} \frac{1}{S(\hat{H})} \prod_{v \in V(\hat{H})} (-g_{e_v(\mu)})$$
$$\cdot \left[\prod_{w \in W^0(\hat{H})} (-\delta m^2_{e_w(\mu)}) \right]$$
$$\cdot \left[\prod_{w \in W^1(\hat{H})} (-\delta a_{e_w(\mu)}) \right] \int \prod_{v \in V \cup W^0 \cup W^1(\hat{H})} dx_v$$
$$\cdot \left[\prod_{h \in \mathbf{D}(\hat{H})_\mu, h \neq \hat{H}} (1 - \tau_h^*) \right] \tau_{\hat{H}} Z_{\hat{H},\mu} \tag{II.4.23}$$

where $\tau_{\hat{H}} Z_{\hat{H},\mu} = Z_{\hat{H},\mu}|_{p=0} = \prod_{l \in H} C^{i_l(\mu)}(x_l, y_l)$ is the integrand for $\hat{H}$ taken at 0 external momenta;

$$\delta m_i^2 = \delta m_{i+1}^2 + \sum_{\substack{\hat{B} \text{ biped}, \mu \leq \rho \\ i_{\hat{B}}(\emptyset) = i+1}} \frac{1}{S(\hat{B})} \prod_{v \in V(\hat{B})} (-g_{e_v(\mu)})$$
$$\cdot \left[\prod_{w \in W^0(\hat{B})} (-\delta m^2_{e_w(\mu)}) \right] \left[\prod_{w \in W^1(\hat{B})} (-\delta a_{e_w(\mu)}) \right]$$
$$\cdot \int \prod_{v \in V \cup W^0 \cup W^1(\hat{B})} dx_v \left[\prod_{h \in \mathbf{D}(\hat{B})_\mu, h \neq \hat{B}} (1 - \tau_h^*) \right] Z^0_{\hat{B},\mu}; \tag{II.4.24}$$

$$\delta a_i = \delta a_{i+1} + \sum_{\substack{\hat{B} \text{ biped}, \mu \leq \rho \\ i_{\hat{B}}(\emptyset) = i+1}} \frac{1}{S(\hat{B})} \prod_{v \in V(\hat{B})} (-g_{e_v(\mu)})$$
$$\cdot \left[\prod_{w \in W^0(\hat{B})} (-\delta m^2_{e_w(\mu)}) \right] \left[\prod_{w \in W^1(\hat{B})} (-\delta a_{e_w(\mu)}) \right]$$

$$\cdot \int \prod_{v \in V \cup W^0 \cup W^1(\hat{B})} dx_v \left[\prod_{h \in \mathbf{D}(\hat{B})_\mu, h \neq \hat{B}} (1 - \tau_h^*) \right] Z^1_{\hat{B},\mu}, \quad \text{(II.4.25)}$$

where $Z^0_{\hat{B},\mu} = \prod_{l \in \hat{B}} C^{i_l(\mu)}(x_l, y_l)$ and $Z^1_{\hat{B},\mu} = \frac{1}{8} |x - y|^2 \prod_{l \in \hat{B}} C^{i_l(\mu)}(x_l, y_l)$, x and y being the border vertices of $\hat{B}$. These last definitions are again consistent with the action of the Taylor operator for a biped as shown in (II.3.44–46).

This theorem is checked by induction exactly as the previous one. A similar theorem for the 1PI or vertex functions Γ_N is left to the reader to formulate, the sums being of course restricted everywhere in this case to one particle irreducible graphs.

Inverting the series g_i we can reexpress them as series (still depending on ρ) in $g_{-1} = g_r$ and substituting them everywhere we obtain the BPHZ renormalized expansion for C_N^ρ. In particular we can obtain renormalized masses and wave function constants $m_r^2 = m^2 + \delta m_{-1}^2$ and $a_r = 1 + \delta a_{-1}$, if m^2 and 1 were the bare values used to build the propagator. These renormalized parameters coincide with the ones of the BPHZ prescription at 0 external momenta, so that $m_r^2 = -\Gamma_2(0)$ and $a_r = -[d\Gamma_2/dp^2](0)$.

For these series we can pass to the ultraviolet limit $\rho \to \infty$ and the generalization of Theorem II.4.2 also holds:

Theorem II.4.4

The effective expansion reexpressed in terms of g_r, m_r and a_r in the limit $\rho \to \infty$ is the same order by order in g_r as the usual BPHZ renormalized series.

It is in principle possible to work out a direct proof of Theorem II.4.2 and II.4.4, i.e., to reshuffle directly the renormalized series into effective ones without ever using an ultraviolet cutoff and the bare expansion. The effective constants come from the resummation of the useless pieces of the counterterms of the renormalized expansion. This approach has been partly implemented in [Ri1] (see also [DFR]), in which such explicit resummations have been introduced for all counterterms of the bubble type (the so called parquet forests). A general resummation rule in the sense of formal power series obviously exists by Theorem II.4.4, but to write it in the form of a simple set of explicit resummation rules seems difficult since forests are now an essential ingredient. This illustrates the fact that path 2 in Fig. II.4.1 is less straightforward than path 1.

Expansion (II.4.20) is not as canonical as (II.4.6). In particular it is rather natural to work out similar versions in which chains of two point insertions in the generalized graphs of (II.4.20) have been resummed. This is possible essentially because two point insertions lead to geometric, explicitly summable series. This leads to an effective expansion with effective coupling constants and effective *propagators*; the advantage is that only ordinary graphs are required, the drawback is that the formula for the effective propagators is somewhat complicated. We refer to [FMRS4] for an example of such a construction.

Finally one may take advantage of the fact that mass renormalization does neither create forest nor renormalon problems, as was shown in Theorem II.3.2; hence one may derive a perturbative expansion which has full mass renormalization and effective coupling and wave function constants; the amplitudes for this expansion are the $A_G^{MR,UR}$ of (II.3.49).

Furthermore in asymptotically free theories which will be discussed soon, it is even useful to consider an expansion with effective coupling constants, bare wave function constant and renormalized mass, since we will see that asymptotic freedom makes in fact the wave function renormalization finite. The corresponding amplitudes are noted $A_G^{MR,CCUR}$ for "mass renormalized, coupling constant usefully renormalized" (and wave function not renormalized...). Such expansions, although of mixed character, are optimal for instance for the construction of the planar $-g\varphi_4^4$ theory which is the goal of the next chapter. They can also be used in the constructive context [FMRS4][FMRS5].

To summarize this section, the multiscale decomposition provides a simple, well defined prescription for writing down discrete flows for the relevant and marginal couplings of a renormalizable theory like φ_4^4; the bare or renormalized parameters simply provide particular boundary conditions for these flows. This formalism is completely rigorous at the level of formal power series, and it leads to effective expansions which are formally equivalent to the bare or renormalized ones but are both ultraviolet finite and free of forest complications and of renormalon effects.

Chapter II.5

Construction of "Wrong Sign" Planar Φ_4^4

In this section we show how to apply the formalism of the effective expansion beyond the sterile level of formal power series, on a simple model which is not a full-fledged field theory but has nevertheless some physical interest.

Even in the effective version without renormalon effects, it is not easy to sum up perturbation theory because of the large number of graphs involved at large order. This divergence of perturbation theory occurs even in 0 dimension, for a single integral $\int e^{-x^2-gx^4}dx$. In the next section we will analyze it in some detail and conclude that we must trade absolute summation for Borel summation at best. However there is a second problem. It is usual in the construction of a theory in the weak coupling regime to take care of combinatoric factors like the constants in the uniform theorems of Sect. II.1–3 by requiring the coupling constant to be small enough. But in the effective expansion of Sect. II.4 there is an infinite set of such coupling constants, and we need therefore them to be all simultaneously small enough. If perturbation theory is asymptotic, the sign of the first term in the recursion relation (II.4.9) or (II.4.23) for $g_i - g_{i+1}$ becomes very important. If this sign is negative, g_{i+1} will be larger than g_i (for small enough g_i) and holding the renormalized constant $g_r = g_{-1}$ fixed to a small value it is very doubtful that all other effective constants g_i can be made simultaneously small.

On the other hand if it is positive, g_{i+1} is smaller than g_i and the goal seems attainable. In this second case the theory is called asymptotically free (or more precisely ultraviolet asymptotically free); indeed in the cases where we can in fact make sense out of the recursion relation (II.4.9) beyond formal power series, g_i tends to 0 as $i \to \infty$, with g_r kept fixed and small.

The first term in (II.4.9) or (II.4.23) corresponds to H (or $\hat{H}$) being the bubble graph. In the $\exp(-g\int \varphi^4)$ theory with the ordinary sign of the coupling constant ($g > 0$), $\tau_H Z_H$ is positive; because of the minus sign in the exponential there is however a minus sign in (II.4.9)–(II.4.23) and the theory is not ultraviolet asymptotically free. This fact of life is mathematically (and perhaps even physically) frustrating because it deprives us of the simplest and most natural model to construct. However by reversing the direction in the index space, a theory has to be either asymptotically free in the ultraviolet or in the infrared direction; the ordinary φ^4 theory is therefore asymptotically free in the infrared regime, which opens up the possibility, exploited in part III, to construct the critical or massless limit of the model with fixed ultraviolet cutoff. Another idea is to change the sign of the coupling constant: we call the corresponding theory, with action $\exp(-g\int \varphi^4)$ and $g < 0$ the "wrong sign" φ^4 (or the "negative coupling" φ^4 theory, but this may lead to some confusion, because $-g$ is positive for $g < 0\ldots$). This model is ultraviolet asymptotically free, but a negative φ^4 potential means that $\varphi = 0$ is an unstable minimum, and in spite of many efforts there is therefore no construction of the model (apart from analytic continuations which are expected not to meet the axiomatic requirements [GK5]).

But there is a restricted version of the wrong sign φ^4, the planar wrong sign φ^4, which offers a nice benchmark to study asymptotic freedom and make sense of the flows and recursion relations of the previous section beyond formal power series. The planar theory was proposed initially by 't Hooft ['tH2] as an approximation for the study of SU(N) gauge theories at large N; it has also been studied in connection with random surface models or string theory. The name planar comes from the fact that the perturbation theory of these models is roughly speaking restricted to graphs which can be drawn on a plane without self-intersections. Constructions of the wrong sign planar φ^4 were performed in ['tH5–7][Ri1][GaNi]; here we try a compromise retaining some of the best aspects of the initial constructions. Because it is not adapted to the constructive versions of part III we do not however retain the nice continuous slicing provided by α parameters and used in [Ri1][Hu]; we invite the reader in search of a "truly optimal" version to write down the necessary modifications as an exercise.

The φ^4 planar model which we consider is the $N \to \infty$ limit of an N by N matrix valued φ^4 model. Hence the field φ is a bosonic field with components $\varphi_{p,q}$, $1 \leq p, q \leq N$. The formal functional measure for this model is similar to (I.3.1) (for a real valued φ):

$$d\nu = \frac{1}{Z} e^{+(g/4!N)\int \mathrm{Tr}(\varphi^t\varphi)^2 - (m^2/2)\int \mathrm{Tr}\,\varphi^t\varphi - (a/2)\int \mathrm{Tr}\,\partial_\mu\varphi^t\partial^\mu\varphi} D\varphi \qquad \text{(II.5.1)}$$

where $D\varphi$ is a product of independent formal Lebesgue measures for each component of φ. This model has global $O(N)$ invariance. Since the measure depends on φ only through the combination $\varphi^t\varphi$, its non-trivial piece is supported by symmetric matrices. One can also consider the complex version with global U(N) invariance, called the hermitian matrix model. The "wrong sign" case in (II.5.1) corresponds to $g > 0$, which avoids minus signs in many of the formulas below.

The Feynman rules of this model are discussed for instance in [GrK]. Propagators carry a matrix index, hence it is convenient to represent them as double lines, one for each matrix index. They are shown together with vertices and some graphs in Fig. II.5.1. Remark the cyclic symmetry of the vertex. To each closed loop there corresponds a sum over possible values of the index flowing through this loop, which gives a corresponding factor N. There is a factor N^{-1} per vertex, hence the overall factor for a graph with k external lines is

$$N^{2-k-2h} \qquad \text{(II.5.2)}$$

where h is the number of handles of the surface on which the graph is a triangulation. This number may be visualized by considering each propagator (double line) as a thin ribbon. All the ribbons of the external legs are tied to a single point at infinity (without tangling them). Then each closed index loop is filled with a flat piece of surface matching with each ribbon on its boundary. In this way a compact surface is generated, and h is simply its genus (its number of holes or of "handles"). The formula (II.5.2) is then a standard homological formula, which can be checked on the examples of Fig. II.5.1.

At fixed k in the limit $N \to \infty$ the leading term surviving in (II.5.2) corresponds to the sum of all graphs with no handles ($h = 0$), and is called the planar theory. Apart from the constant overall factor N^{2-k} in front, the amplitudes are then exactly similar to the ones of the ordinary φ^4 theory, the only difference lying in the planar restriction.

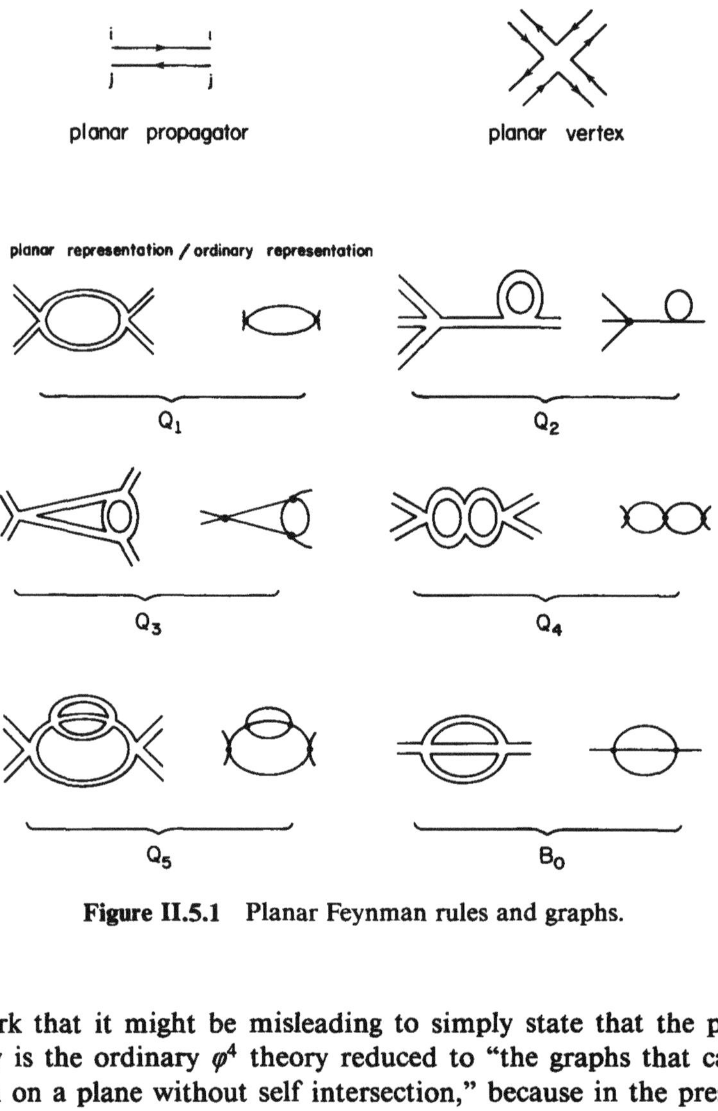

Figure II.5.1 Planar Feynman rules and graphs.

Remark that it might be misleading to simply state that the planar theory is the ordinary φ^4 theory reduced to "the graphs that can be drawn on a plane without self intersection," because in the prescription for computing the value of h above we cannot avoid to consider the double lines. It is true that for any planar graph the collapse of the double propagators to single lines gives an ordinary φ^4 graph that can be drawn on a plane without self crossings. But the combinatoric factors associated to such a graph are usually different in the planar and the ordinary theory; this is already true for the "bubble graph" (and leads for instance to different values of the first term in the β function for the ordinary and the planar theory).

From (II.5.2) only the two point function ($k = 2$) has a non-zero limit, so that the planar theory strictly speaking is a free field theory with a complicated propagator. In fact we will give a non-trivial meaning to the sum over planar graphs for *any* Schwinger function S_k, discarding the constant overall factor N^{2-k} in front. This point of view allows to introduce the renormalized coupling constant as usually as the value of the connected four point function at 0 external momenta, etc.... Nevertheless one should keep in mind that these planar series without the overall factor N^{2-k} are no longer exactly the $N \to \infty$ limit of (II.5.1), and in particular one should not believe that they correspond to a full fledged interacting field theory satisfying the Osterwalder–Schrader axioms.

There is an even simpler model with similar features: it is the $N \to \infty$ limit of the N-vector φ^4 theory, which we pause to discuss briefly. In this model the action is similar to (II.5.1) but φ is an N component vector, so that $\varphi^t\varphi$ is a scalar product, and there is no need for traces in (II.5.1). The Feynman rules and typical graphs for this model are shown in Fig. II.5.2.

The leading terms as $N \to \infty$ for vacuum graphs, two, and four point functions behave as N, 1, and N^{-1}, respectively (some of them are pictured in Fig. II.5.2). Again this limit should be considered a free field. By Wick ordering one may eliminate the tadpoles (lines with both ends at the same vertex). Then the leading graphs for instance for the four point function are simply the bubble chains of Fig. II.5.3.

They form a geometric series which can be summed explicitly and replaced by a wavy line. This point of view allows to reorganize the graphs of the expansion as in Fig. II.5.4.

Again for "wrong sign" g the model is asymptotically free. Nevertheless the "bubble chain" model is too simple to keep some interesting

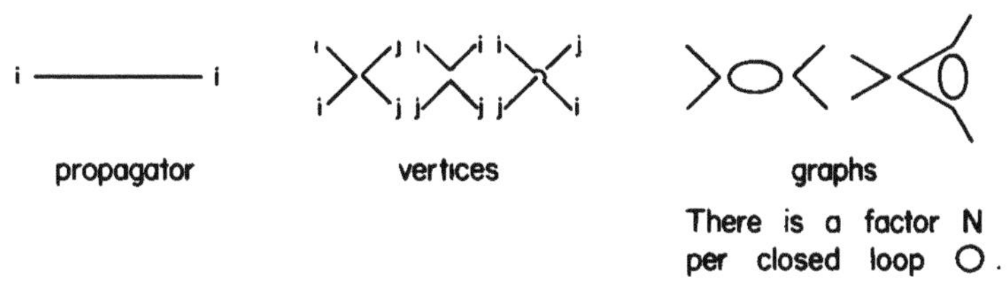

Figure II.5.2 The vector Feynman rules.

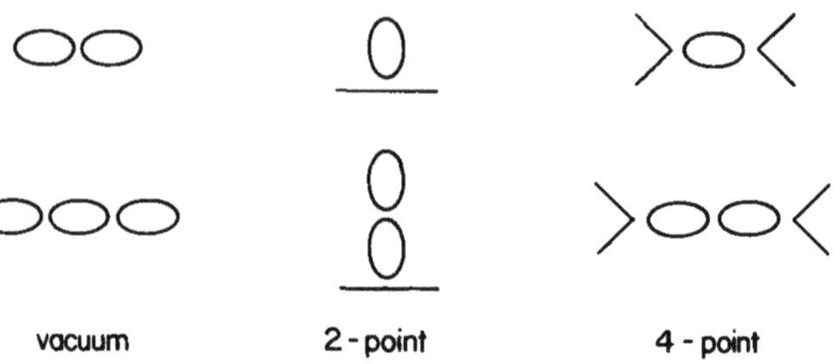

Figure II.5.3 Some leading order graphs.

features of asymptotic freedom like the loglog behavior analyzed below (unless next to leading order in N^{-1} is included). This is why in this section we decided to analyze the less trivial planar series.†

Although not explicitly soluble like the "bubble chain" model, the planar model remains entirely tractable with perturbative methods because of the following main simplification:

Theorem II.5.1

There exists some constant K such that the number of planar Feynman graphs for the φ^4 theory, counted with their proper multiplicity, is bounded at order n by K^n (times a function of the number of their external legs).

Theorem II.5.1 means that the planar restriction trims most of the graphs of ordinary φ^4 theory and that for instance the bare series with a fixed ultraviolet cutoff has a finite radius of convergence.

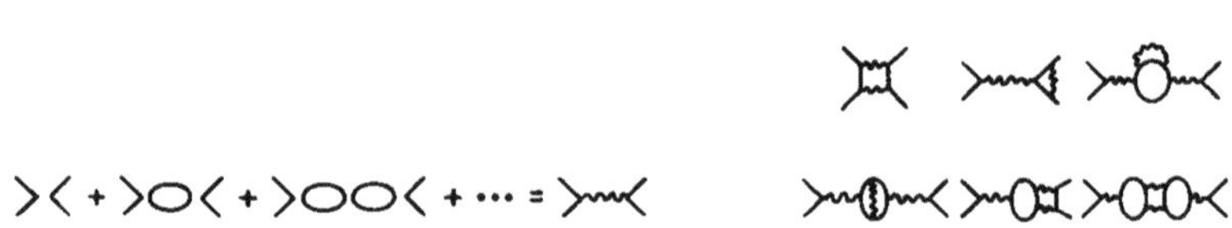

Figure II.5.4 "The bubble chain" and $1/N$ expansion rules.

†As an interesting problem, which is open to the author's knowledge, we suggest to work out these models and their $N \to \infty$ limit for tensor fields φ with more than 1 or 2 indices (there may be no canonical choice for the vertex), and to search for an analogue of Theorem II.5.1.

Counting the number of planar graphs at order n is similar to studying the large N limit of a single site or "zero-dimensional" matrix model similar to (II.5.1) (without any kinetic term). This problem can be fully solved. One of the earliest proofs of Theorem II.5.1 is in [KNN]. Another proof together with an asymptotic analysis of the number of planar diagrams at large order is in [BIPZ]; it uses a saddle point analysis. However the best method to prove the theorem is presumably the method of orthogonal polynomials introduced in [Be]. For a model of hermitian matrices similar to (II.5.1) one can express the measure (II.5.1) solely in terms of the eigenvalues of φ; the integration over the unitary matrix diagonalizing φ factorizes explicitly (this fact would remain true for any lattice matrix model without loops [Me]). The Jacobian of the corresponding change of variables is non-trivial however and adds to the ordinary measure the square of a Vandermonde determinant so that we have to analyze the measure:

$$Z(g,N) \equiv \int \prod_{i=1}^{N} d\mu(a_i) \prod_{i<j} (a_i - a_j)^2 \qquad \text{(II.5.3)}$$

where $d\mu(a) = e^{-(1/2)a^2-(g/N)a^4}da$. For instance with these notations the number of connected vacuum planar graphs at order n is exactly $(-1)^n$ times the coefficient at order n of

$$E(g) - E(0) \equiv \lim_{N\to\infty} \frac{1}{N^2} \log \frac{Z(g,N)}{Z(0,N)}. \qquad \text{(II.5.4)}$$

According to [Me] it is convenient in order to analyze the integral (II.5.3) to introduce the set of orthogonal polynomials P_n for the measure $d\mu$ normalized so that $P_n(a)$ starts with a^n. Using integration by parts a recursion relation for the corresponding normalizations $h_n = \int d\mu P_n^2$ can be found. Since

$$Z(g,N) = N! \prod_{i=0}^{N-1} h_i, \qquad \text{(II.5.5)}$$

this recursion relation can be used to investigate (II.5.4). The limit $N \to \infty$ corresponds under suitable rescalings to the discrete index i between 0 and N becoming a continuous variable between 0 and 1, and the recursion relation becomes an algebraic equation which after some analysis leads to the exact computation:

$$E(g) - E(0) = \frac{1}{24}(A-1)(9-A) - \frac{1}{2}\log A, \qquad \text{(II.5.6)}$$

$$A = \frac{1}{24g}\left(\sqrt{1+48g}-1\right), \tag{II.5.7}$$

where the square root is the one which makes A regular near $g = 0$. From these formulas Theorem II.5.1 follows easily. We can remark also that the radius of convergence of the 0-dimensional φ^4 planar series is exactly $\frac{1}{48}$.

The result (II.5.6–7) can be found already in [BIPZ] but the method of orthogonal polynomials is more powerful, allowing one to compute as well subleading corrections corresponding to graphs that can be drawn on Riemann surfaces with a fixed number of holes. For these series the radius of convergence remains the same as for the planar series (this can be understood intuitively by remarking that for a fixed number of holes and a graph with many vertices "most" of the graph has to be planar). It was recently discovered that there are double-scaling limits both in $N \to \infty$ and $g \to g_c = \frac{1}{48}$ in which the series corresponding to these Riemann surfaces do survive and add up to the planar series in a non-trivial way [BK][DS][GM]. The corresponding topological sum is again analyzed with the help of the orthogonal polynomial recursion relation, but the double scaling limit corresponds then to a non-trivial differential equation rather than an algebraic equation. The solutions of these equations can be interpreted as non-trivial solutions of two dimensional quantum gravity; if analogue phenomena were to exist in higher dimensional matrix models, it might be possible to discover completely new field theories associated to them.

Here we do not need in fact to know the precise value of K in Theorem II.5.1. We will construct only the wrong sign planar φ_4^4 series. Let us simply mention that it seems clear to us that the method could be extended to the construction of the corresponding series on surfaces with a fixed genus; of course it would be fascinating if the corresponding sums would in some double scaling limit also combine in a non-trivial way!

Following the remarks at the end of last section, we use as a starting point the perturbation theory with mass renormalization fully performed and with bare coupling constant g_ρ and bare wave function constant a_ρ. The Schwinger function $C^\rho_{N,\text{planar}}$ are therefore expressed as:

$$C^\rho_{N,\text{planar}} = \sum_{G,\mu\leq\rho} \frac{(g_\rho)^n}{S_{\text{planar}}(G)} A^{MR}_{G,\mu}, \tag{II.5.8}$$

the sum being performed over planar graphs G, and $S_{\text{planar}}(G)$ being their combinatoric weight in the planar theory. From now on we forget the subscript *planar* in the rest of this section. The mass-renormalized amplitudes A^{MR} are defined by:

$$A^{MR}_{G,\mu} \equiv \int \prod_v dx_v \prod_{b \in \mathbf{B}(G)} (1 - \tau_b^{0*}) Z_{G,\mu}. \tag{II.5.9}$$

We pass to an effective expansion only for the coupling constant. Hence we derive an analogue of Theorem II.4.1, rather than of the more complicated Theorem II.4.3. (This will be justified only a posteriori when asymptotic freedom will make finite the apparent divergences due to the lack of wave function renormalization). Therefore we really do not need all the apparatus of Sect. II.3B to treat bipeds. We can stick to the simpler definitions (II.3.4–5) of external and internal indices. We also do not need the lemmas II.3.1 and II.3.2 for the classification of forests. We simply define $\mathbf{D}_\mu$ as the forest of all quadrupeds $q \subseteq G$ which satisfy the almost locality condition $i_q(\mu) > e_q(\mu)$. Then we derive the following analogue of Theorem II.4.1:

Theorem II.5.2

There exist $\rho + 1$ formal power series g_i^ρ, $i = \rho - 1, \ldots, -1$ in g_ρ, such that (II.5.3) is the same power series in g_ρ as

$$C_N^\rho = \sum_{G,\mu \leq \rho} \left[\prod_{v \in G} g^\rho_{e_v(\mu)} \right] \frac{1}{S(G)} A^{MR;CCUR}_{G,\mu}, \tag{II.5.10}$$

where

$$A^{MR;CCUR}_{G,\mu} \equiv \int \prod_v dx_v \prod_{b \in \mathbf{B}(G)} (1 - \tau_b^{0*}) \prod_{q \in \mathbf{D}_\mu} (1 - \tau_q^*) Z_{G,\mu} \tag{II.5.11}$$

and

$$g_i^\rho = g_{i+1}^\rho + \sum_{\substack{H \text{ quadruped}, \mu \leq \rho \\ i_H(\emptyset) = i+1}} \frac{1}{S(H)} \left[\prod_{v \in H} g^\rho_{e_v(\mu)} \right]$$
$$\cdot \int \prod_v dx_v \prod_{b \in \mathbf{B}(H)} (1 - \tau_b^{0*}) \prod_{q \in \mathbf{D}_\mu(H), q \neq H} (1 - \tau_q^*) \tau_H Z_{H,\mu}. \tag{II.5.12}$$

To establish this formula one follows the same path as for Theorem II.4.1, except that the combinatoric of planar Wick contractions is different from the ordinary one, so that the combinatoric aspect of the Bogoliubov induction must be checked again. The key point to notice is that inserting a 4 point subgraph at a particular vertex, one must now preserve the cyclic ordering in the plane of the 4 double lines or "ribbons" of the planar vertex [Ri1]. Apart from that, the proof is just as before.

To go beyond formal power series, we want to choose first g_ρ so that the power series g_i^ρ and (II.5.10) are convergent. They will then define the theory. Using Theorem II.5.1 and the bounds of Sect. II.3, this is relatively easy, provided g_ρ is very small as $\rho \to \infty$. But if g_ρ is too small as $\rho \to \infty$ we end up on an uninteresting theory with $g_r = 0$. Hence the real challenge is to find a clever ρ dependent ansatz for g_ρ so that as $\rho \to \infty$, the renormalized coupling $g_{-1}^\rho \equiv g_r^\rho$ tends to some given small fixed g_r; then the limit constructed in this way is not trivial, but corresponds to this prescribed renormalized coupling.

We start with a heuristic search for the right ansatz, and then prove that it actually works.

The smallest possible graph in the recursion relation (II.5.12) is our friend the bubble; the only other connected four point graph with two vertices is Q_2 in Fig. II.5.1, which is 0 after mass renormalization is performed. To second order (II.5.12) reduces therefore to:

$$g_i^\rho \simeq g_{i+1}^\rho + b \sum_{\substack{i_1,i_2=i+1,\ldots,\rho \\ i_{12}\equiv\inf\{i_1,i_2\}=i+1}} [g_{j_{12}}^\rho]^2 \int d^4y C^{i_1}(x,y)C^{i_2}(x,y). \quad \text{(II.5.13)}$$

By translation invariance the right hand side is independent of x; b is by definition the planar combinatoric coefficient of the bubble, and we define $j_{12} = \sup\{i_1, i_2\}$. For fixed i_{12} after integration over y the sum over $j_{12} \geq i_{12}$ is exponentially decreasing as shown in Sect. II.2; furthermore the mass term in the propagators is also small if $i \gg 1$. More precisely it is a simple exercise to check:

Lemma II.5.1

There exists a numerical constant β_2 such that:

$$\beta_2 \log M = b \lim_{i\to\infty} \lim_{\rho\to\infty} \sum_{\substack{i_{12}=i+1 \\ j_{12}\leq\rho}} \int d^4y C^{i_1}(x,y)C^{i_2}(x,y). \quad \text{(II.5.14)}$$

Furthermore both limits are "exponential," in the sense that for some small enough δ:

$$\left| \beta_2 \log M - b \sum_{\substack{i_{12}=i+1 \\ j_{12}\leq\rho}} \int d^4y C^{i_1}(x,y)C^{i_2}(x,y) \right| \leq e^{-\delta(\rho-i)} + e^{-\delta i}. \quad \text{(II.5.15)}$$

β_2 may be identified with the second order coefficient of the Callan–Symanzik β function [Ca1][Sy1]; in the planar theory it has the particular value $16/(16\pi^2)$; this value, smaller than the standard value $72/(16\pi^2)$ of β_2 for one component φ_4^4 reflects the fact that there are few contraction schemes which respect planarity and cyclic ordering of the planar vertices.

The sum over j_{12} being exponentially decreasing it should also be no problem to replace (at least in the second order approximation we are using) $g_{j_{12}}^2$ by g_{i+1}^2 or even $g_i g_{i+1}$. With all these changes the recursion relation (II.5.13) takes the simpler form (we forget superscripts ρ since $\rho \to \infty$ has been taken in (II.5.14)):

$$\frac{1}{g_{i+1}} - \frac{1}{g_i} \simeq \beta_2 \log M, \qquad \text{(II.5.16)}$$

an equation whose exact solution is, in terms of $g_r = g_{-1}$:

$$g_i = \frac{g_r}{1 + (i+1) g_r \beta_2 \log M} \cdot \qquad \text{(II.5.17)}$$

Let us assume that the approximate "asymptotically free" behavior $g_i \sim \text{const.} i^{-1}$ deduced from (II.5.17) is correct; it has important consequences.

The first consequence is that only third order terms in the recursion (II.5.12) should be relevant in determining the exact form of the ansatz for g_ρ. This is because relation (II.5.16) when generalized to higher orders becomes:

$$\begin{aligned} \frac{1}{g_{i+1}} - \frac{1}{g_i} &= \beta_2 \log M + \text{const.} g_{i+1} + O(g_{i+1}^2) \\ &\simeq \beta_2 \log M + \frac{\text{const.}}{i} + O\left(\frac{\log i}{i^2}\right). \end{aligned} \qquad \text{(II.5.18)}$$

Since $\frac{1}{i}$ is still a divergent series but $(\log i)/i^2$ is not, the correct asymptotic behavior of g_i^{-1}, starting from a fixed value of $g_{-1} = g_r$ should be $(\beta_2 \log M) i + \text{const.} \log i + \text{const.}$ We call β_3/β_2 the constant in front of $\log i$ (indeed β_3 turns out to be the third coefficient in the Callan–Symanzik beta function).

The second consequence is that if we perform the effective analysis for the wave function constant, we should derive a recursion relation of type (II.4.25). The leading term would correspond to the second order biped B_0 in Fig. II.5.1 (since mass renormalization kills the first order tadpole), hence to a contribution in g_{i+1}^2. Therefore we can expect:

$$\delta a_{i+1} - \delta a_i \sim \frac{\text{const.}}{i^2} + O\left(\frac{\log i}{i^3}\right) \qquad \text{(II.5.19)}$$

and $\sum_0^\infty \delta a_i$ should be finite (and even small for small g_r). This means that a bare ansatz in which a_ρ is constant in ρ (e.g., close to 1) is acceptable since it leads to a finite renormalized wave function constant, also close to 1 for small g_r. We summarize this phenomenon by saying that in an asymptotically free theory of this kind the flow of the wave function constant is bounded (and small for small coupling). This is neither true for the flow of the mass nor for the flow of the coupling constant. Another important aspect of this phenomenon is that the apparent logarithmic divergences associated to dangerous bipeds in $\mathbf{D}_\mu$ are spurious. Although the second Taylor subtraction corresponding to wave function renormalization has not been performed for these bipeds, their contribution is nevertheless finite since there are at least two vertices for each them, and $\sum_0^\infty i^{-2}$ converges. But one should be aware that there is some price to pay for that: the index space convergence of the amplitudes in the effective expansion (II.5.10) now requires to use the decay of the vertex factors $\prod_v g_{e_v(\mu)}$, and is no longer exponential in index space, but only power-like when bipeds are present. This requires sometimes additional care, as will happen in the computation of the coefficient β_3 below.

The conclusion of this heuristic analysis of asymptotic freedom is to justify, as announced, our use of a formula with no wave function renormalization, and to tell us that in order to land on a small finite renormalized coupling constant, we should try as an ansatz for a_ρ a constant a close to 1 and for g_ρ the formula:

$$g_\rho = f(\rho, C)^{-1}; \qquad f(\rho, C) \equiv (\beta_2 \log M)\rho + \frac{\beta_3}{\beta_2} \log \rho + C, \tag{II.5.20}$$

where β_2 is defined by (II.5.14), β_3 is another computable constant (defined by (II.5.26-27-29) below) which results from the careful study of the subleading terms in the recursion relation (II.5.12), and C is a (large) constant, whose value is related to the exact value of g_r that one wants to obtain. Hence the constants a and C play the role of two bare parameters which parametrize the two parameter family of theories one is looking for (remember that from the beginning the third parameter, the mass, is the renormalized one).

To formulate the construction of the planar theory as a precise theorem it is convenient to use complex values of a and of C in (II.5.20) and to introduce the half-plane $H_K = \{C \mid \operatorname{Re} C > K\}$ and the disks $D_\delta^g = \{g \mid \operatorname{Re} \frac{1}{g} > \frac{2}{\delta}\}$ and $D_\delta^1 = \{a; |a-1| < \delta\}$ of radius δ and centers

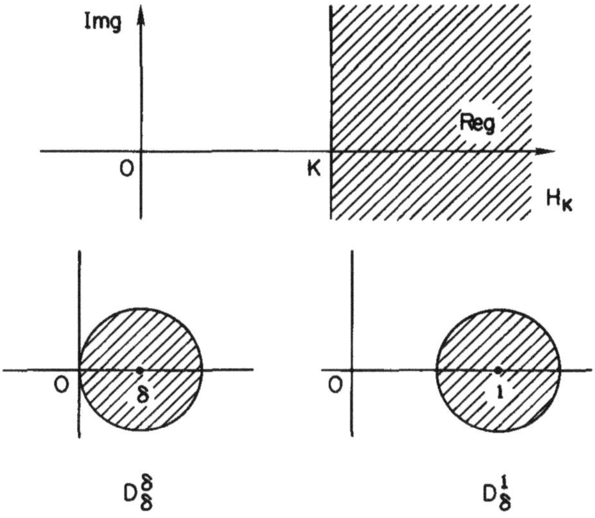

Figure II.5.5

respectively at δ and 1 (see Fig. II.5.5). (Of course $D_\delta^\delta = C_{R/2}$ in the notations of section I.5).

The main result is then summarized by:

Theorem II.5.3

Let g_ρ be given by (II.5.20), with $C \in H_K$ and K large enough, and $a_\rho = a$, $a \in D_\delta^1$, with δ small enough. Then the recursive relation (II.5.12) and the effective expansion (II.5.10) are absolutely convergent, uniformly in $H_K \times D_\delta^1$. The corresponding sums g_i^ρ and C_N^ρ are analytic in $H_K \times D_\delta^1$ and converge uniformly as $\rho \to \infty$ to functions g_i and C_N, therefore also analytic in $H_K \times D_\delta^1$, which define the theory with no cutoff.

In particular there is a doubly analytic map from (C, a) to the renormalized coupling and the renormalized wave function constant (g_r, a_r); this map is from $H_K \times D_\delta^1$ to $D_{\delta'}^{\delta'} \times D_{\delta'}^1$ for some δ'. It can be inverted to a map $(g_r, a_r) \to (C, a)$ from a smaller double disk $D_{\delta''}^{\delta''} \times D_{\delta''}^1$ to $H_K \times D_\delta^1$ for some $\delta'' < \delta'$. In this way the theory can be parametrized by the renormalized parameters g_r, m_r and a_r; it is in particular analytic in g_r for g_r in $D_{\delta''}^{\delta''}$ and the functions $C_N(g_r, m_r, a_r)$ are the Borel sums of the ordinary fully renormalized planar series in g_r with propagator $(a_r p^2 + m_r^2)^{-1}$ (see section I.5 for the definition of Borel summability).

We prove this theorem by an induction in index space from $i=\rho$ to $i=-1$. Let us study first in more detail (II.5.12). The only third order graphs for the four point function to be considered are Q_3 and Q_4 in Fig. II.5.1. Hence to third order (II.5.12) leads to:

$$\begin{aligned} g_i^\rho - g_{i+1}^\rho = b &\sum_{i+1=i_{12}\leq j_{12}\leq\rho} (g_{j_{12}}^\rho)^2 \int C_{xy}^{i_1} C_{xy}^{i_2} d^4y \\ &+ c \sum_{i+1=i_{1234}\leq j_{1234}\leq\rho} g_{j_{12}}^\rho g_{j_{134}}^\rho g_{j_{234}}^\rho \int d^4y d^4z \mathbf{R}_{\text{int}}^U C_{xy}^{i_1} C_{xz}^{i_2} C_{yz}^{i_3} C_{yz}^{i_4} \\ &+ d \sum_{i+1=i_{1234}\leq j_{1234}\leq\rho} g_{j_{12}}^\rho g_{j_{1234}}^\rho g_{j_{34}}^\rho \int d^4y d^4z \mathbf{R}_{\text{int}}^U C_{xy}^{i_1} C_{xy}^{i_2} C_{yz}^{i_3} C_{yz}^{i_4} \\ &+ \text{graphs with at least 4 vertices,} \end{aligned} \tag{II.5.21}$$

where we use shortened notations $C_{xy}^i = C^i(x,y)$, $i_{abc...} = \inf\{i_a, i_b, i_c, \ldots\}$ and $j_{abc...} = \sup\{i_a, i_b, i_c, \ldots\}$. b, c and d are respectively the planar combinatoric coefficients of the graphs Q_1 (the bubble) Q_3 and Q_4 of Fig. II.5.1, and the "useful" internal renormalization operator $\mathbf{R}_{\text{int}}^U$ is by definition $\prod_{h\in \mathbf{D}_\mu(H), h\neq H}(1-\tau_h^*)$. In the second term of (II.5.21) we have therefore:

$$\begin{aligned} &\mathbf{R}_{\text{int}}^U C_{xy}^{i_1} C_{xz}^{i_2} C_{yz}^{i_3} C_{yz}^{i_4} = C_{xy}^{i_1} C_{xz}^{i_2} C_{yz}^{i_3} C_{yz}^{i_4} \text{ if } i_{34} \leq j_{12}; \\ &\mathbf{R}_{\text{int}}^U C_{xy}^{i_1} C_{xz}^{i_2} C_{yz}^{i_3} C_{yz}^{i_4} = C_{xy}^{i_1} [C_{xz}^{i_2} - C_{xy}^{i_2}] C_{yz}^{i_3} C_{yz}^{i_4} \text{ if } i_{34} > j_{12}. \end{aligned} \tag{II.5.22}$$

In the third term, the full amplitude factorizes as a product so that when one internal operator $(1-\tau_h^*)$ is performed, the result vanishes. Therefore:

$$\mathbf{R}_{\text{int}}^U C_{xy}^{i_1} C_{xy}^{i_2} C_{yz}^{i_3} C_{yz}^{i_4} = C_{xy}^{i_1} C_{xy}^{i_2} C_{yz}^{i_3} C_{yz}^{i_4} \tag{II.5.23a}$$

if $i_{12} \leq j_{34}$ and $i_{34} \leq j_{12}$, and

$$\mathbf{R}_{\text{int}}^U C_{xy}^{i_1} C_{xy}^{i_2} C_{yz}^{i_3} C_{yz}^{i_4} = 0 \tag{II.5.23b}$$

otherwise.

To obtain a recursion relation involving solely g_i^ρ and g_{i+1}^ρ at third order, one should replace, in each third order term of (II.5.21) every g_k^ρ with $k \geq i+1$ simply by g_{i+1}^ρ, since they are equal at first order. But in the second order term one should reexpress $g_{j_{12}}^\rho$ in terms of g_{i+1}^ρ, taking into account the third order terms that this operation generates. This results in contributions with a main bubble H corresponding to the initial second order graph (hence with $i_H = i+1$), and a reduction vertex corresponding to the counterterm for another bubble h, generated by the second order recursion relation (II.5.13), which we apply to reexpress $g_{j_{12}}^{\rho}$ in terms of g_{i+1}^{ρ}. This second bubble h may be inserted "transversally" or "longitudinally," in which case we associate this contribution respectively to the graphs Q_3 and Q_4. For Q_3, the corresponding counterterm completes the renormalization of h; namely the inner bubble h satisfies $j_{12} \geq i_{34} > i+1 = i_{12}$: adding its counterterm replaces precisely the operator $\mathbf{R}_{\text{int}}^{U}$ by $\mathbf{R}_{\text{int}} \equiv (1 - \tau_h^*)$, the full internal renormalization operator, except for a subtlety; the counterterm for internal subgraphs h with $i_{34} = i+1$ is missing. We add it and subtract it, so that Q_3 gets full internal renormalization, and there is an exceptional term which is $+\tau_H \tau_h^*$, with $i_H = i_h = i+1$.

Similarly for Q_4, taking into account the symmetry factor of the graph, missing counterterms for h to the right and $j_{12} \geq i_{34} > i+1 = i_{12}$ *and* for h to the left and $j_{34} \geq i_{12} > i+1 = i_{34}$ are generated. They correspond almost exactly to the first case of (II.5.23), so that the contribution for Q_4 becomes 0 as in the second case of (II.5.23), except again for the subtlety that the case $i_{12} = i_{34}$ remains. But combining this term with the exceptional term for the graph Q_3, we reconstruct exactly the product of two independent bubble counterterms $\tau_H \tau_{H'}$ (with their full combinatoric coefficients), and with $i_H = i_{H'} = i+1$. The conclusion of this tedious analysis is that to third order in g_{i+1}, (II.5.21) becomes:

$$\begin{aligned} g_i^{\rho} - g_{i+1}^{\rho} = b & \sum_{i+1=i_{12}\leq j_{12}\leq \rho} (g_{i+1}^{\rho})^2 \int C_{xy}^{i_1} C_{xy}^{i_2} d^4y \\ & + c \sum_{i+1=i_{1234}\leq j_{1234}\leq \rho} (g_{i+1}^{\rho})^3 \int d^4y\, d^4z\, \mathbf{R}_{\text{int}} C_{xy}^{i_1} C_{xz}^{i_2} C_{yz}^{i_3} C_{yz}^{i_4} \\ & + (g_{i+1}^{\rho})^3 \left\{ b \sum_{i+1=i_{12}\leq j_{12}\leq \rho} \int C_{xy}^{i_1} C_{xy}^{i_2} d^4y \right\}^2 \\ & + \text{contributions with at least 4 vertices.} \end{aligned} \qquad \text{(II.5.24)}$$

It remains to check that the "contributions with at least 4 vertices" in (II.5.24) are unimportant. These corrections correspond to graphs with 4 vertices or more in (II.5.21) or are generated by the recursion which changes (II.5.21) into (II.5.24). Taking into account the form of the error terms in (II.5.18) and (II.5.15), we expect a bound of the type $O((\log i)/i^4) + i^{-2} O(e^{-\delta(\rho - i)})$ for the sum of all these terms. Such a bound would not be too hard to prove if one could directly combine the uniform bound on usefully renormalized graphs of Section II.3 with Theorem II.5.1. Indeed taking C large enough, one is certainly inside the convergence radius of the planar power series; the remainder in (II.5.21) for $i+1=\rho$ can be evaluated by $O(g_\rho^4)=O(\rho^{-4})$. By induction, one would remain inside convergence radius of the series in (II.5.21) for smaller i's; the behavior (II.5.18) would be checked inductively and the series of contributions generated from (II.5.21) to (II.5.24) again would be controlled inductively. For this last bound one can simply use the exponential decay in index space generated by phase space analysis to obtain a uniform bound $O((g_{i+1}^\rho)^4) = O(i^{-4})$ for series of contributions of the type $\sum_{j>i+1} e^{-\delta(j-i)} \sum_{n\geq 4}(g_j^\rho)^n a_n$.

This program is basically right except for one subtle point: the amplitudes appearing in (II.5.21) are not exactly the usefully renormalized ones; in particular no wave function renormalization is performed. We need to use the decay of the effective constants to sum over logarithmically divergent mass-renormalized bipeds. This is possible because there are at least two effective constants available for each such biped. Hence the analysis sketched above remains correct, except that graphs H with k bipeds, n vertices and $i_H > i$ should not be evaluated naively by the regular bound $O(i^{-n})$ but rather by $O(i^{-(n-k)})$ since each logarithmic divergence "eats" the decay of one coupling constant. This small change would not be relevant at all, except for the fact that the smallest possible H with $k \neq 0$, the graph Q_5 in Fig. I.5.1, should not be considered as fourth order, but rather promoted to third order, so that it becomes relevant for the correct value of β_3.†

The detailed analysis of the graph Q_5 is easy. The non-trivial biped B_0 corresponds to lines 4,5,6. The logarithmic divergence is fully contained in the case $i_{456} > i_{123}$ and the corresponding (mass-renormalized) contribution is obtained by applying $\tau_{B_0}^{1*}$. Hence if e is the combinatoric coefficient for the planar graph Q_5 the full apparent third order contribution of Q_5 in (II.5.21), up to $O((\log i)/i^2)$, is simply:

$$e(g_{i+1}^\rho)^2 \sum_{i_{123}=i+1} \sum_{k=i_{456}>i_{123}} \int d^4x d^4y d^4z C_{uz}^{i_1} \frac{|y-x|^2}{8}$$
$$\cdot \{C_{xu}^{i_2} \Delta C_{xz}^{i_3}\} \frac{1}{(\beta_2 k \log M)^2} C^{i_4}{}_{xy} C_{xy}^{i_5} C_{xy}^{i_6}. \quad \text{(II.5.25)}$$

---

†Of course the phenomenon discussed here at the level of graphs is well known at the level of the Callan–Symanzik equation; the second order term in (II.5.19) reacts on the third order term of this equation.

We generalize now Lemma II.5.1 to the contributions in (II.5.24–25):

Lemma II.5.2

There exist numerical constants γ_3 and δ_3 such that:

$$\gamma_3 \log M = c \lim_{i\to\infty} \lim_{\rho\to\infty} \sum_{i+1=i_{1234}\le j_{1234}\le\rho} \int d^4y d^4z \cdots$$
$$\cdot \mathbf{R}_{\text{int}} C^{i_1}_{xy} C^{i_2}_{xz} C^{i_3}_{yz} C^{i_4}_{yz}, \tag{II.5.26}$$
$$\frac{\delta_3}{\beta_2} = e \lim_{i\to\infty} i. \lim_{\rho\to\infty} \sum_{i_{123}=i+1} \sum_{k=i_{456}>i+1} \int d^4x d^4y d^4z \cdots$$
$$\cdot C^{i_1}_{uz} \frac{|y-x|^2}{8} \{C^{i_2}_{xu} \Delta C^{i_3}_{xz}\} \frac{1}{(\beta_2 k \log M)^2} C^{i_4}_{xy} C^{i_5}_{xy} C^{i_6}_{xy}. \tag{II.5.27}$$

Furthermore the sum in (II.5.26) converges exponentially and the generalization of (II.5.15) holds, namely differences between the right hand side of (II.5.26) at finite i and ρ and the left hand side are uniformly bounded by $O(e^{-\delta(\rho-i)}) + e^{-\delta i})$.

The limit in (II.5.27) is more complex; up to exponentially small errors in $\rho - i$ and i, we have an integral invariant under translation in index space, hence a contribution const. $\sum_{i+1<k\le\rho} k^{-2} = \text{const.}[i^{-1} - \rho^{-1} + O(i^{-2})]$.

Finally by Lemma II.5.1 the third contribution in (II.5.24) converges simply to $(\beta_2 \log M)^2$ with exponentially small corrections as ρ and i tend to infinity. Therefore we obtain:

Lemma II.5.3

$$g_i^\rho - g_{i+1}^\rho$$
$$= \beta_2 \log M (g_{i+1}^\rho)^2 + \gamma_3 \log M (g_{i+1}^\rho)^3 + (\beta_2 \log M)^2 (g_{i+1}^\rho)^3$$
$$+ \frac{\delta_3}{\beta_2}(g_{i+1}^\rho)^2 \left(\frac{1}{i} - \frac{1}{\rho}\right) + O\left(\frac{\log i}{i^4}\right) + (g_{i+1}^\rho)^2 O(e^{-\delta(\rho-i)}). \tag{II.5.28}$$

Let us define:

$$\beta_3 = \gamma_3 + \delta_3. \tag{II.5.29}$$

The behavior of the recursion relation (II.5.28) is investigated easily under the initial ansatz (II.5.20), even for complex C with $\operatorname{Re} C$ large.

Lemma II.5.4

For $\operatorname{Re} C$ large enough, we have:

$$\left| \frac{1}{g_i^\rho} - (\beta_2 \log M)i + \frac{\beta_3}{\beta_2} \log i + C \right| \leq \operatorname{Re} C. \tag{II.5.30}$$

This is because we may rewrite (II.5.28) as:

$$\frac{1}{g_{i+1}^\rho} - \frac{1}{g_i^\rho} = \beta_2 \log M + \frac{\beta_3}{\beta_2 i} - \frac{\delta_3}{\beta_2 \rho} + O\left(\frac{\log i}{i^2}\right) + O(e^{-\delta(\rho - i)}). \tag{II.5.31}$$

(II.5.30) follows from the uniform summability of $(\log i)/i^2$, $e^{-\delta(\rho-i)}$, and the obvious bound:

$$\sum_{j=i}^{\rho} \frac{\delta_3}{\beta_2 \rho} \leq \frac{\delta_3}{\beta_2} \qquad \forall i. \tag{II.5.32}$$

We can now extend uniform bounds like Theorem II.3 to the amplitudes which appear in the particular expansion (II.5.8):

Lemma II.5.5

There exists a constant K such that:

$$\left| \sum_{\mu} \left[\prod_{v \in G} g_{e_v(\mu)}^\rho \right] A_{G,\mu}^{MR;CCUR} \right| \leq K^{n(G)} (1 + \sup_j |p_j|)^{\hat{N}}. \tag{II.5.33}$$

This bound is obtained by combining the bounds on usefully renormalized amplitudes (Theorem II.3) for the pieces of G without bipeds, the argument that mass insertions do not create renormalon effects (see Theorem II.3.2) and the existence of at least two specific coupling constants of the right scale associated to each dangerous biped; using the decay of these constants as expressed in Lemma II.5.4 the logarithmic divergence for these bipeds not only becomes convergent but does not disturb the uniform nature of the estimate.

It is important to notice that the proof of Lemmas II.5.3–5 is inductive. For each scale i from ρ to -1, Lemma II.5.3 is proved first, then Lemma II.5.4, then the piece of Lemma II.5.5 which deals with the subgraphs G_i^k of G (made of lines with indices $j \geq i$). There is no logical loop, because the uniform bounds on the remainders necessary for Lemma II.5.3 at scale i only depend on the bounds (II.5.31) for effective couplings of scales $j > i$, and on bounds of the type (II.5.33) for subgraphs H with $i_H > i$.

Once the renormalization group "discrete flow" for the coupling constants g_i and the bounds (II.5.33) are established by Lemmas II.5.4–5 for all scales, it is easy to complete the proof of Theorem II.5.3. Using

Lemmas II.5.4, II.5.5 and Theorem II.5.1, the series (II.5.10) are absolutely and uniformly convergent, hence their sum defines the planar theory with cutoff ρ. Furthermore every estimate and the sums over index space being uniform in ρ, the effective constants g_i^ρ have limits g_i as $\rho \to \infty$ which still satisfy Lemmas II.5.4–5. This constructs the theory without cutoff. The analyticity result follows in the straightforward way from uniform convergence of series term by term analytic. The results on the behavior of g_r and a_r as functions of C and a are obtained by considering the recursion relation (II.5.19) for the wave function constant in addition to the flow of the coupling constant expressed in Lemma II.5.3.

(Finally for Borel summability in g_r, one checks directly the hypotheses of Nevanlinna–Sokal theorem (Theorem I.5.1). The region of Fig. II.5.5 is exactly the region necessary to apply this theorem (the true region of analyticity is in fact much larger because the discrete flow (II.5.28) still leads to an asymptotically free theory ($g_\rho \to 0$ as $\rho \to \infty$) for g_r at least in the region pictured in Fig. II.5.6).

The uniform Taylor remainder estimates (I.5.2) are just another exercise in establishing factorial bounds (in this case solely due to the renormalon effects [Ri1]). Since this kind of bound has been studied at length in Sect. II.3, we leave this problem to the reader.

It might be interesting at this stage to compare briefly what we have done to the standard continuous renormalization group flows, limiting ourselves for simplicity to the example of the β function. In the standard definition of the Callan Symanzik or renormalization group equations the key rôle is played by the ultraviolet β_{uv} function which is defined as $[dg_{ren}/dx]\,|_{g_\rho}\,(g_{ren})$, where g_{ren} is the renormalized coupling; g_ρ, the bare coupling, is held fixed, and $x = \log(\kappa/m_{ren})$ is the logarithm of the quotient between the scale κ of the ultraviolet cutoff

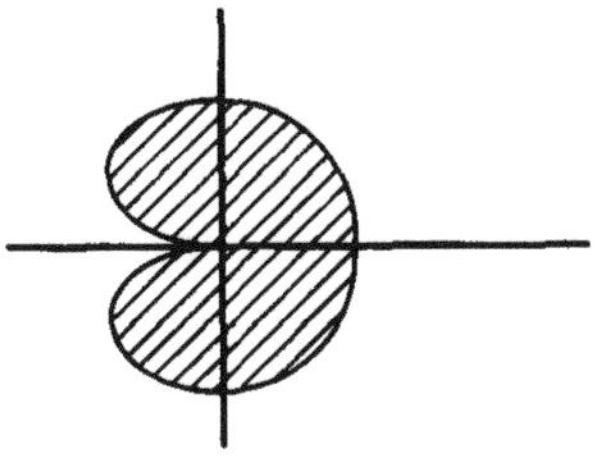

Figure II.5.6

and the renormalized mass, hence in our case where we work with a unit renormalized mass which is fixed, x is equal to $\rho \log M$. In the BPHZ scheme that we use, it is known that there are formulas which relate this ultraviolet β function to renormalized Schwinger functions $\Gamma_{\Delta}^{(N)}$ at zero momentum with one mass insertion Δ on one propagator and the minimal subtraction prescription. This means that two or four point functions which contain the mass insertion Δ should be less subtracted, according to their true (improved) degree of convergence. The formula for the ultraviolet β formal power series is [IZ]:

$$\beta_{uv}(g) = \left[1 + \frac{\partial \Gamma_{\Delta}^{2}(0)}{\partial p^2}\right]^{-1} \left[-2\Gamma_{\Delta}^{(4)}(0) + 4g\frac{d\Gamma_{\Delta}^{2}(0)}{dp^2}\right]. \qquad \text{(II.5.34)}$$

This formula can be used for practical numerical computations of coefficients such as β_2 or β_3. What is its relationship to the method described above? When we compute the difference $g_i^\rho - g_{i+1}^\rho$ we compute clearly a discrete analogue of the ultraviolet beta function with cutoff $\kappa = M^\rho$, but this function is not expressed as a power series in the last (renormalized) coupling g_i^ρ but in the whole sequence of previous effective couplings. Usually this is better than expressing it in terms of a single renormalized constant, an operation which generates useless counterterms and renormalons. However in this particular case, to study the behavior of the effective couplings it is practical to develop at least the first orders of the equation in terms of the last coupling g_i^ρ, as is done above to third order. In this way we see that the coefficients of the usual β series (II.5.34) are not exactly generated both because of the remaining ultraviolet cutoff and of a slice effect due to the fact that our flow is discrete rather than continuous. More precisely the condition that one propagator in our contributions to the flow is in slice i is asymptotically the analogue of the mass insertion in (II.5.34). However since our slices have finite thickness, there are some terms with several legs in the slice (a situation of measure zero for infinitesimal slices hence for continuous flows). These terms are the source of corrections to the β function such as the term in $(\beta_2 \log M)^2$ in (II.5.28) which are characterized by a power of $\log M$ higher than one. In other words if we develop the finite difference equation (II.5.28) up to a given finite order in terms of the last (renormalized) constant g_i^ρ, take $\rho \to \infty$, divide by $\log M$ (since $x = \rho \log M$ and take $M \to 1$, then order by order only the regular contribution of the ultraviolet β function will survive, but in a scheme with an infrared cutoff; in the limit $i \to \infty$ the small corrections to scale invariance due to the mass disappear and we will find exactly the same coefficients than in (II.5.34) (this explains the two limits in (II.5.14)).

In conclusion the discrete flows considered in this section and later in part III are naturally expressed in terms of effective quantities with effective constants; they are the correct way to replace the formal renormalization group functions by well defined ones. The standard formal power series are recovered when cutoffs are removed and effects due to the discretization are removed.

Chapter II.6

The Large Order Behavior of Perturbation Theory

Aussi loin que la science recule ses frontières,
et sur tout l'arc étendu de ces frontières,
on entendra courir encore la meute chasseresse du poète.
—Saint-John-Perse

In this last section on perturbation theory we no longer discuss large order bounds for individual Feynman amplitudes, but consider the more difficult problem of the exact large order *behavior* of the renormalized perturbation series. We will meet again the problem of the large number of graphs in the ordinary φ^4 theory, and the renormalon problem for φ^4_4, and discuss how they shape the large order behavior of the theory, using the convenient mathematical formalism of the Borel transform introduced in section I.5. The rigorous results obtained so far are still fragmentary and in our opinion a lot of interesting work remains to be done in this area.

In the regular φ^4 theory the total number of Wick contractions for graphs (not necessarily connected) with n vertices and a fixed number N of external lines is $(4n+N-1)!!$. The number of connected graphs is of course smaller, but it is rather easy to show that it is more than $(\text{const.})^n n!^2$; indeed we may first build a spanning tree in more than (const.)$^n n!$ different ways (apply Cayley's theorem, Sect. I.4, with co-ordination numbers bounded by 4), then still have (const.)$^n n!$ different contraction schemes for the remaining lines.

In a theory like φ_1^4 (the anharmonic oscillator), all graphs add up at any given order with the same sign. Furthermore it is easy to check that any amplitude of order n satisfies both an upper and a lower bound of the kind (const.)^n. The upper bound is simply Weinberg's uniform theorem, although in this case it may be obtained for instance in a space by simpler arguments. The lower bound is also extremely easy in a space: simply integrate only over $1 \leq a_i \leq 2$ $\forall i$ and use the fact that the number of spanning trees is bounded by $2^{l(G)} \leq 4^{n(G)}$ for an upper bound on the Symanzik polynomial U_G.

Hence in this case of φ_1^4, taking into account the factor $\frac{1}{n!}$ in (I.4.1), the nth order of perturbation theory a_n satisfies some bound of the type:

$$K_1^n n! \leq (-1)^n a_n \leq K_2^n n! \qquad \text{(II.6.1)}$$

for some positive constants K_1 and K_2. As a consequence, the radius of convergence of the perturbative series is 0; we say in short that it diverges. Nevertheless a behavior like (II.6.1), although incompatible with ordinary summability may still allow Borel summability.

It is not easy to prove that the corresponding renormalized series diverge for φ^4 in higher dimensions. BPHZ renormalization in two dimensions is equivalent to Wick ordering, hence simply suppresses the graphs with tadpoles. It has been first proved in [Ja] that enough graphs remain so that the nth order of perturbation series for φ_2^4 still satisfy a lower bound which implies divergence. For φ_3^4 there is a first non-trivial mass renormalization which changes the sign of some amplitudes. Lower bounds for sums with different signs is more difficult because one has to rule out the possibility of systematic cancellations. The divergence for φ_3^4 was proved in [dCR2], but no longer from a lower bound simply on the nth order of the series; the argument already mixes different orders. For φ_4^4 a rigorous proof is still missing; renormalization introduces changes of signs much more difficult to track, so that a proof of divergence seems impossible except as a by-product of a detailed analysis of the large order behavior. Significant progress towards this goal has been made in [MNRS][DFR], using the multiscale representation, and the main goal of this section is to introduce the reader to this approach.

We turn now to a description of the rigorous results and heuristic expectations on the large order behavior of φ^4. We will not comment on the large order behavior of other models, except to notice that theories with fermionic fields have better convergence properties due to the cancellations in the corresponding fermionic determinants; this fact, a direct consequence of Pauli's principle, will be used extensively in the construction of the Gross–Neveu model (Sect. III.4). The study of the corresponding large order behavior of these models is much less advanced than for bosonic theories, and is a beautiful open problem.

The large order behavior of φ^4 is of course best understood in the one dimensional case (anharmonic oscillator) (not to speak of the 0-dimensional case, which is the study of the moments of the measure $e^{-gx^4-ax^2}dx$, for which explicit formulae can been derived in terms of hypergeometric functions [Wig2]). There is a long history of numerical and rigorous results on the anharmonic oscillator, using BKW methods or the functional integral and steepest descent methods, reviewed in [Si2]. But here we intend to put the emphasis on higher dimensions, $d = 3$ and mostly $d = 4$. The first major progress in these cases came from a semi-rigorous extension of the steepest descent to the functional integral of φ^4_d, called the Lipatov method [Lip]. This method was developed and applied in [BGZ]. Let us summarize now its guiding principle.

For simplicity the large order behavior may be investigated on the simplest typical quantity for the theory with non-trivial renormalized perturbation theory. For instance in 1, 2 and 3 dimensions we may use the pressure:

$$p = \lim_{\Lambda\to\infty,\rho\to\infty} \frac{1}{|\Lambda|} \log Z_{\rho,\Lambda} = \sum (-g)^n a_n \tag{II.6.2}$$

but in 4 dimensions the renormalized series for the pressure or for the 2 and 4 point functions at 0 external momenta are trivial, so that the simplest quantity may be the connected 6 point function at 0 external momenta, or the 2 point function at a particular momentum, or the connected 4 point function at some symmetric set of external momenta. If C_N is the quantity of interest we write again:

$$C_N = \sum_n (-g_R)^n a_n^R \tag{II.6.3}$$

with g_R the renormalized constant.

The Lipatov method leads to an asymptotic behavior of a_n or a_n^R at large n which is always of the type:

$$a_n \simeq n!a^n.n^b.c,(1 + O(1/n)) \tag{II.6.4}$$

where a, b and c are some constants. a depends only on the dimension (and may also depend on some parameters of the theory, like the mass for $d < 4$). b may also depend on which particular Schwinger function one is investigating (hence, may depend on N) and c depends on further details of the theory (in particular in 4 dimensions, on the particular renormalization scheme and subtraction scale which is used).

Since we are interested in universal features of the large order behavior which are valid for the perturbative series of any reasonable quantity in the theory, a is the most important constant in (II.6.3). The Lipatov method predicts:

$$a = a_{\text{Lip}} = e^{-\inf_\varphi S(\varphi)+2}, \tag{II.6.5}$$

$$S(\varphi) = \frac{1}{2}\int(\partial_\mu\varphi\partial^\mu\varphi + m^2\varphi^2) - \log\int\varphi^4, \tag{II.6.6}$$

where the infimum in (II.6.5) is taken over the appropriate Sobolev space where the functional (II.6.6) is well defined.

Before providing a heuristic motivation for (II.6.5–6), let us show that the functional $S(\varphi)$ is bounded below in dimensions $d \leq 4$ by virtue of some Sobolev inequality. In dimension $d \leq 4$ indeed there exists a constant K_1 (depending on m for $d < 4$) such that:

$$\int_{\mathbb{R}^d}\varphi^4 \leq \left[K_1\int_{\mathbb{R}^d}(\partial_\mu\varphi\partial^\mu\varphi + m^2\varphi^2)\right]^2, \tag{II.6.7}$$

so that whenever φ belongs to the Sobolev space $H^{1,2}$ in which the right hand side of (II.6.7) is well defined, it belongs also to L^4. We call $K_1(d)$ the infimum of the constants K_1 for which (II.6.7) holds; this infimum is of course a minimum, i.e., (II.6.7) still holds with $K_1(d)$ instead of K_1.

Similarly for C^∞ functions with compact support (or under suitable decay conditions at infinity) we have also for some optimal $K_2(d)$:

$$\int_{\mathbb{R}^d}\varphi^4 \leq \left[K_2(d)\int_{\mathbb{R}^d}(\partial_\mu\varphi\partial^\mu\varphi)\right]^2. \tag{II.6.8}$$

Remark that a in (II.6.5) and $K_1(d)$ are related by:

$$a = (4K_1(d))^2, \tag{II.6.9}$$

since taking $\varphi = af$ with $\int(\partial_\mu f \partial^\mu f + m^2 f^2) = 1$ we may optimize in (II.6.6) and get:

$$\inf_{a,f}\left(\frac{a^2}{2} - \log a^4 \int f^4\right) = \inf_f\left(2 - \log 4^2 \int f^4\right) = 2 - \log(4K_1(d))^2. \tag{II.6.10}$$

In dimension 1, 2 and 3 the smallest constants $K_1(d)$ and $K_2(d)$ for which (II.6.7) or (II.6.8) hold are different; $K_1(d)$ depends on m, and for $m \neq 0$, $K_1(d) < K_2(d)$. But in the critical case $d = 4$ we have equality: $K_1(d) = K_2(d)$. In fact in dimensions 1, 2 and 3 the infimum in (II.6.5) is a minimum, hence is attained for a particular smooth minimizing function φ_0 which by the variational principle is a solution of the differential equation:

$$(-\Delta + m^2)\varphi_0 = \lambda \frac{\varphi_0^3}{\int \varphi_0^4} \tag{II.6.11}$$

for some constant λ; we may find a particular φ_0 with radial symmetry (to break the translation invariance of (II.6.11)) which has fast decrease at infinity, and $S(\varphi_0)$ depends on m. This is also true in 4 dimensions for the finite volume analogue of $S(\varphi)$, $S_{\Lambda,X}(\varphi)$ (one has to specify some boundary conditions X on $\partial\Lambda$), which attains its minimum for a particular smooth $\varphi_{0,\Lambda,X}$, and the corresponding minimum $K_{1,\Lambda,X}(d = 4)$ of S_Λ depends on Λ and m. As $\Lambda \to \infty$, $K_{1,\Lambda,X} \to K_1(d = 4)$; it can be shown however that the minimizing functions do not converge, that $K_1(d = 4) = K_2(d = 4)$, the optimal constant for which (II.6.8) holds, which is of course independent of m, and that the infimum of $S(\varphi)$ is therefore a true infimum (not attained). Furthermore the value of this infimum may be computed exactly: $K_1(d = 4) = K_2(d = 4) = \frac{1}{4\pi}\sqrt{3/2}$ [Aub], hence $a = 3/(2\pi^2)$ in (II.6.5). This change of behavior in dimension 4 is of course due to the marginal character of the Sobolev inequality (II.6.7) in this case.

A crude motivation for the Lipatov prediction of the value (II.6.5) of a in (II.6.4) goes as follows. Let us pretend that an ultra violet cutoff κ_n and a finite volume Λ_n (with boundary conditions X_n) may be imposed on the nth order of perturbation theory in such a way that $\kappa_n, \Lambda_n \to \infty$ as $n \to \infty$, but that as far as the leading order behavior is concerned, a_n^R and $a_n^{X_n,\Lambda_n,\kappa_n}$ (the bare amplitudes with cutoffs and *no* renormalization) are equivalent. (We will see below that this assumption turns out to be justified in dimensions less than 4 but wrong for $d = 4$).

Let us assume also that the restriction of connectedness in a_n is also unimportant as far as leading large order behavior is concerned. Then the leading behavior of $(-1)^n a_n^R/n!$ is the same as the one of

$$b_n = \int \left(\frac{1}{n!}\right)^2 \left(\int_{\Lambda_n} \varphi^4\right)^n d\mu_{X_n,\Lambda_n,\kappa_n}. \qquad \text{(II.6.12)}$$

For conceptual simplicity, let us choose the ultraviolet cutoff to be a lattice with lattice spacing δ_n such that $\lim_{n\to\infty} \delta_n = 0$, and let us choose the dependence in n of Λ_n so that the total number of sites in Λ_n grows only slowly, like n^ε. The Gaussian measure on the lattice may be written in terms of the ordinary Lebesgue measure (see (I.3.13)). Rescaling φ to $\psi = \varphi/\sqrt{n}$ we get:

$$b_n = \left(\frac{n^n}{n!}\right)^2 \int e^{-(n/2)\left\{(\partial_\mu\varphi\partial^\mu\varphi)/n) + m^2(\varphi^2/n) - 2\log\int_{\Lambda_n}(\varphi^4/n^2)\right\}} \prod_{x\in\Lambda_n} d\varphi(x)$$

$$= (\sqrt{n})^{n^\varepsilon} \left(\frac{n^n}{n!}\right)^2 \int e^{-(n/2)\left\{(\partial_\mu\psi\partial^\mu\psi) + m^2\psi^2 - 2\log\int_{\Lambda_n}\psi^4\right\}} \prod_{x\in\Lambda_n} d\psi(x), \qquad \text{(II.6.13)}$$

so that:

$$(b_n)^{1/n} = e^{n^\varepsilon \frac{\log n}{2n}} \left(\frac{n^n}{n!}\right)^{2/n} \|e^{-S(\varphi)}\|_n, \qquad \text{(II.6.14)}$$

which, as $n \to \infty$ should tend to $1.e^2\|e^{-S(\varphi)}\|_\infty$, hence to a by (II.6.5).

To refine this crude prediction, one has to apply the steepest descent method to the functional integral (II.6.12), hence to expand the action around the configurations which minimize $S(\varphi)$. By a rescaling of φ the variational equation (II.6.11) is the same as the equation of motion of the theory $(-\Delta + m^2)\varphi = g\varphi^3$, and these configurations are the non-trivial classical solutions of finite action which are called instantons. Expanding around these solutions to second order, as usual, and performing the corresponding Gaussian functional integral gives an explicit determinant with which one can compute subleading coefficients at large order like b and c in (II.6.4), and in principle a systematic expansion in $1/n$. The Lipatov method in this way analyzes the asymptotic behavior of perturbation theory at large order and relates it to singularities on the negative real axis in the Borel plane which are therefore also called *instanton* singularities. In the case of the anharmonic oscillator these singularities are known to induce a corresponding essential singularity at $g = 0$ corresponding to a cut along the negative real axis for quantities like the ground state energy (II.6.2). We do not develop further this point of view here, and refer to [Zin] for a review on instanton calculus.

However in the standard Lipatov argument as sketched above, the possible effect of renormalization on large order behavior is neglected; more precisely it is assumed simply to change the determinants corresponding to fluctuation around the saddle points into renormalized determinants. This assumption is expected to fail in the case of φ_4^4, where renormalization affects in a major way the large order behavior. As argued by Parisi and 't Hooft [Pa1– 2]['tH3], the factorial behavior of single Feynman graphs that we met and discussed at length in Section II.3 creates corresponding singularities on the right hand side of the real axis in the Borel plane, called *renormalons*. It happens for ordinary φ^4 as well as for vector φ^4 models with N components that the first renormalon singularity on the positive real axis is closer to the origin of the Borel plane than the first instanton singularity on the negative real axis, so that the large order behavior of φ_4^4 is in fact expected to be governed more by renormalization than by the instantons of the Lipatov method. For instance for one component φ_4^4 the position of the first expected renormalon is at $t = 2/\beta_2 = 4\pi^2/9$; in contrast the value $a_{\text{Lip}} = 3/(2\pi^2)$ above for the Lipatov behavior corresponds to an instanton singularity in the Borel plane at $t = -2\pi^2/3$ hence farther from the origin by a factor 3/2.†

The existence of renormalons, if confirmed rigorously, would mean that the φ_4^4 series are not Borel summable, in contrast to what has been proved for $\varphi_{1,2,3}^4$ [GGS][EMS][MS]. This fact is presumably related to the difficulty in defining a non-trivial φ_4^4 theory satisfying the axioms [Aiz][Frö]. We return to this point in Sect. III.4 with a weak coupling triviality theorem; for reviews on triviality, see [Sok2][GaRi].

We recall now briefly the rigorously proved large order results. For φ_1^4, the anharmonic oscillator, Borel summability [GGS] and all of formula (II.6.4) is proven (with expected values of a, b and c) [HS]; the Lipatov method has been justified [Sp3][Bre], and much is known about the analyticity properties of the corresponding sums in the Borel plane, either numerically or rigorously (see [Si2][Wig2]). In two and three dimensions Borel summability has been proved by constructive theory [EMS][MS] and the Lipatov method has been justified basically to leading order only, i.e., up to the computation of the coef-

---

†There is no deep explanation yet for this simple rational factor of 3/2.

ficient a in (II.6.4) [Bre][MR][FR]. Finally in 4 dimensions the results [MNRS][DFR] fall short of proving that the large order behavior is really governed by the first renormalon. Many of these rigorous results are therefore summarized by:

Theorem II.6.1

(a) The perturbation series for $\varphi^4_{1,2,3}$ are Borel summable (in the Watson or in the Nevanlinna–Sokal sense [GGS][EMS][MS]).
(b) For $\varphi^4_{1,2,3}$ there is a disk of analyticity in the Borel plane of radius a^{-1}, where a is defined by (II.6.5) and there is a singularity in the Borel plane at $t = -a^{-1}$ ("instanton") [Bre][MR][FR].
(c) For φ^4_4 there is a disk of analyticity in the Borel plane of radius $2/\beta_2 = 4\pi^2/9$ (the optimal expected disk)[MNRS][DFR].

In dimensions 3 and 4 the comparison between expected and proven results is sketched in Fig. II.6.1.

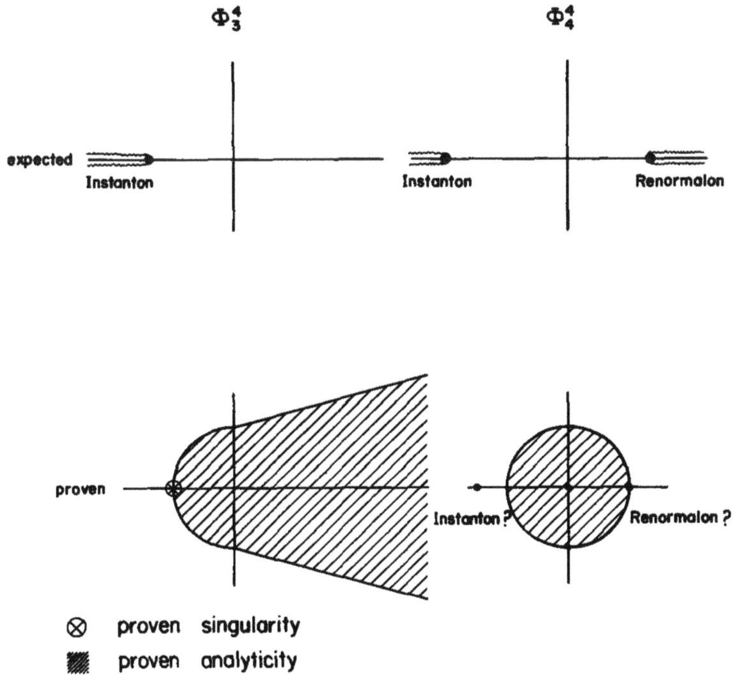

Figure II.6.1 The Borel plane of Φ^4; expected and proven results.

The proof of item a) relies on constructive methods and we will return to it in the next chapter; more precisely we will give a construction and prove a Borel summability result for infrared φ_4^4 which is general enough to apply directly with straightforward modifications to ultraviolet φ^4 in lower dimensions. Such a method is certainly a bit of an overkill for the cases of dimensions 1 and 2, perhaps even for dimension 3, but it is certainly not more complicated than the sum of the specific proofs derived earlier for these lower dimensions, which did not use the full machinery necessary in dimension 4 (multiscale expansion).

Similarly we will not discuss the proof of item b) which concerns superrenormalizable theories because the phase space language is not strictly speaking necessary there and because we prefer to concentrate on the more difficult issue of the four dimensional case. Let us simply state that item b) may be decomposed into the proof of an upper and a lower bound of the Lipatov type. The upper bound [Bre][MR] may be considered a simple corollary of the four dimensional upper Lipatov bound which is discussed below (in dimensions 1, 2 and 3, the usefully renormalized series which appear below, in which mass renormalization may be fully performed as in Theorem II.3.2, coincide with the regular series since there are only a few mass renormalizations). For the lower bound, which implies the existence of a singularity at $t = -a^{-1}$ we refer to [Bre][MR][FR] and notice simply that in three dimensions the changes of signs induced by the single nontrivial mass renormalization require a separate argument in which all orders of perturbation theory are mixed [FR]. In this sense the proven existence of an "instanton" singularity is still a weaker result than strict asymptotics of the Lipatov type for a_n, which remains an open problem in dimension 3, even at leading order. Subleading behavior (with the right constants b and c) remains also an open problem, even for φ_2^4.

We turn now our attention to item c) of Theorem II.6.1 which we rewrite in more detail as:

Theorem II.6.2

There exists a function $\varepsilon(n)$ which tends to 0 as $n \to \infty$ such that

$$|a_n^R| \leq n! \left[\frac{\beta_2}{2}\right]^n (1 + \varepsilon(n))^n, \qquad \text{(II.6.15)}$$

where a_n^R is the nth order of perturbation theory for the φ_4^4 model, and $\beta_2 = 9/2\pi^2$ is the one loop coefficient of the β function (see Sect. II.4–5).

The proof of course generalizes to the N-component φ_4^4 theory, in which case $\beta_2 = (N+8)/2\pi^2$.

Our goal is to describe in some detail the proof of this theorem. We will skip many of the technicalities, for which the reader is invited to look to the original articles, but we will try to explain clearly the structure of every important argument. We start by the following bound, proved in [MNRS]:

Theorem II.6.3: Upper Lipatov bound

There exists a function $\varepsilon(n)$ which tends to 0 as $n \to \infty$ such that

$$|a_n^{UR}| \leq n![a_{\text{Lip}}]^n(1+\varepsilon(n))^n \tag{II.6.16}$$

and a_n^{UR} is the sum of all usefully renormalized amplitudes (the theorem applies also to $a_n^{MR,UR}$, the sum of the mass-renormalized, usefully renormalized amplitudes of Theorem II.3.2).

Upper bounds of the Lipatov type are indeed very natural because of one key simple observation: the critical fields which minimize the functional (II.6.6) also saturate the corresponding Sobolev inequality (II.6.7) (see (II.6.10)). Therefore using the Sobolev inequality on the vertices $\int \varphi^4$ in (II.6.12) should lead to an upper bound of the Lipatov type with correct value of a. In contrast a lower bound of the Lipatov type typically requires a more complicated analysis on the speed at which, at large order, functional integrals like (II.6.13) become peaked around these minimizing configurations.

However using a Sobolev inequality in (II.6.12) replaces a φ^4 local vertex by two (disconnected) φ^2 vertices (with derivative couplings). There is a loss of connectivity which prevents one from applying the key observation above in a too naive manner. It is here that phase space analysis becomes useful. The outline of the strategy is as follows. Either the graphs contributing to a_n or b_n are spread over a large number p of cubes of the series of scaled lattices naturally associated to the multiscale decomposition, or they are concentrated in a few such cubes. The transition is somewhat arbitrary but may be taken at $p \simeq n/(\log n)^\delta$ for some small $\delta > 0$. In the first case, using the by now familiar horizontal and vertical decay associated to phase space (recall that such decay requires the use of "usefully renormalized" expansions) we should prove that the corresponding contributions do not contribute at all to the leading behavior at large n. In the second case, one should think of the vertices as densely packed into small areas of phase space. In this case one can apply the Sobolev inequality; the loss of connectivity is harmless (as far as leading large order behavior is concerned) because the total volume is then small.

Returning to the language of Sect II.3, we introduce the triplets $(G, \mathbf{F}, \mu)$ made of a graph, an assignment, and a forest $\mathbf{F}$ which is safe for μ, and the corresponding integrands

$$Z_{G,\mathbf{F},\mu} \equiv \prod_{g\in\mathbf{F}}(-\tau_g^*) \prod_{g\in\mathbf{H}_\mu(\mathbf{F})} (1-\tau_g^*) Z_{G,\mu} \tag{II.6.17}$$

where $Z_{G,\mu}$ is as in (II.4.8) the ordinary Feynman integrand for the assignment μ. Useful renormalization corresponds to the case $\mathbf{F} = \emptyset$, hence the bound (II.6.16) may be written as:

$$\sum_{(G,\mathbf{F},\mu);n(G)=n,\mathbf{F}=\emptyset} \int \prod_{v\in G} dx_v |Z_{(G,\mathbf{F},\mu)}| \leq n![a_{\mathrm{Lip}}]^n(1+\varepsilon(n))^n. \tag{II.6.18}$$

Next we decompose the (usefully renormalized) perturbation theory according to the multiscale slicing, as in Sect. II.3. The crucial point is to decompose the series in two according to whether the graphs and assignments (G, μ) spread over a large region of phase space or not. To measure the size of a region in phase space, it is natural to consider that distance scales are the inverse of momentum scales.

For each scale i, $i = 0, 1, 2 \ldots$ this leads one to introduce the scaled lattice $\mathbf{D}_i$ made of cubes of side M^{-i}, and define $\mathbf{D} = \cup_i \mathbf{D}_i$ (see Fig. II.6.2). Now at a vertex v of a graph sitting at x_v there are at most 4 propagators which meet, with scales $i_1(v)$, $i_2(v)$, $i_3(v)$ and $i_4(v)$, the maximum being $e_v(\mu)$ by (II.4.4), and we may associate to v the set of the four cubes $\Delta_1, \ldots, \Delta_4$, $\Delta_j \in \mathbf{D}_{i_j(v)}$ to which x_v belongs. If we repeat this for each vertex of a graph, we obtain a region which is the natural domain in $\mathbf{D}$ of the contribution associated to (G, μ). This domain $X_0^{\mathrm{vert}}(G, \mu) \subset \mathbf{D}$ is simply obtained by coloring the cubes of $\mathbf{D}$ which contain the meeting of a line at a vertex in our standard representation Fig. II.1.2 of phase space (see Fig. II.6.3). It will be called the "vertex domain" of the contribution in phase space, and a good measure of its size is simply the total number $x_0^{\mathrm{vert}}(G, \mu)$ of cubes in it.

However there is a problem with this simple approach. Remember that our strategy is to bound crudely the contributions spread over large domains of phase space (they should be small, anyway), and to apply the Sobolev inequality only to the sum over all graphs spread over a small domain of phase space. The Sobolev inequality (II.6.8) in a finite volume X_0 applies with the same constant as in infinite volume

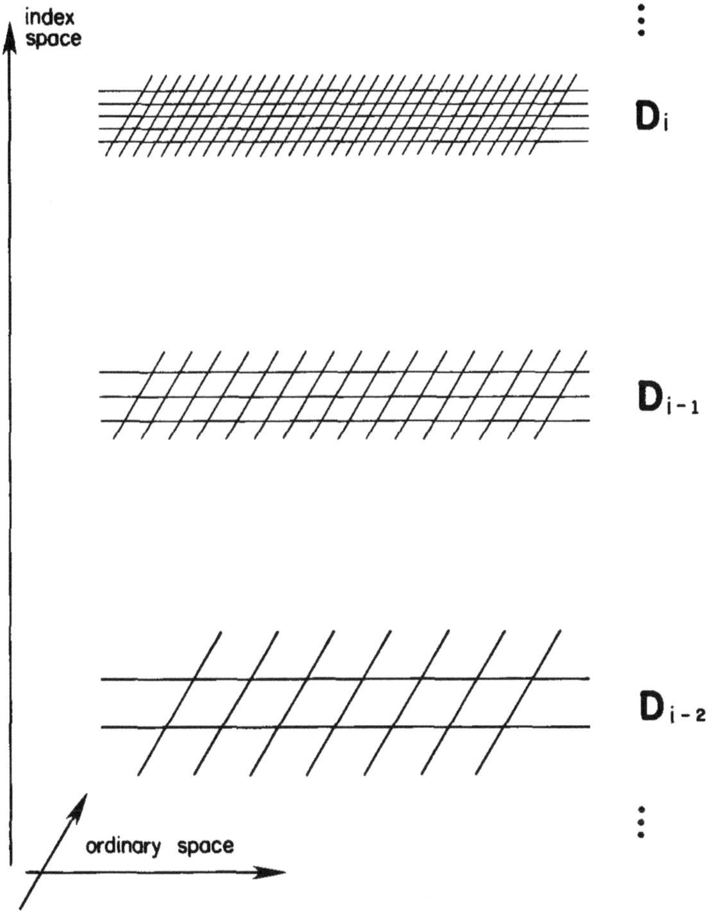

Figure II.6.2 The scaled lattices $\mathbf{D}_i$.

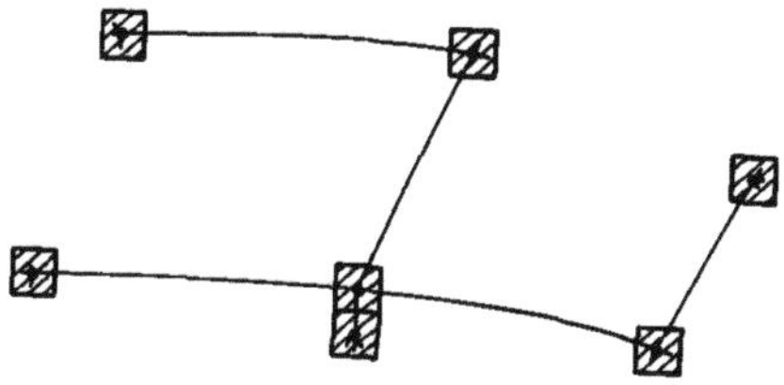

Figure II.6.3 The vertex domain.

only for fields in $H_0^1(X_0)$, the Sobolev space of functions φ with square integrable gradients which vanish on the boundary of X_0, which is a natural subset of $H_0^1(\mathbb{R}^4)$. But the sum of graphs that we are studying is related to an integral $(\int_{X_0} \varphi^4)^n d\mu_C(\varphi)$ where the Gaussian measure $d\mu_C(\varphi)$ corresponds to the propagator C. Using this propagator creates a problem because the sample fields for $d\mu_C$ are not in $H_0^1(X_0)$. This is not a regularity problem since in a finite region of phase space the propagator has an ultraviolet cutoff of type (I.3.7) so that the sample fields are very smooth, but the problem is that these sample fields have no reason to vanish on ∂X_0. This would be however the case for instance if we could use a propagator in which the Laplacian entering the propagator's definition has Dirichlet boundary conditions on X_0 (for the definition of Gaussian measures and their support properties, see [Er1][Si1]). For such a task it is convenient to introduce the Wiener path representation of the propagator [GJS][FO]. Fortunately this representation is fully compatible with the parametric representation, hence with cutoffs (I.3.7) or (II.1.3), since a is simply the proper time of the path. It reads:

$$C^i(x,y) = \int_{M^{-2i}}^{M^{-2(i-1)}} dt e^{-m^2 t} \int P_t(x,y) d\omega, \tag{II.6.19}$$

where $P_t(x,y)d\omega$ is the Wiener measure on the sets of all paths starting at x at time 0 and ending at y at time t.

Our goal of using the Sobolev inequality then leads us to consider a larger domain in phase space called the "vertex and propagator" domain. In [MNRS] this domain arises in a natural way as the result of an inductive cluster expansion of the Glimm–Jaffe Spencer type [GJS]. Since we postpone the definition of cluster expansions to the next section, we will give an equivalent global (set-theoretic) definition of this domain.

We decompose first each propagator in the perturbative expansion as in (II.6.19), so we rewrite the perturbative expansion as a sum over triplets (G,μ,Ω) where G and μ are a graph and an assignment, as before, and Ω is a set of $l(G)$ paths $\omega_1, \ldots, \omega_{l(G)}$; the first two sums are discrete, but the sum over Ω means a product of Wiener integrals $\prod_{l=1}^{l(G)} \int P_{t_l}(x_l, y_l) d\omega_l$. For this generalized multiscale representation, the straight lines which represent propagators in the standard picture (Fig. II.1.2) should be replaced by the more complicated paths of Fig. II.6.4.

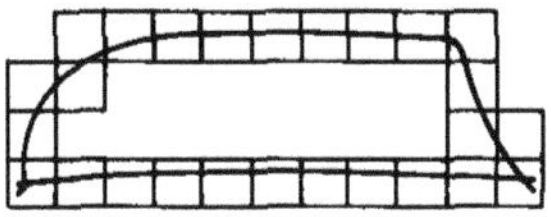

Naïve propagator domain for a graph with 2 propagators.

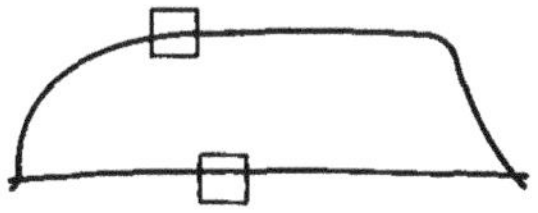

A true propagator domain for the same graph.

Figure II.6.4

Then it may seem natural to define the "full propagator domain" as the union over l of all the cubes of $\mathbf{D}_{i_l(\mu)}$ visited by the path ω_l. However this naive domain is again not the correct concept in this case, because we look for a definition such that when the propagator domain is large (compared to $n/(\log n)^{\delta}$), the corresponding contributions are small (in the sense of Lemma II.6.2 below). This will not be the case with the above naive definition, for instance when the vertex domain is small and the "full propagator domain" is large only because of a *single* propagator whose path extends over a large set of cubes. In fact a definition adapted to our purpose is the following:

Let us consider a subset $X = \{\Delta_1, \ldots, \Delta_j, \ldots, \Delta_q\}$ of $\mathbf{D}$. We say that this subset has the property HZ (for "horizontally connected") with respect to (G, μ, Ω) if and only if for any $\Delta_j \in X \cap \mathbf{D}_i$ (hence for any cube of X of scale i) there is a *distinct* line l_j of the same scale ($i_{l_j}(\mu) = i$) such that the corresponding path ω_{l_j} visits Δ_j. (The important fact is that all the lines l_j have to be different: $l_j \neq l_{j'}$ for $j \neq j'$).

Then we define the propagator domain $X_0^{\mathrm{prop}}(G, \mu, \Omega)$ of (G, μ, Ω) as one (arbitrarily chosen) subset of maximal cardinality $x_0^{\mathrm{prop}}(G, \mu, \Omega) = |X_0^{\mathrm{prop}}(G, \mu, \Omega)|$ among all subsets which have the property HZ. Associated to this propagator domain, there is therefore a particular set of distinct lines l_j, $j = 1, \ldots, x_0^{\mathrm{prop}}(G, \mu, \Omega)$, such that each ω_{l_j} visits the corresponding cube Δ_j of $X_0^{\mathrm{prop}}(G, \mu, \Omega)$. We choose a particular set of such lines, $L(G, \mu, \Omega)$, and call it the set of "horizontal connections" of the propagator domain. It is important to notice that for $l \in L(G, \mu, \Omega)$

the path ω_l does not necessarily stay in $X_0^{\text{prop}}(G,\mu,\Omega)$, but for l not in $L(G,\mu,\Omega)$, the path ω_l must be confined in the cubes of $X_0^{\text{prop}}(G,\mu,\Omega)$; otherwise $x_0^{\text{prop}}(G,\mu,\Omega)$ would not be maximal (see Fig. II.6.4 for examples of the natural naive propagator domain and of the propagator domain as defined above).

That this definition is an appropriate one for our purpose will become clear below; but it is not the unique one possible,† and it requires some arbitrary choices, which is unfortunate on an aesthetic level, but does not seem to be easily avoidable: arbitrary choices appear in most cluster expansions, often in the form of an arbitrary ordering of some finite set of geometric objects.

We define now:

$$X_0(G,\mu,\Omega) \equiv X_0^{\text{vert}}(G,\mu) \cup X_0^{prop(G,\mu,\Omega)}$$

and $x_0(G,\mu,\Omega) \equiv |X_0(G,\mu,\Omega)|$.

The proof of (II.6.18) is decomposed into two steps. Fix $\delta > 0$, and write:

$$a_n^{UR} = a_{n,\text{small}}^{UR} + a_{n,\text{large}}^{UR} \tag{II.6.20}$$

where $a_{n,\text{small}}$ is the sum over contributions with $x_0(G,\mu,\Omega) \leq n(\log n)^{-\delta}$, and $a_{n,\text{large}}$ is the complement. We want to bound $a_{n,\text{large}}^{UR}$. We may again distinguish two subcases. Let $\delta' > 0$ be such that $\delta \ll \delta'$. The first subcase is when $x_0^{\text{vert}}(G,\mu) \leq n(\log n)^{-\delta'}$, and we call the sum of the corresponding contributions $a_{n,\text{large prop}}^{UR}$. In this first subcase, we remark indeed that we must have $x_0^{\text{prop}}(G,\mu,\Omega) > (n/2)(\log n)^{-\delta}$ (since $\delta \ll \delta'$). In the second subcase, when $x_0^{\text{vert}}(G,\mu) > n(\log n)^{-\delta'}$, the sum of contributions is called $a_{n,\text{large vert}}^{UR}$.

Then our first goal is to explain, without entering all technicalities, the following bounds:

Lemma II.6.1

For some ε (depending on δ):

$$|a_{n,\text{large prop}}^{UR}| \leq C^n \frac{n!}{e^{n(\log n)^{\varepsilon}}}\,; \tag{II.6.21a}$$

$$|a_{n,\text{large vert}}^{UR}| \leq C^n \frac{n!}{e^{n(\log n)^{\varepsilon}}}\,. \tag{II.6.21b}$$

---

†To test its understanding of the problem we suggest that the reader tries to invent an other one, for instance with at most *two* cubes associated to distinct propagators.

Let us start with (II.6.21a). The reason for which $a^{UR}_{n,\text{large prop}}$ is small is that many propagators are longer than their typical decay scale. More precisely since there at most const.r^4 cubes in $\mathbf{D}_i$ at a distance less or equal to rM^{-i} of any given cube of $X_0^{\text{vert}}(G,\mu)$, there must be at least half of the cubes of the "propagator domain" which are far (in the relevant scale) from any cube of the "vertex domain" of the same scale. More precisely, at least half of the cubes Δ of $X_0^{\text{prop}}(G,\mu,\Omega)$ must satisfy an inequality:

$$\text{dist}(\Delta, X_0^{\text{vert}}(G,\mu)\cap \mathbf{D}_i) \geq cM^{-i}(\log n)^{(\delta'-\delta)/4} \qquad \text{if} \quad \Delta \in \mathbf{D}_i. \quad \text{(II.6.22)}$$

The full propagator $C^i(x,y)$ satisfies a bound like (II.1.6), but the conditioned propagator:

$$C^i_\Delta(x,y) = \int_{M^{-2i}}^{M^{-2(i-1)}} dte^{-m^2t} \int P_t(x,y)\chi_\Delta(\omega)d\omega \qquad \text{(II.6.23)}$$

where $\chi_\Delta(\omega)$ is the characteristic function forcing ω to visit Δ, satisfies a more detailed bound:

$$C^i_\Delta(x,y) \leq KM^{2i}e^{-\eta M^i[\text{dist}(x,\Delta)+\text{dist}(\Delta,y)]} \qquad \text{(II.6.24)}$$

where η is a small number (the notation δ like in (II.1.6) would be confusing).

This bound is intuitively obvious if we recall that the travel time on paths acts as an infrared cutoff. To prove it is a standard exercise in using the "additive" Markovian structure of the Wiener measure, and decomposing the path ω into two pieces by introducing the first hitting time of ω with Δ (see e.g., [GJS]). (Actually, remark that in the literature it is standard to use as primary objects the faces which make the boundary of the cubes rather than the cubes themselves; this point of view helps to write a more systematic cluster expansion).

We may collect half of the decay (II.6.24) from the propagators of the horizontal connections $L(G,\mu,\Omega)$ and combine it with the information (II.6.22) to extract a factor

$$\prod_{j=1}^{(n/4)(\log n)^{-\delta}} e^{-\eta(\log n)^{(\delta'-\delta)/4}} = e^{-\eta' n(\log n)^{(\delta'-5\delta)/4}}. \qquad \text{(II.6.25)}$$

(Remark that here it is crucial to know that each cube of the propagator domain is associated to a distinct line of the graph).

We may take $\delta' - 5\delta > 0$. After extraction of the factor (II.6.25), propagators still satisfy the bound (II.1.6) (with a different constant for the decay). Hence (now forgetting the restrictions introduced by the condition $_{\text{large prop}}$) we can apply Theorem II.3.1–2, and bound $|a_{n,\text{large prop}}|$ by $K^n n! e^{-n(\log n)^\varepsilon}$ for some $\varepsilon > 0$; roughly speaking K^n comes from the estimate of Theorem II.3.1–2 for single amplitudes, $n!$ comes from the number of graphs, and $e^{-n(\log n)^\varepsilon}$ from the factor (II.6.25). This establishes (II.6.21a).

The second subcase, $a^{UR}_{n,\text{large vert}}$, is bounded according to a slightly different idea: it is no longer the value of each individual amplitude which is small because of propagators longer than usual, but it is "statistically" that the sum of amplitudes is not as large as $c^n n!$, as should be expected from the number of Wick contractions divided by the vertex symmetry factor, whose leading behavior is $4n!!/n! \simeq n!$. Let us illustrate first on a simpler example this idea, which one may call the "local factorials" principle.

Lemma II.6.2: the local factorial principle

Let us consider the φ^4 theory (in any dimension) with fixed cutoff $\kappa = 1$ (so only the first slice $i = 0$ is kept). Let $n(\Delta)$ be a family of integers associated to each cube $\Delta \in \mathbf{D}_0$ such that $\sum_\Delta n(\Delta) = n$. We call $a_n^{\{n(\Delta)\}}$ the sum of the contributions of the graphs which have exactly $n(\Delta)$ vertices in each Δ. There exists a constant K such that:

$$|a_n^{\{n(\Delta)\}}| \leq K^n \prod_\Delta n(\Delta)!. \tag{II.6.26}$$

The proof of this local factorial principle uses the exponential decay of the propagator, which suppresses strongly the Wick contractions which join distant cubes. The result has the same form (up to a large value of K) as if in fact all Wick contractions between *distinct* cubes were suppressed, which means that the usual factorial $n!$ reflecting the number of Wick contractions of $4n$ fields divided by the vertex symmetry factor $n!$ is replaced by a product of local such factorials in each cube. To prove the lemma we start with n vertices. The multinomial factor $n!/[\prod_\Delta n(\Delta)!]$ allows to distribute them in the cubes. Then we build Wick contractions by choosing for a given field, first the cube containing the field to which it contracts, then the particular field to which it contracts, and iterating this process until all the fields are exhausted. The first choice (of the cubes) lead to a sequence of sums, each of which is controlled by the decay of the corresponding propagators, and leads to a constant per sum hence to c^n. The second choice (the field in the cube) leads to a factor $c'^n \prod_\Delta n(\Delta)!^2$, provided at each step j, $j = 1, \ldots, 2n$, we contract a field in a cube Δ_j with contains a maximal number $n_j(\Delta)$ of remaining fields not yet contracted; hence the choice of the field in Δ'_j to which it will contract will cost a factor $n'_j \leq \sqrt{n_j.n'_j}$ and the total process will cost a factor bounded by $\sqrt{\prod_\Delta n(\Delta)^{4n(\Delta)}}$ as announced. Taking the multinomial coefficient into account achieves the proof of the Lemma.

We remark that when the number p of occupied cubes gets large, the behavior of $\prod n(\Delta)!$ may become significantly smaller than $n!$; it is certainly bounded by $n!/p!$. This motivates the following stronger result, in which the domain of occupied cubes and the numbers $n(\Delta)$ are no longer given, but summed up:

Lemma II.6.3: Large "vertex domain" bound

Let $p \leq n$ be an integer. For the one slice model, the sum of all perturbative contributions to a connected Schwinger function which contain vertices in exactly p different cubes, $a_{n,p}$ satisfies:

$$a_{n,p} \leq K^n \frac{n!}{p!}. \tag{II.6.27}$$

To prove this result it is natural to use a (single-scale) cluster expansion like the one introduced in the next section (III.1). Roughly speaking, the convergence of the Brydges–Battle–Federbush tree cluster expansion, Theorem III.1.1, applied to this problem, means that we can choose the cubes containing the vertices and build a tree of Wick contractions connecting them at the cost of only c^p (hence without any factorial factor). This tree eats up $2p$ fields, and the remaining Wick contractions create only a factor of order $(4n - 2p)!! \simeq 2n!/p!$, which makes the result plausible. The true proof, however, is more complicated because there is some book-keeping of vertex symmetry factors to do, and we give it in section III.1, as an example of application of the (single slice) cluster expansion.

We understand now that when the contributions $a^{UR}_{n,\text{large vert}}$ are restricted to the 0th slice, the desired bound (II.6.21b) holds, since $n!/[(n(\log n)^{-\delta'})!] \leq c^n(n!/e^{n(\log n)^\varepsilon})$ for some ε. Hence to extend this bound to $a^{UR}_{n,\text{large vert}}$ is the same thing as to extend the local factorial principle and lemma II.6.3 from a single slice model to the general phase space situation. This can be done because the vertical and horizontal exponential decay of usefully renormalized perturbation theory is the correct generalization to phase space of the horizontal "spatial decay" of the single slice model. However this requires the generalization to phase space of the single-slice (standard) cluster expansion (which is what we call a "multiscale cluster expansion"), and even with this tool the details of the combinatoric ([MNRS, section 3 and Appendix B) remain complicated. Since the next chapter of this book is devoted to cluster expansions and their use in constructive theory, we suggest that after studying this chapter, the interested reader returns to (II.6.21b) and builds up its own proof for it, using the simplest possible multiscale expansion (of the "pair of cubes" type) rather than the one of [GJS][MNRS] (to use section 3 of [MNRS] for some clues is of course allowed!). Indeed for a proof of (II.6.21b) a cluster expansion which localizes only vertices is clearly enough; it is only for (II.6.21a) and Lemma II.6.5 below that a cluster expansion which localizes both vertices and propagators is required.

It remains now to apply the Sobolev inequality when the total vertex and propagator domain in phase space is small. We follow the same approach as for the large vertex domain, namely we start with an easy lemma as a motivation:

Lemma II.6.4 Upper Lipatov bound in finite volume

Let us consider the φ_4^4 model with fixed ultraviolet cutoff κ. Let X_0 be a region of total volume $|X_0| \leq n/(\log n)^\delta$ with $0 < \delta < 1$. There exists $\varepsilon(n)$ such that $\lim_{n\to\infty} \varepsilon(n) = 0$ and:

$$\frac{1}{n!} \int d\mu_\kappa(\varphi) \left[\int_{X_0} \varphi^4(x) dx \right]^n \leq (1 + \varepsilon(n))^n n!\, a_{\text{Lip}}^n \qquad \text{(II.6.28)}$$

where we recall that a_{Lip}, the Lipatov constant, is defined by (II.6.5), hence $a = (4K_{\text{Lip}})^2$, K_{Lip} being defined by (II.6.7–8) for $d = 4$.

Proof Applying the Sobolev inequality (II.6.7) we obtain:

$$\text{[left hand side of (II.6.28)]} \leq \frac{1}{n!} K_{\text{Lip}}^{2n} \int d\mu_\kappa(\varphi) \left[\int_{X_0} \partial_\mu\varphi\partial^\mu\varphi + m^2\varphi^2 \right]^{2n} . \qquad \text{(II.6.29)}$$

Integrating over $d\mu_\kappa(\varphi)$, we get a sum over closed loops, each propagator being $(-\Delta + m^2)C_\kappa$, where C_κ, the cutoff propagator (I.3.7–8) is smaller than $(p^2 + m^2)^{-1}$ in Fourier space and decays exponentially in direct space (over lengths of order κ^{-1}). Therefore each closed loop contribution is bounded by $C|X_0|$, where C is a constant depending on κ (the single factor $|X_0|$ takes into account the single translation invariance of the connected loop). Hence:

$$\int d\mu(\varphi) \left[\int_{X_0} \partial_\mu \varphi \partial^\mu \varphi + m^2 \varphi^2 \right]^{2n}$$

$$\leq \sum_{p=1}^{2n} \frac{(2n)!}{p!} (C|X_0|)^p \sum_{\substack{t_1,\dots,t_p \\ \sum_{j=1}^{p} t_j = 2n}} \frac{(2t_1-1)!!\dots(2t_p-1)!!}{t_1!\dots t_p!}$$

$$\leq \sum_{p=1}^{2n} \frac{(2n)!}{p!} (C|X_0|)^p 2^{2n} C_{2n}^p \tag{II.6.30}$$

where C_{2n}^p is the binomial coefficient, p is the total number of closed loops, and $t_1, \dots, t_p$ are the number of vertices in each loop. The combinatorial factors arise from the number of possible Wick contractions for each loop, which is easy to compute. Since $C|X_0| < n/(\log n)^\delta$, we have obviously:

$$\sum_{p=1}^{2n} C_{2n}^p \frac{(C|X_0|)^p}{p!} \leq e^{Cn(\log n)^{-\varepsilon}} \quad \text{for some} \quad \varepsilon. \tag{II.6.31}$$

Remembering $(2n)! \leq 2^{2n}(n!)^2$, this proves the lemma.

It remains to extend this lemma to the phase space context:

Lemma II.6.5

$$|a_{n,\text{small}}^{UR}| \leq n![a_{\text{Lip}}]^n(1+\varepsilon(n))^n. \tag{II.6.32}$$

Several difficulties arise when one tries to extend Lemma II.6.4 to Lemma II.6.5. First in Lemma II.6.4 the region X_0 is given, but in Lemma II.6.5 we know only that it is small and we must sum over all possibilities. A similar difficulty separates Lemma II.6.2 from II.6.3; and we said that the solution requires a kind of cluster expansion. This is true also here, and this cluster expansion must be a multiscale one which localizes both propagators and vertices. Again here we provide simply minimal guiding remarks, in order to avoid overlap with the next chapter. A cluster expansion is, roughly speaking, an algebraic machinery to define a domain and select a particular explicit subset of connections between the different regions (here, cubes) of the domain, which ensure sufficient decay between these regions, so that one can sum upon the position and shape of the domain. Concretely this is usually done by Taylor expansions in interpolating ("decoupling") parameters. But in perturbation theory (which is free of the stringent positivity requirements of constructive theory) these interpolating parameters are not strictly speaking necessary; in [MNRS] a mixed approach is used in which for the horizontal (in the sense of Fig. II.2.1) cluster expansions, interpolating parameters are used, but for the vertical connections, a set of vertices with fields in different slices (vertical dotted lines in our standard multiscale representation) is selected in a set-theoretic way (i.e., without interpolating parameters and Taylor expansions). Here since we define the full "vertex and propagator domain" in a set-theoretic way it is more natural to select the set of horizontal connections also in a set theoretic way, so that no interpolating parameters and Taylor formulae are ever needed; this set is simply $L(G,\mu,\Omega)$, defined as above. This simplifies some details in [MNRS].

All together this process selects particular propagators and vertices which connect together the vertex and propagator domain, and contain enough convergence factors to realize the vertical and horizontal exponential decay between its cubes, hence to solve the problem of summing over all such possible domains. Roughly speaking, for convergent almost local subgraphs, it is enough to select vertices in them with at least 5 external fields of lower momenta. When divergent subgraphs of the almost local type appear in some places, the corresponding useful counterterms are associated to them. We may select all the external vertices of these subgraphs and let the corresponding subtractions act on their external lines. In both cases vertical decay is generated, the main problem being the burden of notations. The price to pay is simply that all fields hooked to selected vertices cannot be included in the set of remaining fields to which the Sobolev inequality is applied. However an important point is that, as a whole, the number of fields contracted into horizontal connections or hooked to selected vertices remain bounded by const.$x_0(G,\mu,\Omega)$, hence by const.$(n/(\log n)^{\delta})$. Hence the remaining fields are vastly the majority.

There are still implicit restrictions on the Wick contractions of these remaining fields (for instance the ones who tell that G itself must be connected). We can now lift all these restrictions, (hence producing an overestimate for $a^{UR}_{n,\mathrm{small}}$). Then the remaining fields may be written as a single functional integral of the type $\int (\int_{X_0} \varphi^4)^{n'} d\mu_{X_0}(\varphi)$ and *no longer* developed into Feynman graphs. Remark that we use the word "field" for simplicity, but since Feynman graphs rather than fields were our starting point, one should more precisely say that the remaining "pieces of Feynman graphs" are written as a single functional integral. Also our notations are very loose. In this integral $\int(\int_{X_0}\varphi^4)^{n'}d\mu_{X_0}(\varphi)$, for simplicity we did not decompose the field φ as the sum of fields associated to each slice (using (II.1.5)) as should be done in fact. Also for simplicity a single notation X_0 recalls the restrictions over the range of integration for vertices coming from the definition of the vertex domain, and the restrictions on the propagator, hence on the Gaussian measure corresponding to the definition of the propagator domain. The important point to stress is that this definition is chosen so that although the paths of the propagators l_j of the horizontal connections do not necessarily lie in the "vertex and propagator domain," all the paths of the other propagators do. This implies that the Gaussian measure $d\mu_{X_0}$ with which these remaining fields are integrated is supported on fields which have their support inside this domain and vanish, together with their gradient, at the boundary ∂X_0. Hence these fields do obey the (infinite volume) Sobolev inequality. Therefore we can apply the Sobolev inequality in the manner of Lemma II.6.4 to this functional integral. Since the number of cubes in phase space turns out to be, as expected intuitively, the correct factor which generalizes the volume $|X_0|$ in Lemma II.6.4, Lemma II.6.5 follows. This completes our sketch of the proof of Theorem II.6.3.

Finally we will sketch how the Lipatov upper bound on usefully renormalized series combines with an analysis of the recursion relation for effective coupling constant in order to obtain Theorem II.6.2 [DFR]. We will need in fact the slightly more detailed corollaries of the analysis above:

For any $\delta > 0$ there exists a function $\varepsilon(n)$ with $\lim_{n\to\infty} = 0$ and

$$\sum_{(G,\mathrm{F},\mu);n(G)=n,\mathrm{F}=\emptyset} \int \prod_{v\in G} dx_v |Z_{(G,\mathrm{F},\mu)}| e^{(2-\delta)i_{\max}(\mu)} \leq n![a_{\mathrm{Lip}}]^n(1+\varepsilon(n))^n, \tag{II.6.33}$$

where $i_{\max(\mu)}$ is the maximum index in μ. This bound will result from the fact that in index space the exponential decay rate is at least 2, since the worst superficial convergence degree is at least 2 after useful renormalization has been performed. After the factor $e^{(2-\delta)i_{\max}(\mu)}$ of (II.6.33) is included, there remains indeed a vertical exponential decay with rate at least δ.

We need also related bounds for the 4 point counterterms:

$$\sum_{\substack{(G,\mathbf{F},\mu)\\ n(G)=n,N(G)=4,\mathbf{F}=\emptyset\\ k\leq i_G(\mathbf{F})\leq e}} \int \prod_{v\in G} dx_v |Z^0_{(G,\mathbf{F},\mu)}| e^{(2-\delta)[i_{\max}(\mu)-i_G(\mathbf{F})]}$$
$$\leq (e-k+1)n![a_{\mathrm{Lip}}]^n(1+\varepsilon(n))^n, \quad \text{(II.6.34)}$$

where Z^0 is the coupling constant counterterm obtained by applying τ_G instead of $(1-\tau_G)$ in (II.6.17). Remark that now we have only exponential decay between the maximal index $i_{\max}(\mu)$ and the minimal index $i_G(\mathbf{F})$ (as defined by (II.3.34)) in G. There is one global translation invariance in vertical index space for G corresponding to the logarithmic divergence of a coupling constant counterterm. These features explain the form of (II.6.34).

Similarly:

$$\sum_{\substack{(G,\mathbf{F},\mu)\\ n(G)=n,N(G)=2,\mathbf{F}=\emptyset\\ k\leq i_G(\mathbf{F})\leq e}} \int \prod_{v\in G} dx_v |Z^0_{(G,\mathbf{F},\mu)}| e^{(2-\delta)[i_{\max}(\mu)-i_G(\mathbf{F})]}$$
$$\leq M^{2(e-k)}n![a_{\mathrm{Lip}}]^n(1+\varepsilon(n))^n, \quad \text{(II.6.35)}$$

where Z^0 is the mass counterterm obtained by applying τ_G^0 instead of $(1-\tau_G)$ in (II.6.17). This bound reflects the quadratic divergence of a mass counterterm (which in practice in the case of useless counterterms is always compensated by the quadratic convergence of one of the external legs of G).

In [DFR] the renormalized perturbation series is then recast into a form which is not exactly the effective expansion of Sect. II.4, but an expansion intermediate between this one and the renormalized expansion. More precisely only the useless counterterms for the subgraphs isomorphic to the bubble are resummed into effective constants, and the other useless counterterms for all subgraphs except the ones isomorphic to the bubble are kept in the expansion. This may seem complicated and not very natural, but it has the advantage of leading to a very simple recursion rule for effective constants which is simply a second order polynomial [DFR]:

$$g^i(g_r) = g^{i-1}(g_r) + Y(i-1)(g^{i-1}(g_r))^2 \quad \text{(II.6.36)}$$

where $Y(i)$ is the value of the bubble graph with minimal index i:

$$Y(i) = b \sum_{\min\{j_1,j_2\}=i} \int d^4x C^{j_1}(0,x) C^{j_2}(x,0), \tag{II.6.37}$$

b being the symmetry factor of the bubble. The advantage is that this simple recursion leads to an easy bound on the Borel transform of the product of effective couplings associated to a contribution $(G,\mu,\mathbf{F})$. This product $DR_{(G,\mu,\mathbf{F})} = \prod_{v\in G} g^{e_v(\mu,\mathbf{F})}$, using (II.6.36–37) satisfies indeed:

$$|B'(DR_{(G,\mu,\mathbf{F})})(b)| \leq \frac{|b|}{(n-1)!} e^{\sum_{0\leq j<i_{\max}(\mu)} Y(j)|b|}$$
$$\leq \frac{|b|^{n-1}}{(n-1)!} e^{|b|\beta_2 i_{\max}(\mu)+const.}, \tag{II.6.38}$$

where $\beta_2 = 9/2\pi^2$ is the coefficient in Theorem II.6.2, and the modified Borel transform (easily related to the ordinary one) is simply

$$B'\left(\sum_{n\geq 1} a_n g_r^n\right) \equiv \sum_{n\geq 1} \frac{a_n b^n}{(n-1)!}. \tag{II.6.39}$$

Then in [DFR] some analysis of the useless counterterms is performed by combining the bounds (II.6.34–35) with some tedious combinatoric arguments. The outcome is that as long as $|b| \leq 2/\beta_2$ the bound (II.6.38) essentially tells us that there is still vertical index space decay and (II.6.33) leads to analyticity in the Borel plane, because the Lipatov constant is bigger than $2/\beta_2$ by a factor 3/2. Theorem II.6.2 then follows.

The use of a modified effective expansion is a technical device which is not very appealing. Also Theorem II.6.2 remains unsatisfactory. We would like to rule out possible ("miraculous") cancellations and have a proof of the existence of the first renormalon singularity at $b = \beta_2/2$, by showing that some derivative of $B(b)$, the Borel transform of the renormalized series blows up at this point. This result would be the first direct proof that the φ_4^4 series diverge and are not Borel summable. Ecalle's theory of resurgent functions [Ec] (see also [GKT]) may be relevant for this difficult problem, because it gives information on the asymptotic behavior of the Borel transform of recursive relations much more general than (II.6.36). However the problem is difficult, because the natural approach would be to define rigorously the full β function. It is not yet clear whether this is possible in the BPHZ scheme [Kop].† From [FMRS5] we know that the β function for infrared φ^4 is Borel summable in terms of the *bare* coupling; transcribed to the ultraviolet problem these results only mean that the β function of the massive theory with an ultraviolet cutoff M^ρ is Borel summable in terms of the renormalized coupling in a disk which shrinks as ρ^{-1} as $\rho \to \infty$. If we cannot use the full ultraviolet β function the only available approach seems to improve on the explicit resummations of [DFR]; if they could be extended to third order graphs, so as to reconstruct the first two terms of the β function rather than the first one, a proof of existence of the first φ_4^4 renormalon might become possible, in particular in the case of a vector model with large number of components, in which remainder terms might be bounded in $\frac{1}{N}$ so as to exclude the possibility of miraculous cancellations.

This completes our review of the mathematical problem of φ_4^4 large order behavior. Although our guided tour of [MNRS]-[DFR] is no mathematical substitute for more rigorous proofs, we hope that it may provide some help in understanding Theorem II.6.2. Also we think that a study of this problem, in particular of the Lipatov upper bound, provides a natural introduction to the main theme of constructive theory, hence to the next part of this book.

Indeed although at first sight the large order behavior of φ^4 is a purely perturbative problem, we see that for this problem the language of Feynman graphs reaches its limits, and functional integration becomes unavoidable, precisely at the point where the Sobolev inequality is used to gain information on the collective behavior of the graphs. Very similar phenomena occur in constructive theory. The main theme in constructive theory, as we shall see now, is indeed exactly to sort out explicitly some critical connections between regions of phase space (by a cluster expansion), but to treat the rest of the theory as a functional integral. In constructive theory the goal is no longer to apply a Sobolev inequality to this rest, but it is still to take the functional form into account to gain some information which is hidden at the level of Feynman graphs, for instance the positivity used in "domination" or large fields bounds.

---

†In the minimal subtraction scheme based on dimensional regularization there is some belief that the β function should be Borel summable [BDZ], but a proof may have to wait until constructive theory finds a non-perturbative way to define dimensional regularization....

PART III

CONSTRUCTIVE RENORMALIZATION

Chapter III.1

Single Scale Cluster and Mayer Expansions

Cluster and Mayer expansions are key tools in many areas of mathematical physics. Introduced in constructive field theory by Glimm, Jaffe and Spencer to complete the construction of φ_2^4 [GJS], they have been improved or generalized over the years, in particular by Brydges, Battle and Federbush [BrFe][BaF1–3][Bat]. Unfortunately these expansions had for a while a reputation of being heavy to handle. In this section we try to dispel this impression by underlying the main facts behind their convergence. We do not try to present the best techniques or the optimal bounds. Our goal is simply to introduce beginners to the paradise of expansions; for a more complete review we suggest the reading of [Bry].

In this section we work always in a single momentum slice, for instance the slice with index 0 in phase space. In other words we have both fixed infrared and ultraviolet cutoff. Nevertheless we have in mind to use the single scale expansion as the building block for the multiscale expansions of the next section. This gives us some flexibility which we would not have if we were forced to treat as far as possible models with a single scale expansion. For instance to some extent the way the slicing is done and the form of these cutoffs may be chosen at will. In particular we can always require the sliced propagator to have fast decay at infinity, for instance exponential decay.

Taking advantage of this simplifies some arguments. For instance once the propagator decays sufficiently fast, we can use what we call the "volume argument." This means simply that in a *finite dimensional* lattice of cubes, there are not many cubes close to a given one. This simple observation, combined with the rapid decay of the propagator, simplifies often the combinatoric which arises within the expansion.

We describe first how to recast a Gaussian measure perturbed by a small stable interaction in the form of a polymer system with hard core interaction: the very name of polymer comes from this "excluded volume" effect, hence from the hard core interaction. This step is what we specifically call the cluster expansion. Then we show how to remove the hard core interaction and compute normalized quantities (or the pressure of the system). This is what we call the Mayer expansion, and it allows us to control the thermodynamic (or infinite volume) limit. This Mayer expansion is really a systematic way to compute the logarithm of a grand canonical partition function. It would be an interesting exercise, which we leave to the reader, to apply this expansion formalism to ordinary perturbation theory (which may be considered as a grand canonical Bose gas of vertices). With this formalism one could for instance recover the reasons for which connected functions are sums of connected graphs and for which the combinatoric of the Bogoliubov recursion works.

The typical situation we study is a massive Gaussian measure $d\mu(\varphi)$ with an ultraviolet cutoff, perturbed by a $g\varphi^4$ interaction with g small. The measure could also be massless but with both infrared and ultraviolet cutoffs (remember that it should correspond to a momentum *slice*.) This seems a somewhat trivial situation, but remember that even in the finite volume the perturbation expansion would diverge because of the large number of graphs. Moreover, even for a massive theory defined in a finite volume Λ, the thermodynamic limit $\Lambda \to \infty$ is not trivial to define directly, and that is precisely what a single cluster and Mayer expansion will easily do (the problem of the large number of graphs is bypassed because the expansion automatically develops only a piece (typically small) of the interaction).

The partition function in a volume Λ is

$$Z(\Lambda) = \int d\mu(\varphi) e^{-g \int_\Lambda \varphi^4(x) dx}, \tag{III.1.1}$$

the pressure is

$$P(\Lambda) = \frac{1}{\Lambda} \log Z(\Lambda), \tag{III.1.2}$$

the unnormalized Schwinger functions are

$$S^u_{N,\Lambda}(z_1,\ldots,z_N) = \int \varphi(z_1)\ldots\varphi(z_N)d\mu(\varphi)e^{-g\int_\Lambda \varphi^4(x)dx} \qquad \text{(III.1.3)}$$

(remember that they are distributions which should be smeared by test functions $\chi(z_1)$, ..., $\chi(z_N)$). The normalized Schwinger functions are

$$S_{N,\Lambda}(z_1,\ldots,z_N) = \frac{1}{Z(\Lambda)} S^u_{N,\Lambda}(z_1,\ldots,z_N). \qquad \text{(III.1.4)}$$

The interaction $\exp(-g\int_\Lambda \varphi^4)$ could be generalized, but we should then assume that it remains stable, that is has a small constant like g in front, and that it is local in the sense that if we cut Λ in several regions Λ_i, it factorizes as a product of functions of the field in Λ_i: in our case this is just $\exp(-g\int_\Lambda \varphi^4) = \prod_i \exp(-g\int_{\Lambda_i} \varphi^4)$.

The dimension d of space time has to be finite (for the volume argument below) but is not necessarily 4 at this stage.

A The cluster expansion

We consider a normalized Gaussian measure $d\mu$ whose covariance is a symmetric positive definite operator in x-space $C(x,y)$ which has good decrease at infinity. It could be for instance the covariance C^0 (first slice) of (II.1.4), but in this section we drop the superscript 0 for simplicity. We assume that its decay is either exponential:

$$|C(x,y)| \le O(1)e^{-|x-y|} \qquad \text{(III.1.5)}$$

or power-law with a very large (adjustable) power rate r:

$$|C(x,y)| \le O(1)\left(\frac{1}{1+|x-y|}\right)^r. \qquad \text{(III.1.6)}$$

(We have rescaled the mass or the unit of length so that there is no scale coefficient in (III.1.5) or (III.1.6)). Then we divide our volume Λ into hypercubes of side size unity, which form a lattice $\mathbf{D}$. On $\mathbf{D}$ we adopt for convenience an arbitrary order: $\mathbf{D} = \{\Delta^1,\ldots,\Delta^{|\Lambda|}\}$. We adopt also the same notation Δ for a cube and for its characteristic function (also called sharp characteristic function): $\Delta(x) = 0$ if $x \notin \Delta$, $\Delta(x) = 1$ if $x \in \Delta$. We can write

$$\Lambda(x) = \sum_{\Delta\in\mathbf{D}} \Delta(x) \qquad \text{(III.1.7)}$$

and we might also use C_0^∞ versions of the characteristic functions Δ (which we call smooth characteristic functions) so that (III.1.7) would then be a smooth partition of unity over Λ. In both cases the notation

$\int_\Delta f(x)$ means really $\int_\Lambda \Delta(x) f(x)$. Working within volume Λ means that we consider $\Lambda(x)C(x,y)\Lambda(y)$ as our covariance.

Since coupling between different cubes, which prevents factorization of (III.1.1) over the cubes of **D**, comes solely from $d\mu$, hence from the covariance C, we want to interpolate directly in C between an uncoupled situation and the coupled one. But we would like the interpolated covariance to preserve the positivity of C (as an operator) so that it still corresponds to an interpolated Gaussian measure. To weaken off-diagonal elements (which generate the unwanted couplings) in a discrete, finite dimensional positive symmetric matrix C, one could multiply these off-diagonal elements C_{ij} and C_{ji} by an interpolating parameter s_{ij} for each pair i,j, $i \neq j$. This point of view leads naturally to what is probably the simplest cluster expansion, the "pair of cubes" cluster expansion of Fig. III.1.1.

However this process does not preserve positivity in general. For instance $\begin{pmatrix} 1 & 1 & 1 \\ 1 & 2 & 2 \\ 1 & 2 & 3 \end{pmatrix}$ is positive, but $\begin{pmatrix} 1 & 1 & 0 \\ 1 & 2 & 2 \\ 0 & 2 & 3 \end{pmatrix}$ is not! Multiplying off-diagonal elements by parameters between 0 and 1 preserves positivity only if the diagonal elements are quite big compared to the off-diagonal ones. Hence to be used the pair of cubes expansion may require that we take a lattice of rather *large* cubes compared to the unit scale. This in effect reinforces the diagonal piece of C viewed as a matrix between the cubes of **D** and allows preservation of positivity (see [FMRS5]). For the moment we do not want to use the (somewhat complicated) trick of enlarging our cubes. Then we can remark that in the particular case of a 2 by 2 matrix, positivity is preserved by damping the (single!) off-diagonal piece.

This suggests an inductive expansion in which at each step one tests the coupling between a set of cubes and the complement. This

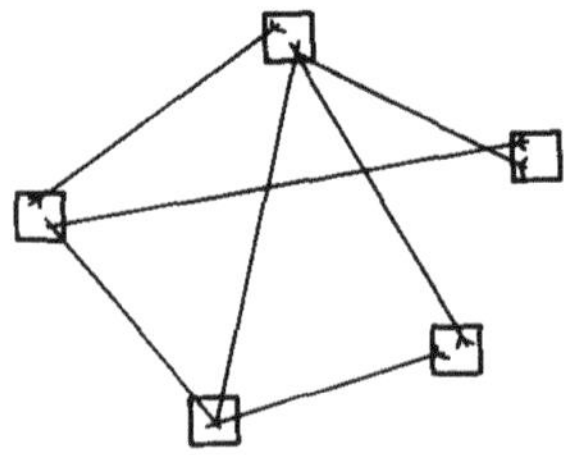

Figure III.1.1 A connected contribution in the "pair of cubes" cluster expansion.

automatically preserves positivity and generates the tree-like expansion of Fig. III.1.2. It is this powerful point of view that we explain now; its only drawback is that it is inductive, hence its outcome is difficult to capture in a single formula.

For the tree expansion, we start with the first labeled cube in **D**, say $\Delta_1 = \Delta^1$ (the upper index is reserved for the arbitrary order on **D**), and we introduce a first parameter s_1 which tests the coupling between Δ_1 and the rest of **D** noted $\bar{\Delta}_1$ (a bar indicates the complement in **D**)

$$\begin{aligned} C(s_1,x,y) &= s_1 C(x,y) + (1-s_1)[\Delta_1(x)C(x,y)\Delta_1(y) + \bar{\Delta}_1(x)C(x,y)\bar{\Delta}_1(y)] \\ &= \Delta_1 C\Delta_1 + \bar{\Delta}_1 C \Delta_1 + s_1 \left[\Delta_1 C\bar{\Delta}_1 + \bar{\Delta}_1 C\Delta_1\right]. \end{aligned} \tag{III.1.8}$$

The first expression is a convex combination of functions of positive type, hence is of positive type, and proves that there is no problem to define the corresponding Gaussian measure $d\mu_{s_1}$; the second form is also useful to check that the s dependence is indeed on the off-diagonal terms. We follow this interpolation by a first order Taylor expansion. For instance for the partition function we write:

$$\begin{aligned} Z(\Lambda) &= \int d\mu_{s_1}(\varphi) e^{-g\int_\Lambda \varphi^4}\Big|_{s_1=1} \\ &= \int d\mu_{s_1} e^{-g\int_\Lambda \varphi^4}\Big|_{s_1=0} + \int_0^1 ds_1 \frac{d}{ds_1} \int d\mu_{s_1}(\varphi) e^{-g\int_\Lambda \varphi^4} \\ &= Z(\Delta_1)Z(\bar{\Delta}_1) + \sum_{\Delta_2 \neq \Delta_1} Z_{\Delta_1\Delta_2}(\Lambda), \end{aligned} \tag{III.1.9}$$

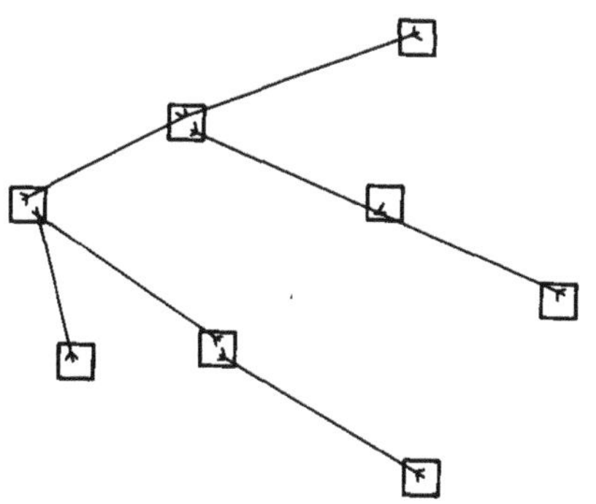

Figure III.1.2 A connected contribution in the tree cluster expansion.

with

$$Z_{\Delta_1\Delta_2}(\Lambda) = \int_0^1 ds_1 \int d\mu_{s_1}(\varphi) \int_{\Delta_1} dx \int_{\Delta_2} dy \cdots$$

$$\cdots C(x,y)\left[\frac{\delta}{\delta\varphi(x)}\frac{\delta}{\delta\varphi(y)} e^{-g\int_\Lambda \varphi^4}\right] \quad \text{(III.1.10)}$$

(here it is necessary to distinguish between upper and lower indices because in general Δ_2 is not Δ^2, the second cube in our list!). To check formulae (III.1.9)–(III.1.10) we recall that at $s_1 = 0$ we can consider $\Delta_1\varphi$ and $\bar{\Delta}_1\varphi$ as two independent fields, and since the interaction factorizes, Z also factorizes. The advantage here is that we do not really need to worry about boundary conditions like Dirichlet, Neumann, etc.... which were important technicalities in former formalisms. Formula (III.1.10) follows from

$$\frac{d}{ds_1} C(s,x,y) = \sum_{\Delta_2 \neq \Delta_1} [\Delta_1(x)C(x,y)\Delta_2(y) + \Delta_2(x)C(x,y)\Delta_1(y)] \quad \text{(III.1.11)}$$

and from integration by parts with respect to φ [GJ2, Ch. 9].† It creates both a propagator between points localized in Δ_l and Δ_2, whose decay becomes available to perform the sum over Δ_2 at Δ_1 fixed in (III.1.9), and vertices hooked to this propagator, which are created by the derivatives $\delta/(\delta\varphi(x))$ and $\delta/(\delta\varphi(y))$ acting in (III.1.10) on the exponential. This is also useful because these vertices have precious factors g attached to them.

The definition of our second interpolation depends on which term in (III.1.9) we look at, and that's why this expansion is inductive. For the first term, Δ_1 is decoupled from the rest and we should start again with **D** now restricted to $\bar{\Delta}_1$, hence pick the next cube (which, this time, is Δ^2) and test its coupling with $\overline{\Delta_1 \cup \Delta^2}$. But for the $Z_{\Delta_1\Delta_2}$ term we will introduce an interpolating parameter s_2 which tests the coupling of the *union* of Δ_1 and Δ_2 to $\overline{\Delta_1 \cup \Delta_2}$ (remember the former remarks about positivity). Hence in this case the interpolated propagator is a natural iteration of (III.1.8). Writing Δ_{12} for $\Delta_1 \cup \Delta_2$, $\bar{\Delta}_{12}$ for its complement

---

†This functional integration by parts can be checked first for a polynomial integrand using the explicit rules of Gaussian integration for this case; it can be extended to more general C^∞ functions by a continuity argument.

in **D**, and using $\Delta_1\Delta_{12} = \Delta_1$, $\Delta_1\bar{\Delta}_{12} = 0$, etc...., we get:

$$\begin{aligned}
&C(s_1, s_2, x, y) \\
&\quad = s_2 C(s_1, x, y) + (1 - s_2)(\Delta_{12} C(s_1, x, y)\Delta_{12} \\
&\qquad + \bar{\Delta}_{12} C(s_1, x, y)\bar{\Delta}_{12}) \\
&\quad = s_1 s_2 C + s_1(1 - s_2)[\Delta_{12} C\Delta_{12} + \bar{\Delta}_{12} C\bar{\Delta}_{12}] \\
&\qquad + (1 - s_1)s_2[\Delta_1 C\Delta_1 + \bar{\Delta}_1 C\bar{\Delta}_1] \\
&\qquad + (1 - s_1)(1 - s_2)[\Delta_1 C\Delta_1 + \Delta_2 C\Delta_2 + \bar{\Delta}_{12} C\bar{\Delta}_{12}] \\
&\quad = \Delta_1 C\Delta_1 + \Delta_2 C\Delta_2 + \bar{\Delta}_{12} C\bar{\Delta}_{12} + s_1[\Delta_1 C\Delta_2 + \Delta_2 C\Delta_1] \\
&\qquad + s_2[\Delta_2 C\bar{\Delta}_{12} + \bar{\Delta}_{12} C\Delta_2] + s_1 s_2[\Delta_1 C\bar{\Delta}_{12} + \bar{\Delta}_{12} C\Delta_1]
\end{aligned} \tag{III.1.12}$$

and the Taylor expansion in s_2 gives:

$$\begin{aligned}
Z_{\Delta_1\Delta_2}(\Lambda) &= Z_{\Delta_1\Delta_2}(\Delta_{12}) Z(\bar{\Delta}_{12}) \\
+ \sum_{\Delta_3 \neq \Delta_1, \Delta_2} &\int_0^1 ds_1 \int_0^1 ds_2 \int d\mu_{s_1, s_2}(\varphi) \int_{\Delta_1} dx_1 \int_{\Delta_2} dy_1 \int_{\Delta_{12}} dx_2 \int_{\Delta_3} dy_2 \\
\cdots C(x_1, y_1) C(s_1, x_2, y_2) &\left[\frac{\delta}{\delta\varphi(x_2)} \frac{\delta}{\delta\varphi(y_2)} \frac{\delta}{\delta\varphi(x_1)} \frac{\delta}{\delta\varphi(y_1)} e^{-g\int_\Lambda \varphi^4} \right].
\end{aligned} \tag{III.1.13}$$

If we want to have spatial integrals over unit cubes, we see that we need to break (III.1.13) according to whether x_2 is in Δ_1 or in Δ_2, so we get a term indexed by the set $\{\Delta_1, \Delta_2, \Delta_3\}$, *and* by the two possible trees $\{(\Delta_1\Delta_2), (\Delta_1\Delta_3)\}$ or $\{(\Delta_1\Delta_2), (\Delta_2\Delta_3)\}$ corresponding to the first and the second case respectively. We can iterate this process until the finite set **D** is exhausted. Clearly the factorized contributions that we obtain, once they are decomposed so that each spatial integral is over a single cube, are not only indexed by their content or support in terms of cubes, but also by all possible tree structures that can be built on them and generated by the expansion. After reading part I, we should no longer be afraid of trees anyway!

For summation we use a formalism of ordered sequences rather than sets because the former allows more easily the repetition of a single element. We should remark that in formula (III.1.9) for the first term $Z(\Delta_1)Z(\bar{\Delta}_1)$ there are no $\delta/(\delta\varphi)$'s, so no small constant generated. In fact the partition function of a single cube $Z(\Delta_1) = Z(\Delta)$ is simply $\int d\mu_\Delta(\varphi) \exp(-g \int_\Delta \varphi^4)$, where $d\mu_\Delta$ is the normalized Gaussian measure corresponding to the covariance $\Delta(x)C(x, y)\Delta(y)$, and we expect $Z(\Delta)$ to be close to 1, since g is small. To have really 1 we could do a first order Taylor expansion in g but it would complicate the picture. We prefer to compute a slightly different partition function:

$$\tilde{Z}(\Lambda) = Z(\Lambda)/Z(\Delta)^{|\Lambda|} \tag{III.1.14}$$

($|\Lambda|$ is of course the number of cubes in Λ, also equal to its volume since the cubes are of unit size). This corresponds to a finite shift of the pressure $p(\Lambda)$ by the constant $\log Z(\Delta)$. With this choice, isolated cubes automatically cancel against the corresponding denominators in (III.1.14). The result of the cluster expansion is then simply:

$$\tilde{Z}(\Lambda) = \sum_{\substack{Y_1,\ldots,Y_q \subset \mathbf{D} \\ Y_1,\ldots,Y_q \text{ disjoint}}} \frac{1}{q!} A(Y_1)\cdots A(Y_q), \tag{III.1.15}$$

where $Y_1, \ldots, Y_q$ are polymers (called also vacuum- or $\emptyset$-polymers), i.e., sets of at least 2 cubes of $\mathbf{D}$ (isolated cubes or "monomers" are excluded), and the amplitude corresponding to a polymer $Y = \{\Delta_1, \Delta_2, \ldots, \Delta_p\}$ (where Δ_1 is the first cube of Y in the arbitrary order in $\mathbf{D}$ and $p = |Y|$) is:

$$A(Y) = \sum_{T'} \frac{1}{Z(\Delta)^{|Y|}} \int_0^1 \cdots \int_0^1 ds_1 \ldots ds_{p-1} M_{T'}(s) \int d\mu_{\{s\}}(\varphi)$$
$$\cdot \left\{ \left[\prod_{k=1}^{p-1} \int_{\Delta_{i_k}} dx_k \int_{\Delta_{j_{k+1}}} dy_k C(x_k, y_k) \frac{\delta}{\delta\varphi(x_k)} \frac{\delta}{\delta\varphi(y_k)} \right] e^{-g\int_Y \varphi^4(x)dx} \right\} \tag{III.1.16}$$

where to be precise, the sum is over ordered rooted trees T' with root 1, in the sense of Section I.4. Recall that these trees are ordered sequences of links $l_k \equiv (\Delta_{i_k}, \Delta_{j_{k+1}})$ $k = 1, \ldots, p-1$, connecting all the cubes of Y such that if we define $\Delta_{j_1} \equiv \Delta_1$ we have, for any $k \geq 1$, $\Delta_{i_k} \in \{\Delta_{j_1}, \Delta_{j_2}, \ldots, \Delta_{j_k}\}$ (see Fig. I.4.6). This ensures that for any k with $1 \leq k \leq p$, $Y_k(T') = \{\Delta_{j_r}, 1 \leq r \leq k\}$ has exactly k distinct cubes which are connected together by the subset $T'_{k-1} = \{l_1, \ldots, l_{k-1}\}$ of the $k-1$ first links of T'; in particular $Y_p(T') = Y$. As in Sect. I.4 it is convenient to consider the ancestor function a associated to T' such that the index $a(k) \leq k$ is the one such that $\Delta_{i_k} = \Delta_{j_{a(k)}}$. The Gaussian measure $d\mu_{\{s\}}$ corresponds to the symmetric propagator $C(s_1, \ldots, s_{p-1}, x, y)$ which is simply $C(x, y) \prod_{j=k'}^{k-1} s_j$ when y belongs to Δ_{j_k} and x belongs to $\Delta_{j_{k'}}$, with $k' \leq k$; when y or x are out of Y the covariance corresponding to $d\mu_{\{s\}}$ is defined to be zero. Also by definition $A(\emptyset) = 1$.

At an intermediate stage of the expansion, we may define

$$C(s_1,\dots,s_k,x,y) = C(s_1,\dots,s_{p-1},x,y)\,|_{s_{k+1}=\dots=s_{p-1}=1}\,.$$

There is therefore a polynomial s dependence of the propagators derived by the inductive expansion; by definition this polynomial dependence has been gathered in the factor $M_{T'}(s)$. It is easy to check that this factor is nothing but:

$$M_{T'}(s) = \prod_{k=1}^{p-1}\prod_{j=a(k)}^{k-1} s_j. \tag{III.1.17}$$

(III.1.15) expresses the partition function as the one of a gas of polymers swirling in Λ, with hard core interaction (the condition of disjointness). From the decay of C and the smallness of g we expect this gas to be dilute, and typical polymers to be small.

When external fields $\varphi(z_1)\cdots\varphi(z_N)$ are present we introduce the notion of an external polymer, which is a set Y of p cubes of $\mathbf{D}$ together with a subset $\zeta\subset\{z_1\cdots z_N\}$ of external variables; it is then more precisely called a ζ-polymer. Typically these external variables z_j are smeared against test functions $\chi_j(z_j)$, and one should keep in mind that if the support of these smearing functions does not overlap with the cubes of Y the corresponding amplitude of the polymer will be zero. Remark that we do not require ζ to contain all the external variables which are localized in Y (i.e., have smearing functions with support in the cubes of Y), a fact which will be convenient later.

Again we forbid the trivial case of a single isolated *empty* cube $p=1$, $\zeta=\emptyset$, but $p=1$ and $\zeta\neq\emptyset$ is allowed. Then the truncated unnormalized Schwinger functions $\widetilde{S^u_{N,\Lambda}}(z_1,\dots,z_N) = Z(\Delta)^{-|\Lambda|}S^u_{N,\Lambda}(z_1,\dots,z_N)$ are also expressed by the cluster expansion as a gas of polymers with hard core interaction:

$$\widetilde{S^u_{N,\Lambda}}(z_1,\dots,z_N) = \sum_{\substack{Y_1,\dots,Y_q\subset\mathbf{D}\\ Y_1,\dots,Y_q \text{ disjoint linear in } \{z\}}} \frac{1}{q!}A(Y_1)\cdots A(Y_q), \tag{III.1.18}$$

where the condition of "linearity in the external variables" means by definition that for each z_i, $1\le i\le N$ there is exactly one polymer Y_j, $j\in[1,\dots,q]$ which contains it. The amplitude $A(Y)$ is exactly similar to (III.1.16) but with $\exp(-g\int_Y\varphi^4(x)dx)$ replaced by

$$\left[\prod_{z_j\in\zeta}\int \chi_j(z_j)dz_j\varphi(z_j)e^{-g\int_Y\varphi^4(x)dx}\right]$$

if Y is a ζ-polymer. The next step in the analysis is to perform a Mayer expansion that will take care of the hard core interaction in (III.1.15) or (III.1.18). Anticipating what follows, let us state that this expansion has been shown to converge, for such a polymer gas, provided the amplitudes satisfy the bound [Bry]:

$$\sum_{Y \ni 0} |A(Y)| e^{|Y|} < 1. \tag{III.1.19}$$

The condition $Y \ni 0$ is there to break translation invariance, and if (III.1.19) holds uniformly in Λ, it has to hold in an infinite volume as well. Before continuing with the Mayer expansion we will show why we can achieve estimates even stronger than (III.1.19), using the smallness of the coupling constant, and the fast decay of the propagators. Hence let us prove:

Theorem III.1.1: The polymer bound

Let K be any fixed constant. For g small enough we have:

$$\sum_{Y \ni 0} |A(Y)| K^{|Y|} < 1 \tag{III.1.20}$$

where the sum in (III.1.20) is over all finite polymers in an infinite volume (because the estimate (III.1.20) in finite volume Λ is independent of the volume of Λ).

We explain in detail the proof, restricting for simplicity to vacuum polymers.† A crucial first step is to go from ordered trees to regular, unordered trees. Let us call $T(T')$ the regular (unordered) tree associated to T'. Battle and Federbush realized that at fixed T there is nothing in (III.1.16) allowing to sum over the various orderings of the trees T' with $T(T') = T$, except the small factor associated to the monomial (III.1.17) in the s parameters. The following lemma is inspired by their work [BaF1]:

†For polymers with external variables the proof is similar except the position of the external variables can be used to break translation invariance, hence there is no need to include a condition like $Y \ni 0$ in (III.1.20).

Lemma III.1.1

For any (regular, unordered) tree T we have, summarizing all the s integrals in (III.1.16) by the notation $\int ds$:

$$\sum_{T',T(T')=T} \int M_{T'}(s)ds = 1. \tag{III.1.21}$$

This lemma is often stated as the fact that $\sum_{T',T(T')=T} M_{T'}(s)ds$ is a probability measure $dp_T(s)$ [Bry].

Proof Let us introduce a parameter $\varepsilon_{i,j}$ for each pair of cubes Δ_i, Δ_j in Y. We can apply the inductive analysis above to the function $F(\varepsilon) = \prod_{(i,j)}(1+\varepsilon_{i,j})$. More explicitly starting from the root $i_1 = 1$ we introduce a first interpolating parameter:

$$\varepsilon_{i,j}(s_1) = \varepsilon_{i,j}[s_1 + (1-s_1)[\delta_{1,i}\delta_{1,j} + (1-\delta_{1,i})(1-\delta_{1,j})]] \tag{III.1.22}$$

and perform a first order Taylor expansion $F(1) = F(0) + \int_0^1 (dF/ds_1)ds_1$, and so forth. The polynomial s dependence generated by this process is again $M_{T'}(s)$ for an ordered tree T'; but by comparing powers of $\varepsilon_{i,j}$ we can now compute that the left hand side of (III.1.21) is exactly the coefficient of $\prod_{(i,j)\in T} \varepsilon_{i,j}$ in F, hence is 1.

Remember that isolated cubes are excluded, so if $Y = \{\Delta_1, \dots, \Delta_p\}$ and $p > 1$, we compute now the action of the $\prod(\delta^2/\delta\varphi(x_k)\delta\varphi(y_k))$ in (III.1.16). This action is a bit complicated to write down, but remember that $\delta/(\delta\varphi(y_k))$ acts in $\Delta_{j_{k+1}}$, which is $Y_{k+1}(T') - Y_k(T')$ with previous notation; hence it has to derive a vertex from the exponential by the formula

$$\frac{\delta}{\delta\varphi(y)} e^{-g\int_\Delta \varphi^4} = -4g\varphi^3(y)e^{-g\int_\Delta \varphi^4}. \tag{III.1.23}$$

This remark ensures that in the bound for $A(Y)$ we can extract a factor of at least g^p, and since g is as small as we want we have a "small factor per cube;" in particular this remark takes care of the large constant K in Theorem III.1.1. More precisely we claim that:

Lemma III.1.2

For a propagator decaying like (III.1.6) there exists some large constant $K' = \text{const}.K$ such that

$$K^p|A(Y)| \le (K')^p g^p \sum_T \prod_k \left[\frac{1}{1+\text{dist}(\Delta_{i_k}, \Delta_{j_{k+1}})}\right]^{(3r/4)} \prod_{i=1}^p (d_i!)^{3/2} \tag{III.1.24}$$

where d_i is the coordination number of the (regular, unordered) tree T at cube Δ_i, i.e., the number of propagators of the tree T which hook to a vertex within Δ_i. In the case of exponential decay (III.1.5), simply replace $\prod_k \left[1/(1 + \text{dist}(\Delta_{i_k}\Delta_{j_{k+1}}))\right]^{(3r/4)}$ by $\exp(-(3/4)\sum_k \text{dist}(\Delta_{i_k}, \Delta_{j_{k+1}}))$.

Indeed after performing the $\delta/(\delta\varphi)$ functional derivatives we obtain a sum over procedures P of functional integrals of the type

$$\int \prod_j \varphi_j e^{-g\int_Y \varphi^4} d\mu_{\{s\}}(\varphi);$$

the product $\prod \varphi_j$ is a short notation for a product

$$\prod_{\Delta\in Y} \prod_{k=1}^{3} [\int_\Delta dx \varphi^k(x)]^{n_k}$$

of integrals over Δ of products of one up to three fields which were produced by the $\delta/(\delta\varphi)$ functional derivatives. We commute the spatial integrals dx and the functional integral $d\mu$ and evaluate the latter by a Schwarz inequality:

$$\int \prod_j \varphi_j e^{-g\int_Y \varphi^4} d\mu_{\{s\}} \leq \cdots$$

$$\cdots \left[\int \prod_j \varphi_j^2 d\mu_{\{s\}}(\varphi)\right]^{1/2} \left[\int e^{-2g\int_Y \varphi^4} d\mu_{\{s\}}(\varphi)\right]^{1/2}$$

$$\leq \left[\int \prod_j \varphi_j^2 d\mu_{\{s\}}(\varphi)\right]^{1/2} \leq \left[\int \prod_j \varphi_j^2 d\mu(\varphi)\right]^{1/2}. \qquad \text{(III.1.25)}$$

The last inequality is true because both sides can be computed with Wick's theorem, and $C(s,x,y)$ is bounded pointwise by $C(x,y)$. This is because it is a convex combination of covariances pointwise bounded by C (see (III.1.8) and (III.1.12)), and these covariances are pointwise positive by (I.3.6–7). Remark that it is a nice coincidence that C is both pointwise positive and of positive type; in general it is *not* true that a convex combination of operators B_i (like (III.1.8)), each of which is bounded by B in the operator sense, remains bounded in the operator sense by B.

The measure $d\mu$ has both ultraviolet and infrared cutoff and the number q_i of fields φ_i localized in a cube Δ_i has to be linear in d_i, in fact bounded by $3d_i$ (each derivation step produces at most in fact 3 new fields). At an intermediate stage, after j steps ($1 \leq j \leq d_i$ of the cluster expansion, the number of fields in Δ_i is $q_{i,j} \leq q_i$. At step $j+1$ a new functional derivative in Δ_i can apply to a field already produced, or it can derive a new vertex. In the first case we have to pay a factor $q_{i,j} \leq q_i$, but $q_{i,j+1} = q_{i,j} - 1$; in the second case there is only a factor 1 (coupling constants g have been already taken into account), but $q_{i,j+1} = q_{i,j} + 3$. In the end the Gaussian functional integral (III.1.25), by an analogue of the local factorial principle (Lemma II.6.2), gives a factor $\prod_i (q_i!)^{3/2}$. To use Lemma II.6.2 we need a (summable) piece of the spatial decay of the propagator, e.g., one fourth of the initial decay (III.1.6), which explains the factor $3r/4$ in (III.1.24).

Summing over all possibilities for the functional derivatives $\delta/(\delta\varphi)$, called also *procedures* P, the final bound is therefore similar to the worst scenario where each propagator derives a new vertex, hence is $(\text{const})^n \cdot \prod_i (d_i!)^{3/2}$. Using Lemma III.1.1 (since the s dependence is now factorized) we can perform the s integrals and change the sum over T' into a sum over regular trees T. This completes the proof of (III.1.24) (the term $(Z(\Delta))^{-p}$ is absorbed in the constant of (III.1.24)).

Constant powers of factorials like $(d_i!)^{3/2}$ can be beaten by the decrease of the propagators (in fact with only a piece, say half of it) because of a phenomenon which we call the volume effect: it relies simply on the fact that the d_i cubes hooked to Δ_i by the tree T have to be all *distinct*, hence when d_i gets large, since we are in a finite dimensional space, many of these cubes have to be quite far from Δ_i.

Lemma III.1.3 (volume effect)

For any constant c, taking r large enough (depending on c) we have

$$\prod_{k=1}^{p-1} \left[\frac{1}{1 + \text{dist}(\Delta_{i_k}, \Delta_{j_{k+1}})} \right]^{r/4} \prod_{i=1}^{p} (d_i!)^c \leq (K'')^p \tag{III.1.26}$$

for some constant K''.

Proof At least half of the d_i distinct cubes hooked to Δ_i have to be at distance at least $(a \cdot d_i)^{1/d}$, where a is a numerical constant and d is the space time dimension. From the corresponding $\prod \left(1/(1 + \text{dist}(\Delta_{i_k}, \Delta_{j_k}))\right)^{r/4}$ we can therefore extract a factor

$$\prod_i \left(\frac{1}{1 + (ad_i)^{1/d}} \right)^{rd_i/16} \tag{III.1.27}$$

(the factor 16 is because we take only half of the d_i cubes into account, and a propagator hooks to two cubes). Obviously for r large enough this leads to (III.1.26).

Theorem III.1.1 follows then easily from one further lemma:

Lemma III.1.4

For $r' > d$, there exists a constant $\hat{K}$ such that:

$$A = \sum_{\substack{Y \ni 0 \\ |Y|=p}} \sum_{T} \prod_{i=1}^{p} d_i! \prod_{k} \left[\frac{1}{1 + \text{dist}(\Delta_{i_k}, \Delta_{j_{k+1}})} \right]^{r'} \leq \hat{K}^p. \qquad \text{(III.1.28)}$$

Proof We can interchange the sum over Y and T. By translation invariance we can require the root of the tree to be the cube containing the origin. Knowing the structure of T we can sum over the positions of the cubes of Y, using the decay in (III.1.28); this would even be true with any summable decay in (III.1.28) (not necessarily power-law with a large power). The result is bounded by $c^p/p!$ because the set Y gets counted $p!$ times in the independent summation over its elements. Using Cayley's theorem (section I.4), we can perform the sum over trees T with given coordination numbers and get:

$$A \leq \sum_{\{d_1,\ldots,d_p\}, d_i \geq 1, \sum d_i = 2p-2} c_1^p \leq \hat{K}^p \qquad \text{(III.1.29)}$$

with $\hat{K} = 4c_1$.

Combining Lemmas III.1.1–4 we have finally to sum in Theorem III.1.1 a series bounded by the geometric series $\sum_{p>1}(c'K\hat{K}g)^p$, and taking g small enough achieves the proof of (III.1.20).

Remark that a sloppier version of Lemma III.1.4 without factors $d_i!$ in (III.1.28) (and without Cayley's theorem) would suffice at this stage. Lemma III.1.3 in fact takes care of any factorials of the coordination numbers when a large decay is available. However it is a legitimate question to ask whether Theorem III.1.1 remains true under weaker assumptions than (III.1.5–6). The typical rule of thumb is that a cluster expansion usually does not require more than a summable propagator in x-space. In particular let us sketch the proof of a more powerful theorem:

Theorem III.1.2: Generalization of Theorem III.1.1

Theorem III.1.1 also holds under the assumption (III.1.6) provided only that $r > d$ (summable decay).

With such a limited decay, we have no analogue of Lemma III.1.3, and we must be careful not to consume any decay in Lemma III.1.2, since all the decay of the propagator should be kept for the equivalent of Lemma III.1.4. Comparing Lemmas III.1.2 and III.1.4 it seems that we are going to loose the game anyway because there is a power 3/2 instead of 1 for the local factorials. But there is one point on which we can improve: the Schwartz inequality (III.1.25) is not optimal. We can instead use a Hölder inequality, which we state only in the "worst case:"

$$\int_\Delta \varphi(x)^3 dx C(x,y) \leq \left[\int_\Delta \varphi^4(x) dx\right]^{3/4} \left[\int_\Delta dx C(x,y)^4\right]^{1/4} . \qquad \text{(III.1.30)}$$

The q vertices in Δ are transformed in this way. The integrated propagators $[\int_\Delta dx C(x,y)^4]^{1/4}$ can be used to sum over Δ just as well as $C(x,y)$ if (III.1.6) holds with $r > d$. Hence the analog of Lemma III.1.4 remains true. But we can now use the fact that $x^n e^{-x} \leq n!$ to improve our factorials: in the worst case there are d_i vertices to which (III.1.30) is applied in Δ_i. They are bounded using the inequality $g^{d_i/4} \cdot [\int_{\Delta_i} g\varphi^4]^{3d_i/4} \exp(-g \int_{\Delta_i} \varphi^4) \leq c \cdot g^{d_i/4} (d_i!)^{3/4}$, and (in contrast with Lemma III.1.2) this does not consume any fraction of the propagator's decay, because this bound is completely local (works separately in each cube Δ_i). Comparing to Lemma III.1.4 we win now the game provided g is taken still smaller than before. We leave to the reader to fill in the details, in particular to check that the sum over all possibilities is of the same order as the "worst case" considered. The idea of using the interaction to improve on Gaussian integration is the first example we meet of "domination," a technique to be discussed at length in the next chapter, hence we do not develop it here in full detail.

Before going on to the Mayer expansion, let us apply the cluster expansion formalism to the proof of Lemma II.6.3, which we postponed until now.

Proof of Lemma II.6.3 We apply the cluster expansion no longer to $\exp(-g \int_Y \varphi^4)$ but to $(1/n!)(\int_Y \varphi^4)^n$. The number of occupied cubes (vertex domain in the language of section II.6) is still p. We repeat exactly the same analysis, including the analogue of the Schwarz inequality (III.1.25) to separate the fields hooked to propagators derived by the $\delta/(\delta\varphi)$ operators from the remainder, which is necessarily of the form $(1/q!)(\int_Y \varphi^4)^q$ for some q with $0 \leq q \leq n - p$. The fields of the first kind are localized in particular cubes of Y and again their Wick contractions give only factorials of the coordination number, which are bounded as in Lemma III.1.3. Hence the only difference is in the second factor of the first line of (III.1.25), which is no longer bounded by 1, but by $K^q q!$. This, together with the obvious inequality $q! \leq \frac{n!}{p!}$ for $q \leq n-p$ achieves the proof of Lemma II.6.3.

B The Mayer expansion

The Mayer expansion starts with formulas (III.1.15) or (III.1.18). and allows us to compute the correct quantity for a thermodynamical limit, namely normalized Schwinger functions or the pressure. Let us call a finite ordered sequence of polymers such as $Y_1, \ldots, Y_q$ a configuration M, of length q (the terminology "Mayer graph" in [FMRS4] is not very appropriate). When there are external variables in $Y_1, \ldots, Y_q$, the union of which is $\{z\} = \{z_1, \ldots, z_N\}$ we call M a $\{z\}$-configuration. We introduce also the set $\mathbf{P}(M)$ of all pairs (i,j), $1 \leq i < j \leq q$. We say that the configuration is disjoint if $Y_i \cap Y_j = \emptyset$ for every $(i,j) \in \mathbf{P}(M)$. Conversely it is called connected if for every $(i,j) \in \mathbf{P}(M)$ we can find a chain $(Y_{i_1} = Y_i, Y_{i_2})(Y_{i_2} Y_{i_3}) \cdots (Y_{i_{k-1}}, Y_{i_k} = Y_j)$ of overlapping polymers in the configuration which join Y_i to Y_j, i.e., which are such that $Y_{i_\ell} \cap Y_{i_{\ell+1}} \neq \emptyset$. Defining the amplitude of the configuration as $A(M) = \frac{1}{q!} \prod_{i=1}^{q} A(Y_i)$ and defining the two body hard core interaction $V(Y, Y')$ as 0 if Y and Y' are disjoint and $+\infty$ if they overlap, we can rewrite (III.1.15) or (III.1.18) as

$$
\begin{aligned}
\tilde{Z}(\Lambda) &= \sum_{\text{disjoint } \emptyset\text{-configurations M}} A(M) \\
&= \sum_{\emptyset\text{-configurations } M} A(M) \prod_{(i,j) \in \mathbf{P}(M)} e^{-V(Y_i, Y_j)}, \qquad \text{(III.1.31)}
\end{aligned}
$$

$$
\begin{aligned}
\widetilde{S^u_{N,\Lambda}}(z_1, \ldots, z_N) &= \sum_{\text{disjoint } \{z\}\text{-configurations } M} A(M) \\
&= \sum_{\{z\}\text{-configurations } M} A(M) \prod_{(i,j) \in \mathbf{P}(M)} e^{-V(Y_i, Y_j)}. \qquad \text{(III.1.32)}
\end{aligned}
$$

(III.1.31) is the grand canonical partition function of a gas of polymers with hard core interaction: an ideal gas of polymers would have no such interaction. The Mayer expansion is an expansion for the logarithm of (III.1.31) around the ideal gas situation. In our case it consists simply in writing $e^{-V(Y_i, Y_j)} = (e^{-V(Y_i, Y_j)} - 1) + 1$. The $+1$ term corresponds to the free "ideal gas" where the two polymers Y_i and Y_j

are summed independently (this intuitively restores for Y_i the "translation invariance" which was broken by the forbidden region Y_j); and the $(e^{-V(Y_i,Y_j)} - 1)$ which is -1 if Y_i and Y_j overlap and 0 otherwise is called a "Mayer link" between polymers; it plays indeed the role of a connection somewhat similar to the propagators which link two cubes between a given polymer.

Just as for the cluster expansion we could write a systematic expansion of the "pair of cubes" type, namely write

$$\prod_{(i,j)\in \mathbf{P}(M)} e^{-V(Y_i,Y_j)} = \sum_{J\subset \mathbf{P}(M)} \prod_{(i,j)\in J} (e^{-V(Y_i,Y_j)} - 1). \tag{III.1.33}$$

The result is then factorized over maximal subsequences of M which are connected by the bonds of J. Such subsequences can be considered again as configurations which have to be connected. Their connected amplitude is simply:

$$A^T(M) = T(M)\cdot A(M) \qquad \text{with} \quad T(M) = \sum_{J\in J^T(M)} \prod_{(i,j)\in J} (e^{-V(Y_i,Y_j)} - 1) \tag{III.1.34}$$

where $J^T(M)$ is the set of all subsets of $\mathbf{P}(M)$ connecting M into a single component. Using simple multinomial identities we obtain that

$$\tilde{Z}(\Lambda) = \sum_{n=0}^{\infty} \frac{1}{n!} \left\{ \sum_{\emptyset\text{-configurations } M} A^T(M) \right\}^n \tag{III.1.35}$$

so that

$$\tilde{p}(\Lambda) = \frac{1}{|\Lambda|} \log \tilde{Z}(\Lambda) = \frac{1}{|\Lambda|} \sum_{\emptyset\text{-configurations } M} A^T(M). \tag{III.1.36}$$

The $1/n!$ in (III.1.35) comes from the number of partitions of the q elements of the sequence $Y_1, \ldots, Y_q$ into n subsequences. In (III.1.34) or (III.1.36) only connected configurations contribute, otherwise $A^T(M) = 0$. Convergence of (III.1.36) is not easy to prove in this direct "brute force" expansion because naive estimates which do not take into account the sign cancellations in the factor $T(M)$ in (III.1.34) fail (see e.g. [Se]). Again the "pair of cubes" is not the minimal process to reach factorization. This minimal process is rather of the tree type, and we explain it now.

The pair of cubes formula (III.1.33) has the same algebraic structure as the binomial expansion

$$\prod_{i=1}^{n} (a_i + b_i) = \sum_{J\subset\{1,\ldots,n\}} \prod_{i\in J} a_i \prod_{i\notin J} b_i,$$

and it expands a product of n terms as a sum of 2^n monomials. But if we are chiefly interested into the presence or absence of, say, the b factors, there is a more economical way to expand, which unfortunately is no longer canonical in the sense that the outcome depends on a particular ordering of the n terms. It is the formula $\prod_{i=1}^{n}(a_i + b_i) = \prod_i a_i + \sum_{i=1}^{n} b_i \prod_{j<i} a_j \prod_{k>i}(a_k + b_k)$. The right hand side of this formula contains only $n+1$ terms instead of 2^n. The outcome depends on the ordering of the terms in the product, and for short we will say that this formula amounts to expand the b factors with priority to index 1, then priority to index 2 and so on.

Let us apply this economical expansion scheme to the product on the left hand side of (III.1.33), which is a product of terms $1 + (e^{-V(Y,Y')} - 1)$. The factors 1 play the rôle of the factors a_i and the terms $e^{-V(Y,Y')} - 1$ play the rôle of the b_i factors in the algebraic examples above. These non-trivial factors are also called the overlap constraints, because they are 0 except if Y and Y' overlap. We will expand these factors in the product in the left hand side of (III.1.33) but with an inductive rule on the "priority" with which they are expanded, in the sense above.

Starting from a configuration $Y_1, \ldots, Y_q$, we first give priority to the $q-1$ overlap constraints involving the first polymer Y_1 and the other polymers, in their natural ordering $Y_2, \ldots, Y_q$. Hence we write $\prod_{j>1} e^{-V(Y_1,Y_j)}$ in the left hand side of (III.1.33) as:

$$\prod_{j>1} e^{-V(Y_1,Y_j)} = 1 + \sum_{j_1 \neq 1} (e^{-V(Y_1,Y_{j_1})} - 1) \prod_{k>j_1} e^{-V(Y_1,Y_k)}. \qquad \text{(III.1.37)}$$

As is the rule with a tree expansion the next step depends on which term is selected in the sum (III.1.37). The term 1 frees Y_1 from every hard core interaction with other polymers, in which case we can simply continue, which means to replace Y_1 by Y_2 and to apply an analogue of (III.1.37) to $\prod_{j>2} e^{-V(Y_2,Y_j)}$. But for one of the $q-1$ terms corresponding to some value of j_1 in the sum (III.1.37), we should consider that Y_1 and Y_{j_1} are linked through a "Mayer link," and we should treat them as a single block from now on. Therefore the next step is to expand all the remaining constraints relative to Y_1 and Y_{j_1}, to test whether $Y_1 \cup Y_{j_1}$ is free from hard core interaction with the rest of the configuration or not. This is done with priority to the remaining constraints involving Y_1, then to the constraints involving Y_{j_1}, i.e., we write

$$\prod_{k>j_1} e^{-V(Y_1,Y_k)} \prod_{\substack{1<k' \\ k'\neq j_1}} e^{-V(Y_{j_1},Y_{k'})}$$
$$= 1 + \sum_{j_2>j_1} (e^{-V(Y_1,Y_{j_2})} - 1) \prod_{k>j_2} e^{-V(Y_1,Y_k)} \prod_{\substack{1<k' \\ k'\neq j_1}} e^{-V(Y_{j_1},Y_{k'})}$$
$$+ \sum_{j_2\neq 1,j_1} (e^{-V(Y_{j_1},Y_{j_2})} - 1) \prod_{\substack{j_2<k' \\ k'\neq j_1}} e^{-V(Y_{j_1},Y_{k'})}. \qquad \text{(III.1.38)}$$

The first term frees Y_1 and Y_{j_1} from the rest, the second term corresponds to the tree $\{(Y_1, Y_{j_1}), (Y_1, Y_{j_2})\}$, the third one to the tree $\{(Y_1, Y_{j_1}), (Y_{j_1}, Y_{j_2})\}$. We can continue this process until all polymers are exhausted, but there is still some arbitrariness in the expansion rule because there are several different possible strategies for defining priorities, in the sense defined above. Among the most natural strategies we could define a strategy called "push each branch as far as possible" (also called "turn around the tree") and a strategy called "let the tree grow layer by layer." The second one is perhaps more natural, so let us define it in detail. It means that we develop in priority all the constraints of Y_1 with other polymers, until the process stops by the choice of a factor 1. Then we have built a "first layer" of polymers Y_{j_1}, ..., Y_{j_k} linked to Y_1 by Mayer links, with $j_1 < j_2 < \cdots < j_k$. We expand then in priority the constraints of Y_{j_1} with all polymers other than Y_1, Y_{j_2}, ..., Y_{j_k}, constructing the piece of the "second layer" of the tree linked to Y_{j_1}. Then we expand in priority the hard core constraints of Y_{j_2} with all other remaining polymers, constructing a second piece of the second layer, and continue in this way until all the second layer has been built. Then in this second layer we select the polymer with lowest index in the initial ordering, and expand in priority its constraint with the remaining ones, and continue until all the third layer is built, and so on.

In this way each time the process stops by selecting a term 1 in the sums similar to (III.1.38) for all the polymers of the last layer, we have factorized a particular subsequence. When all polymers are exhausted we end up with independent sums over factorized subsequences connected by Mayer links just as in (III.1.35). Therefore the two expansions must be identical in the sense that they express finally

$\tilde{Z}(\Lambda)$ or $\tilde{p}(\Lambda)$ by the same series of the same connected amplitudes $A^T(M)$ for configurations M; the advantage of the second point of view lies in the fact that it gives an explicit rewriting of the factor $T(M)$ as a sum with much less terms than (III.1.34) and for which individual terms still have absolute values less than 1. In other words there are tremendous cancellations between terms with values plus and minus 1 in (III.1.34) which are not explicit at all in (III.1.34); the inductive expansion process described above is one among many possible ways to pack together many terms in (III.1.34) to display explicitly these cancellations; there is some degree of arbitrariness in the way this packing is done, which corresponds to the choice of a strategy for priorities.

How are we going to index mathematically the outcome of this expansion scheme, which seems at first sight very complicated and inductive? The solution is in fact very simple. If we study with care (III.1.38) and its generalization to arbitrary layers we remark that the scheme generates exactly once each tree built on the configuration; it does not generate ordered trees as the former cluster expansion because at a given layer the constraints are always developed in the order provided by the initial ordering $Y_1, \ldots, Y_q$ of the configuration; we see for instance that the tree $\{(Y_1, Y_{j_1}), (Y_1, Y_{j_2})\}$ is generated once, not twice, because of the condition $j_2 > j_1$ in (III.1.38). Obviously the total factor associated to a particular tree is in absolute value bounded by 1; this was *not* the case for $T(M)$ in (III.1.34). However the precise description of this factor is a bit cumbersome. It is the product over all pairs of polymers of the tree of:

- a Mayer link $e^{-V} - 1$ if the polymers are joined by a line of the tree,
- a hard core constraint e^{-V} if the polymers Y_i and Y_j belong to the same layer of the tree or if they belong to adjacent layers, e.g., respectively layer k and $k+1$, and the ancestor Y_j' at level k to which Y_j is hooked has index smaller than Y_i,
- a factor 1 otherwise

We conclude that the process we describe ((III.1.37–38) and its generalization) is an algebraic way of reorganizing the sum over $J^T(M)$ (III.1.34) as a sum over trees, by grouping together many connecting subsets J having a particular tree in common, according to a particular rule. The details of the rule ("layer by layer") are partly arbitrary. The gain lies in the fact that although many J's are grouped together, the corresponding large sum of factors $\prod(e^{-V} - 1)$ is still bounded by

1, so we have effectively taken into account the sign cancellations in (III.1.34).

Having clarified this point, we proceed with the evaluation of the series (III.1.36). We have to bound

$$\sum_{Y_1,\ldots,Y_q} \frac{1}{q!} \sum_{T \text{ connecting the } Y\text{'s}} \prod_{j=1}^{q} A(Y_i). \tag{III.1.39}$$

We can use Cayley's theorem (Sect. I.4) on the number of tree graphs with fixed incidence numbers $d_1, \ldots, d_q$: this gives

$$\begin{aligned} \left|\sum_T (\cdot)\right| &= \left|\sum_{d_1,\ldots,d_q} \sum_{T,\{d_i\} \text{ fixed}} (\cdots)\right| \\ &\leq \sum_{d_1,\ldots,d_q} \frac{(q-2)!}{\prod_{j=1}^{q}(d_i-1)!} \sup_{T,\{d_i\} \text{ fixed}} |(\cdot)|. \end{aligned} \tag{III.1.40}$$

Taking into account the $\frac{1}{q!}$ in (III.1.39), it remains, for a fixed tree connecting the Y_i's, to sum over the Y_i's, not forgetting the important $\left[\prod_{i=1}^{q}(d_i-1)!\right]^{-1}$ from (III.1.40). We sum over the Y_i's starting from the end branches of the tree (again, Y_1 being the last polymer to be summed, or the "root" of the tree). If Y_q is such an end branch, let Δ_q be a cube in the non-empty intersection $Y_j \cap Y_q$ of Y_q with its immediate ancestor Y_j in the tree (remember that Y_j and Y_q have to overlap since they are joined by a tree branch, which corresponds to a "Mayer link"). We sum over Y_q, holding Δ_q fixed. This produces, by translation invariance, a factor

$$\sum_{Y \ni 0} |A(Y)| = \sum_{Y \ni 0} A(Y)|Y|^{d_q-1} \tag{III.1.41}$$

since $d_q = 1$. Now we can sum over Δ_q in Y_j, obtaining a factor $|Y_j|$. Iterating this process, we can perform inductively the sums over Y's, progressively stripping off the branches of the tree. A sum over a given Y_i gives rise to a factor

$$\sum_{Y \ni 0} |A(Y)|\; |Y|^{d_i-1} \tag{III.1.42}$$

where $|Y|^{d_i-1}$ arises because at the time Y_i is summed one has already summed over the $d_i - 1$ which had Y_i as their ancestor in the tree. We take into account the crucial factors $1/((d_i-1)!)$ by writing $\sum_{d_i}[|Y|^{d_i-1}/((d_i-1)!)] = e^{|Y|}$. Finally in the last sum over Y_1, the last reference cube Δ_1 can be anywhere in Λ, hence we get the estimate:

$$\sum_{\emptyset\text{-configurations } M} |A^T(M)| \leq |\Lambda| \cdot \sum_q \left[\sum_{Y \ni 0} |A(Y)| e^{|Y|} \right]^q \tag{III.1.43}$$

hence using Theorem III.1.1 we conclude that for g small enough the series (III.1.36) for the pressure are absolutely convergent.

We can extend easily this analysis to the normalized Schwinger functions. The main difficulty is that the external variables must then be part of the definition of the polymer, so that strictly speaking the polymer is now a set of cubes (its support) plus a set of external variables, (the corresponding amplitude being zero if the support of the smearing functions for the external variables does not intersect the support of the polymer). One difference is that in the cluster expansion generating the polymers some integrations by parts may "hook" to the external fields instead of deriving a vertex by formula (III.1.23); this is no problem for convergence if the number of external variables is bounded, as is the case here. Another subtlety is that one should take into account the fact that polymers which contain external variables are indexed by them, so the corresponding sums have no longer to be symmetrized by $1/(n!)$ factors as for vacuum polymers. Therefore:

$$\widetilde{S^u_{\Lambda,N}} \{z_1, \ldots, z_N\} = \sum_{\omega=(\omega_1,\ldots\omega_m)} \prod_{i=1}^m \sum_{M\,\omega_i\text{-configuration}} A^T(M) \cdot \sum_{n=0}^{\infty} \frac{1}{n!} \left\{ \sum_{M\emptyset\text{-configuration}} A^T(M) \right\}^n \tag{III.1.44}$$

where the sum is taken over partitions ω of all external variables into m subsets $\omega_1, \ldots, \omega_m$. In (III.1.44) we recognize that $\widetilde{Z}(\Lambda)$ is factorized, so that the normalized functions are given by:

$$S_{\Lambda,N}(z_1, \ldots, z_N) = \sum_{\omega=\{\omega_1,\ldots,\omega_m\}} \prod_{i=1}^m \sum_{M\,\omega_i\text{-configuration}} A^T(M) \tag{III.1.45}$$

and the normalized *truncated* Schwinger functions are given by the simpler expansion:

$$S^T_{\Lambda,N}(z_1, \ldots, z_N) = \sum_{M\{z_1,\ldots,z_N\}\text{-configuration}} A^T(M). \tag{III.1.46}$$

The series (III.1.45) and (III.1.46) are shown to be absolutely convergent exactly as the one for the pressure, except that now the external variables (which for simplicity are localized by the smearing test functions within given cubes of Λ) break translation invariance and can be used to perform the last summation over Y_1: for the pressure this last summation was free, hence the dividing factor $1/|\Lambda|$ had to be included.

Remark that if we are interested only into the normalized (*not truncated*) Schwinger functions, we may expand only the hard core constraints between pairs of polymers for which at least one is a vacuum polymer, keeping the constraints between polymers with external variables unexpanded. This will be enough to factorize $\widetilde{Z}(\Lambda)$ as in (III.1.44). We may also derive intermediate versions, in which e.g. the constraints involving at least one vacuum, two or four point polymer are expanded, but not the others. This results in partly truncated amplitudes, which are useful when renormalization of two and four point functions is involved (recall that in part II we learned that renormalization is best expressed at the level of connected functions). It is a version of this kind which will be used in the next chapters.

The convergence of these various expansions does not lead to any particular problem when the number of external fields is fixed. However in the context of the phase space expansion which we are going to introduce, low momentum fields at a given scale must be considered as external variables. Therefore the number of such variables is no longer bounded at intermediate stages, even if we compute a fixed N-point Schwinger function. This leads to a subtlety: if we were to expand fully all hard core constraints involving not only vacuum polymers but also polymers with e.g. two and four external legs, we would generate Mayer configurations in which an arbitrarily large number of external low momentum fields may accumulate at the same place. From the "local factorial principle" we know that this would lead typically to divergent expressions. Therefore the Mayer expansions we use in phase space are slightly modified to avoid this effect: they keep the cubes containing the external legs of a two or four point function (called their external cubes) fixed and non-overlapping. In this way the truncation is almost completely performed, and the renormalization cancellations can be performed, but a large number of low momentum fields still cannot accumulate at the same place. This process is described in Section III.3D. Remark that this problem is a truly constructive one, which has no analogue in perturbation theory.

Altogether, we have achieved the proof of:

Theorem III.1.3 Convergence of the Mayer expansion

For g small enough, the series (III.1.36) and (III.1.45)–(III.1.46) are absolutely convergent, uniformly in Λ. As $\Lambda \to \infty$ their limit is therefore still absolutely convergent and can be used to define rigorously the massive weak coupling φ^4 theory (with ultraviolet cutoff) in any dimension.

The proof of course works for complex g small enough with $\mathrm{Re}\, g \geq 0$; with straightforward estimates on Taylor remainders in g at order n in the spirit of section II.1 one can verify the bound (I.5.2) and conclude that the infinite volume quantities constructed in Theorem III.1.3 are the Borel sum of the corresponding perturbative expansions. This gives a quick check that the limit is indeed independent of all the particular details of the construction such as the form of the boxes Λ and so on (although of course this can be also proved directly).

Let us gather now some further remarks on various aspects of these expansions.

C Further topics

Our first comment is on the similarities and differences between the cluster and Mayer expansions. In both cases trees emerge as the central structure around which the expansion is best organized. But for the Mayer expansion there are no positivity requirements to preserve, and the algebraic process (III.1.38) directly generates trees. Ordinary perturbation theory which is nothing but a gas of vertices interacting through propagators, shares these features; there are no positivity requirements, and it is possible to design an "algebraic" or "set-theoretic" cluster expansion based on selecting particular propagators, as sketched in section II.6. But when functional integrals are involved, positivity requirements occur and interpolating parameters seem necessary. The corresponding cluster expansions generate ordered trees rather than trees; however by the Battle–Federbush theorem, the summation from ordered to ordinary trees is controlled by the integration over these interpolating parameters.

Another difference is that in the cluster expansion for the functional integral, strong decay of the propagator and the volume argument allow sloppy estimates on the factorials of coordination numbers of the tree. This is no longer true in the Mayer expansion where one has to use with care the single factorial of this type delivered by Cayley's Theorem. To get some intuition of this, we may compare the polymers of the cluster expansion to fermions because their hard core interaction is somewhat reminiscent of the Pauli principle. In contrast, the "Mayer configuration," made of ordinary polymers with Mayer links, have no longer hard core interactions; they behave like bosons which can pile up in arbitrary numbers at a given place. Hence we may compare our remark with the known fact that perturbation theory for fermions is more convergent than for bosons, and that keeping track of correct factorial factors for fermions is less important [FMRS4][IM2,Appendix].

Our last comment is on the difference between cluster expansions à la Brydges–Battle–Federbush, used here, and à la Glimm–Jaffe–Spencer as in [GJS]. In section II.6 we wrote the propagator as a Wiener sum over random paths. The difference is summarized by the statement that the GJS cluster expansion localizes both vertices and the paths building the propagators, when the BBF cluster expansion only localizes the vertices, hence is simpler. However to compute the BBF expansion it is necessary to perform explicitly functional derivatives like (III.1.23). For a polynomial interaction like $\exp(-\int \varphi^4)$ this is no problem, but in more complicated situations, it might be difficult or impossible; in such cases one may have to return to GJS cluster expansions (see [Bry] for an example with "large field holes" where the interaction is non-polynomial).

The cluster and Mayer expansions explained above are sufficient for the purpose of proving existence of the thermodynamic limit. They provide information equivalent to connectivity at the graphical level. But sometimes we are interested in more detailed information about the system, in which case one can push these expansions further. A first extension in order to reach information equivalent to graphical one-particle irreducibility was developed in [FMRS4–5] in order to perform full (rather than *useful*) mass renormalization of the models considered there. This point of view has been made more systematic in [IM1–2] for the purpose of multiparticle structure analysis, under the name of pth order expansions. (Of course the story of the subject is a long one, with some landmarks like [Sp2] or [CFR]). We underline here the basic ideas of the construction, and refer the reader to the very clear sections 3 and 4 of [IM1] for a detailed exposition.

The first idea for pth order cluster expansions is that nothing (at the level of convergence) prevents us to push the Taylor expansions in the interpolating s parameters further than to first order. The price to pay however is that the new structures obtained are no longer indexed by ordered trees but by more general "graphs and procedures." More precisely if we return to interpolation (III.1.8), we start as before with a first order Taylor expansion similar to (III.1.9):

$$Z(\Lambda) = Z(s_1 = 0) + \int_0^1 ds_1 \sum_{\Delta_{1,1} \neq \Delta_1} Z_{\Delta_1, \Delta_{1,1}}(s_1). \qquad \text{(III.1.47)}$$

In the term $Z(s_1 = 0)$, $d\mu$ decouples Δ_1 from its complement $\Lambda - \Delta_1$, and as before we turn our attention to the next cube of $\Lambda - \Delta_1$ and repeat the process. But for the remainder term remark that we call $\Delta_{1,1}$ the cube (previously called Δ_2) linked by a propagator to Δ_1. Since the guiding idea is that a single propagator is not $(p-1)$-particle irreducible for $p > 1$, we are not satisfied and want to push further the expansion, trying to know whether $\Delta_{1,1}$ is linked to Δ_1 through more propagators. It is not enough to simply push the Taylor expansion in s_1 up to order p, because the cubes linked to Δ_1 by each s_1 derivation may be all different. (This defect would not occur if the cluster was of the pair of cubes type, but we know that this one has a positivity problem). Hence the correct procedure is again very inductive. We introduce a second Taylor formula (hence a new parameter s_1' interpolating between 0 and s_1), writing:

$$Z_{\Delta_1, \Delta_{1,1}}(s_1) = Z_{\Delta_1, \Delta_{1,1}}(0) + \int_0^{s_1} ds_1' \sum_{\Delta_{1,2} \neq \Delta_1} Z_{\Delta_1, \Delta_{1,1}, \Delta_{1,2}}(s_1'). \qquad \text{(III.1.48)}$$

In the remainder term Δ_1 is linked by a propagator both to $\Delta_{1,1}$ and $\Delta_{1,2}$, which of course may coincide. For this remainder term the procedure is pursued until either one variable $s_1^{(r)}$ is taken at 0, or p squares among $\Delta_{1,1}, \ldots, \Delta_{1,r}$ coincide. Two typical outcomes of the process are pictured in Fig. III.1.3 (for $p = 4$), one in which at most $p-1$ squares among $\Delta_{1,1}, \ldots, \Delta_{1,r}$ coincide and Δ_1 is decoupled in the measure $d\mu$ from $\Lambda - \Delta_1$, and the other in which $\Delta_{1,r} \equiv \Delta_2$ coincides with $p-1$ previous squares among $\Delta_{1,1}, \ldots, \Delta_{1,r-1}$.

In the first case we choose a new square Δ_2 in $\Lambda - \Delta_1$ and continue the expansion, with an important caveat: the new expansion in s_2, s_2', $\ldots$, $s_2^{(r_2)}$ is no longer pushed until some s_2 parameter is set to 0 or p squares linked to Δ_2 coincide, but until some s_2 parameter is set to 0 or Δ_2 is linked in a $(p-1)$-particle irreducible way to some set of other cubes. This can arrive earlier than for Δ_1, because the propagators already created in the first expansion in s_1, $\ldots$, $s_1^{(r=r_1)}$ may help, as

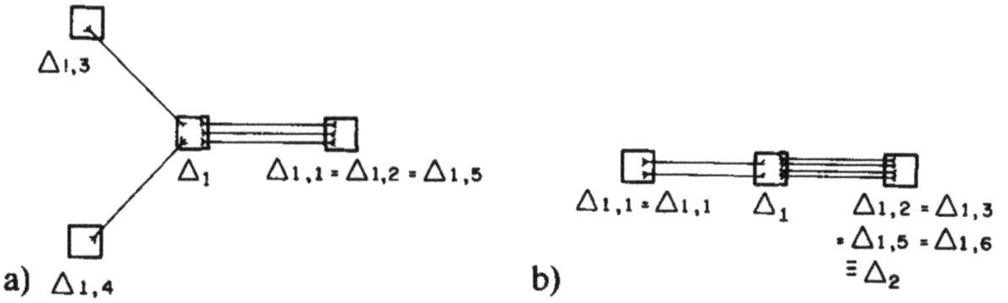

Fig. III.1.3 a) A decoupled term. b) A remainder term with Δ_1 and Δ_2 linked.

shown in Fig. III.1.4a, which is a possible continuation of the process shown in Fig. III.1.3a. Then one considers this whole set of cubes linked in a $(p-1)$-particle irreducible way as a single new block and so on.

In the second case (Fig. III.1.3b) we should consider Δ_1 and Δ_2 as linked in the same $(p-1)$-particle irreducible object and they are treated as a single block in the rest of the expansion, just as were Δ_1 and Δ_2 in the ordinary cluster expansion. This means that we introduce an interpolating parameter s_2 by formula (III.1.12). We make first order Taylor expansions in $s_2, s_2', \ldots, s_2^{(r_2)}$ until either one of these parameters is set to 0 or some square Δ_3 is linked in a $(p-1)$-particle irreducible way to the block $\Delta_1 \cup \Delta_2$, as in Fig. III.1.4b.

This process is continued and larger and larger $(p-1)$-particle irreducible blocks are formed, until the process ends, an issue guaranteed by the finiteness of Λ. The result is an expansion in terms of sets of disjoint cubes $Y_1, \ldots, Y_q$ or polymers (again it is better to factorize the functional integral for an isolated cube, passing to $\tilde{Z}$ as in (III.1.14)).

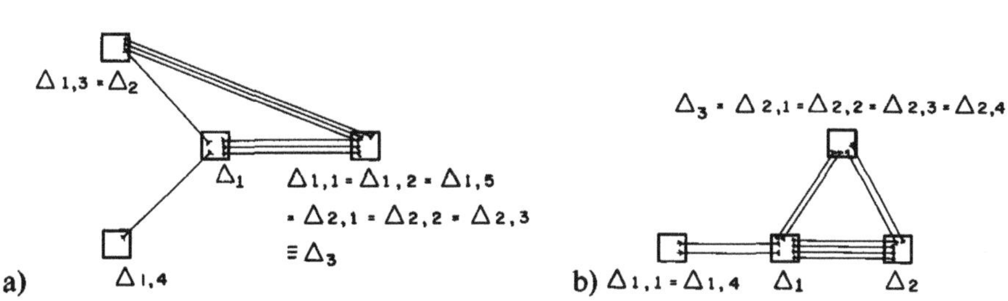

Fig. III.1.4 a) A term with Δ_2, Δ_1 and Δ_3 linked. b) A remainder term with Δ_1, Δ_2 and Δ_3 linked.

For each polymer there is a sum over graphs G made of lines connecting these cubes. These graphs are no longer trees, but there are some restrictions on these graphs, because they never connect subsets of cubes in a p-particle irreducible way. Finally there is also a sum over procedures P leading to these graphs, which is a generalization of the sum over orderings of the trees in the ordinary cluster expansion. The sum over procedures leading to the same graph is again controlled by the integration over interpolation parameters, using a generalization of the Battle–Federbush result [IM1, Lemma 1]. The sum over graphs built on the support of a polymer may also seem more difficult than the sum over trees, but the condition that there is no p-particle irreducible structures is in fact very restrictive, so that Theorem III.1.1 remains valid. Therefore pth order cluster expansion followed by an ordinary Mayer expansion can be used for the computation of thermodynamic limits and normalized quantities. But of course their real interest lies in the fact that in these expansions the k-particle analysis becomes possible. For instance an expansion with $p = 3$ makes explicitly visible all the chains of one-particle irreducible two-point subgraphs in a polymer (this is not the case for $p = 2$, see Fig. III.1.5).

This opens the possibility of performing full non-perturbative mass renormalization in φ_4^4-like models, i.e., to set the renormalized mass of the theory directly at the desired value. Indeed in this renormalization, the one particle irreducible two point functions must be subtracted at 0 external momenta. For the same reason as in perturbation theory, the "useless" part of mass renormalization is not the source of any non-summable effect of the renormalon type. However just as renormalization of the 4 point function requires at least the computation of connected four point structures, hence an ordinary Mayer expansion, the renormalization of such one particle irreducible subgraphs requires a generalized version of the Mayer expansion. Indeed to remove hard core constraints between connected polymers would

Figure III.1.5 This single block (left) in the $p = 2$ expansion reveals a chain (right) of IPI 2-point subgraphs in the $p = 3$ expansion.

Figure III.1.6 The Bethe–Salpeter chain.

not be enough, because inside a connected graph a one particle irreducible two point subgraph has still hard core constraints with the rest of the graph (hence its 0-momentum value is not "background independent," but depends on the rest of the expansion, so that it cannot cancel exactly with a universal, background-independent counterterm). The necessary generalized Mayer expansion removes not only the hard core constraints between connected polymers, but also the hard core constraints of each one particle irreducible 2-point subgraph inside each connected polymer with the rest of the polymer, by algebraic formulas similar to (III.1.37–38). Here again, to display convergence one has to be careful and to proceed in an inductive way, in the natural order provided by the maximal chains of such one particle irreducible subgraphs [FMRS5]. Although appealing at the conceptual level, this method leads to some rather intricate technicalities, so we decide in section III.3 not to use it and to perform a more standard "fixed point" computation of the renormalized mass.

One can continue along these lines and define still more general "Mayer expansions." For instance Iagolnitzer and Magnen have considered, in the analysis of the Bethe–Salpeter equation, the removal of hard core constraints between two-particle irreducible kernels along the chain pictured in Fig. III.1.6 [IM4]. Here again the linear character of the chain is used to organize the expansion. To our knowledge a completely general theory of such expansions remains to be written.

As shown in [IM1–4], pth order expansions can be used for a better understanding of multiparticle structure and may be of help for a proof of asymptotic completeness which is of course one of the most important open mathematical problems in field theory. For a review on this subject we refer to [Ia].

D The φ_2^4 theory

A single slice model like the one considered above may seem somewhat artificial. We could without difficulty remove the ultraviolet cutoff for the φ_1^4 theory (the anharmonic oscillator) and apply the previous formalism to construct its weak coupling infinite volume limit. This is still not very exciting. Hence we prefer to conclude this section by a more interesting exercise which gives a first flavor of the ultraviolet problem in constructive field theory. The construction of the $P(\varphi)_2$ model by Nelson, Glimm–Jaffe and Guerra–Rosen–Simon [Er1] was the birth act of constructive theory, hence no book on this subject should omit it. Here we essentially rephrase Nelson's probabilistic argument in our language. Instead of using his hypercontractive estimates (as in [Si1]), we prefer to use the method of part II for graphical estimates. Combining this method with a truncated perturbative expansion which is organized so as to develop more and more at higher and higher energies, it is easy to prove the existence of the ultraviolet limit of the theory in a unit cube. Although we give below a proof only for φ_2^4 the extension to any $P(\varphi)_2$ theory (with P semi-bounded) is straightforward. Furthermore although Nelson's probabilistic formulation is very appealing, we feel that the method of partially expanding perturbation theory is more consistent with the rest of this book and is a better introduction to the vertical expansion that are necessary for more divergent models.

The only divergent graphs in the φ_2^4 theory are due to loops $\int[d^2p/(p^2+m^2)]$ made of a single line, also called tadpoles. With a glance at (I.4.7) we can indeed check that the only connected divergent graphs are those with a single vertex. There are only two possibilities for such graphs: the two-point tadpole graph and the vacuum graph made of two tadpoles. The Wick ordering, or normal ordering of a polynomial with respect to a Gaussian measure is precisely the mathematical operation which prevents the formation of such subgraphs, so that the perturbative expansion of Wick ordered quantities does not contain any tadpoles [Si1]. More precisely, for any random variable φ with finite moments, we can define normal ordered or Wick ordered powers by the recursive rules:

$$:\varphi^0:=1,\langle:\varphi^n:\rangle=0,\frac{\partial}{\partial\varphi}:\varphi^n:=n:\varphi^{n-1}:. \tag{III.1.49}$$

In the case of a Gaussian random variable φ of variance $\langle\varphi^2\rangle$, the moments of φ are all related to the variance so that in this case one gets the explicit equation:

$$:\varphi^n:=\sum_{m=0}^{[n/2]}\frac{n!}{m!(n-2m)!}\varphi^{n-2m}\left(-\frac{1}{2}\langle\varphi^2\rangle\right)^m \tag{III.1.50}$$

When integrating products of Wick ordered monomials with respect to the corresponding measure, as is done in perturbation theory (see (I.4.1)), the result is expressed as a sum over contraction schemes which do not pair together two variables of the same monomial. Therefore the corresponding diagrams have no tadpoles. In the case of the φ^4 theory, we have:

$$g : \varphi^4 := g(\varphi^4 - 6\varphi^2 \langle \varphi^2 \rangle + 3 \langle \varphi^2 \rangle^2) \qquad \text{(III.1.51)}$$

where expectation value is taken with respect to the corresponding Gaussian measure. Of course we can interpret (III.1.51) along the lines of section II.3, as containing a mass renormalization and a vacuum energy renormalization, which are particularly simple. However we cannot consider (III.1.51) as well defined for a Gaussian measure without ultraviolet cutoff, because as we remarked, $\langle \varphi^2 \rangle$ is the value of the tadpole graph, which is logarithmically divergent.

Therefore we have to return to the theory with an ultraviolet cutoff, define Wick ordering with respect to the corresponding Gaussian measure with cutoff, and see whether this cutoff theory has a limit as the cutoff is removed. We will limit ourselves to prove the convergence of the normalization; the proof for Schwinger functions is similar and left to the reader. We assume also that the coupling constant g is complex with a positive real part.

Introducing our favorite cutoff (II.1.2–4), we consider the Gaussian measure $d\mu_\rho$ with covariance C_ρ. The normalization of the corresponding φ_2^4 theory in a unit cube is defined by

$$Z_\rho = \int d\mu_\rho(\varphi) e^{-g \int_\Delta :\varphi^4:_\rho}. \qquad \text{(III.1.52)}$$

The subscript ρ after the Wick dots means that the Wick ordered product is computed as in (III.1.51) with the expectation value taken with respect to the measure $d\mu_\rho$. Explicitly we have:

$$g : \varphi^4 :_\rho = g(\varphi^4 - 6c_\rho \varphi^2 + 3c_\rho^2), \qquad \text{(III.1.53)}$$

with $c_\rho = c \cdot \rho + O(1)$, and c some constant; this behavior expresses the fact that c_ρ, which is the value of the tadpole graph, diverges logarithmically, hence linearly in ρ.

In (III.1.52) Δ is a unit square in $\mathbb{R}^2$, and for completeness we should choose some boundary conditions on Δ. In fact the proof works for any such boundary conditions, and for instance as in the

BBF formalism we can simply multiply $C_\rho(x,y)$ as defined in (II.1.2–4) by characteristic functions of the cube Δ for x and y. Boundary conditions become important only when one computes the thermodynamic limit of the φ_2^4 theory in the two phase region. The cluster and Mayer expansion considered above correspond to the weak-coupling, high temperature phase, and they do require a small coupling constant in order to converge. In contrast the ultraviolet limit of a theory in a finite cube constructed below is really valid for any g positive, not only for small g, hence it could be the starting point for a study of the low temperature, strong-coupling, two phase region, a problem that however we will not consider in this book.

Instead of proving convergence of Z_ρ, let us prove first that it is uniformly bounded as $\rho \to \infty$; an easy adaptation of the method will then show convergence. A constant lower bound on Z_ρ is provided in a cheap way by Jensen's inequality. In this particular case it gives $Z_\rho \geq 1$ since:

$$\int d\mu_\rho(\varphi) e^{-g\int :\varphi^4:_\rho} \geq e^{-g\int d\mu_\rho(\varphi)\int :\varphi^4:_\rho} = 1. \qquad \text{(III.1.54)}$$

Indeed by (III.1.49) the integral of a Wick ordered vertex vanishes.

The main problem is to prove a uniform upper bound on Z_ρ. Indeed $:\varphi^4:$ is no longer positive for all values of φ. From (III.1.53) we derive easily the bound

$$:\varphi^4:_\rho \geq -6c_\rho^2 = -6c^2\cdot\rho^2 + O(1) \geq -K\rho^2 \qquad \text{(III.1.55)}$$

for some positive constant K (with $K \simeq 6c^2$).

The naive upper bound on Z_ρ obtained by replacing the exponential of the interaction by its supremum in (III.1.52) therefore explodes as $e^{K\cdot\rho^2 \operatorname{Re} g}$; certainly it does not allow to prove the existence of the ultraviolet limit. Nelson refined this crude analysis by showing that the probabilistic measure (with respect to $d\mu_\rho$) of the set of fields where this upper bound for the exponential of the interaction is reached is very small, in fact smaller and smaller as the cutoff increases and that this effect more than compensates the explosion of this upper bound. He could then conclude that Z_ρ converges [Er1].

Equivalently we can test the presence of the interaction at higher and higher energies and when it is there, we will earn a very small factor from graphical estimates; this is the step which corresponds in

Nelson's language to the probabilistic measure of some set of fields being small.

We use our standard splitting of the propagator as $C = \sum_{i=0}^{\rho} C^i$ as in (II.1.2–4). Recall that there is a corresponding decomposition of the Gaussian measure and of the field (II.1.5). We have to distinguish what is the highest energy scale in a given vertex. Therefore we write

$$:\varphi_\rho^4:_\rho = \sum_{i=0}^{\rho} :\varphi^{i,4}: \tag{III.1.56}$$

$$:\varphi^{i,4}: \equiv :\varphi_i^4:_i - :\varphi_{i-1}^4:_{i-1}, \tag{III.1.57}$$

where we recall that $\varphi_i \equiv \sum_{j=0}^{i} \varphi^j$. (III.1.56–7) are consistent because the Wick ordering of a monomial which does not contain any field of index higher than i with respect to $d\mu_\rho$ is the same as its Wick ordering with respect to $d\mu_i$ (the fields φ^i are independently distributed according to the independent Gaussian measures $d\mu^i$ in (II.1.5)). We can consider $:\varphi^{i,4}:$ as a convenient notation for a Wick ordered vertex whose highest field index is exactly i. This highest field index is the exact analogue of the upper index $e_v(\mu)$ of a vertex of a graph as defined in (II.4.4).

Now we apply, for each i, $i = \rho, \ldots, 0$, a Taylor expansion in $:\varphi^{i,4}:$ up to order $n_i = 1 + a \cdot i$, where a is some large integer to be fixed later.†
This means that we write:

$$Z_\rho = \int d\mu_\rho(\varphi_\rho) e^{-g\int_\Delta \sum_{i=0}^{\rho} t_i :\varphi^{i,4}:}|_{t_i=1\,\forall i} = \prod_i (I^{n_i} + R^{n_i}) Z_\rho(\{t_i\}) \tag{III.1.58}$$

where the operator I^{n_i} selects the Taylor expansion in t_i of Z up to order $n_i - 1$ at $t_i = 1$, and R^{n_i} selects the corresponding Taylor remainder at order n_i:

$$I^{n_i} f(t_i) = \sum_{k=0}^{i-1} \frac{1}{k!} \frac{d^k f}{dt_i^k}(t_i = 0); \tag{III.1.59}$$

$$R^{n_i} f = \int_0^1 dt_i \frac{(1-t_i)^{n_i-1}}{(n_i-1)!} \frac{d^{n_i} f}{dt_i^{n_i}}(t_i). \tag{III.1.60}$$

In fact we start from $i = \rho$ down to $i = 0$ and stop expanding the product in (III.1.58) as soon as a remainder term is produced. This

---

†The factor 1 is there so that even in the slice $i = 0$ an expansion is performed.

means that we write

$$\prod_i (I^{n_i} + R^{n_i}) = \prod_i I^{n_i} + \sum_{i=0}^{p} R^{n_i} \prod_{j>i} (I^{n_j}). \tag{III.1.61}$$

Applying this expansion to Z_p, we obtain a sum of $p+2$ terms, which can be written explicitly as (by convention we label by -1 the first term in (III.1.61)):

$$Z_p = \sum_{i=-1}^{p} Z_{p,i} \tag{III.1.62}$$

where:

$$Z_{p,-1} = \sum_{0\le k_j \le n_j -1, j=0,\ldots,p} \int d\mu_p \prod_{j=0}^{p} \frac{(-g : \int_\Delta \varphi^{j,4} :)^{k_j}}{k_j!} \tag{III.1.63}$$

and

$$Z_{p,i} = \sum_{0\le k_j \le n_j -1, j=i+1,\ldots,p} \int d\mu_p e^{-g:\int_\Delta \varphi_i^4:_i}$$

$$\int_0^1 dt_i \frac{(1-t_i)^{n_i-1}}{(n_i-1)!} (-g : \int_\Delta \varphi^{i,4} :)^{n_i} \prod_{j=i+1}^{p} \frac{(-g : \int_\Delta \varphi^{j,4} :)^{k_j}}{k_j!} \tag{III.1.64}$$

for $i \ge 0$.

Remark that in $Z_{p,i}$ the remaining exponential involves only fields with indices bounded by i, because every parameter t_j with $j > i$ has been put to 0 by the action of the corresponding I^{n_j}. Therefore this exponential satisfies an upper bound similar to (III.1.55) but with p replaced by i:

$$|e^{-g\int_\Delta :\varphi_i^4:_i}| \le K_i, \qquad K_i \equiv e^{K\cdot i^2 \cdot \mathrm{Re}\, g}. \tag{III.1.65}$$

(for $i = -1$ we have no exponential at all, which means that we can take $K_{-1} = 1$.

We can describe formulas (III.1.63–4) by saying that for $j > i$ the derivations in the I^{n_j} operators produce $k_j < n_j$ Wick-ordered vertices with highest leg at scale j, and for $i \ne -1$ the R^{n_i} operator produce $n_i = 1 + a \cdot i$ vertices with highest leg at scale i.

We want to insert the upper bound (III.1.65) into (III.1.63–4) and perform the remaining Gaussian integrations. This cannot be done directly because the integral to bound is of the type $\int d\mu(\varphi) F(\varphi) e^{-G(\varphi)}$ and the function F is not necessarily positive (the Wick orderings have introduced subtractions). In such cases one separates first the two pieces to bound by a Schwarz inequality:

$$\int d\mu(\varphi)F(\varphi)e^{-G(\varphi)} \leq \left(\int d\mu(\varphi)F^2(\varphi)\right)^{1/2}\left(\int d\mu(\varphi)e^{-2G(\varphi)}\right)^{1/2} \tag{III.1.66}$$

and then one inserts the bound (III.1.65) on $G(\varphi) = g\int_\Delta :\varphi_i^4:_i$. The other piece of (III.1.66) contains only F^2, and if F is a polynomial (which is the case here) it can be integrated by Wick's theorem. In this process the number of vertices doubles, but because of the power 1/2 the final bound will not be much different. Let us define $k_j' = 2k_j$ and $k' = \sum_j k_j'$, $n = k' + 2n_i$. We are led to the following inequality (which is an overestimate):

$$|Z_p| \leq \cdot \sum_{i=-1}^{p} K_i \left(\sum_{\{0\leq k_j' < 2+2a\cdot j\}} (16|g|)^n (n!)^2 \sup_{G \text{ restricted}} \sum_{\mu \text{ restricted}} A_{G,\mu,\Delta}\right)^{1/2} \tag{III.1.67}$$

where the restricted supremum in (III.1.67) is taken over vacuum φ^4 graphs with n vertices which have no tadpoles, and the restricted sum in (III.1.67) is performed over all momentum assignments which satisfy the constraints that for $j = i+1, \ldots, p$ the number of vertices v whose highest index $e_v(\mu)$ equals j is exactly k_j', and the number of vertices v with $e_v(\mu) = i$ is exactly $2n_i$ (0 if $i = -1$). In other words the graphs are really built from the vertices which appear in (III.1.63–4). Amplitudes $A_{G,\mu,\Delta}$ are defined as in (II.1.8), except for the fact that the vertices are only integrated over the unit cube Δ, as indicated by the subscript Δ.

(III.1.67) is a sloppy estimate in which for simplicity we throw away factorials like $(k_j!)^{-1}$ and $(n_i - 1)!^{-1}$ in (III.1.63–4), and we do not keep the restrictions that the k_j' are even. We have also bounded the total number of graphs with n vertices by $16^n(n!)^2$, a clear overestimate.

The amplitudes $A_{G,\mu,\Delta}$ are now bounded with the universal method of part II. In section II.1 we remarked that for the convergent graphs of a superrenormalizable theory (which is the case considered here, since the divergent tadpoles are forbidden), after spatial integrations have been performed, we obtain some vertical exponential decay in index space between the highest scale of each vertex and the 0th scale, hence some factor $\prod_{v\in G} e^{-\varepsilon e_v(\mu)}$, for some positive ε.† This is different from the just renormalizable convergent case, for which vertical exponential decay occurs only between the highest and the lowest scale of each vertex. Using a piece of this vertical exponential decay we can perform the sum over momentum assignments μ. We are careful to keep another piece of this decay to bound the K_i factor and the sum over k_j's. By the restriction on μ this piece of decay is at least:

$$\prod_{v\in G} e^{-\varepsilon' e_v(\mu)} \leq e^{-\varepsilon\cdot i\cdot n_i} \prod_{j=i+1}^{p} e^{-(\varepsilon/2)j\cdot k_j'} \qquad \text{(III.1.68)}$$

(with by convention the term involving n_i missing if $i = -1$, and $\varepsilon' \equiv \varepsilon/2$). Apart from this factor, the estimate for the supremum over amplitudes with n vertices in (III.1.67) is simply a constant to the number of vertices according to Theorem II.1.1. We conclude that there exists a small constant ε' and a large constant C_1 such that:

$$|Z_p| \leq \sum_{i=-1}^{p} K_i \cdot L_i \left(\sum_{0\leq k_j'<2+2a\cdot j, j=i+1,\ldots,p} (|g|\cdot C_1)^n (n!)^2 \prod_{j>i} e^{-\varepsilon'\cdot j\cdot k_j'} \right)^{1/2} \qquad \text{(III.1.69)}$$

where we recall that $K_i = e^{K\cdot\mathrm{Re}\, g\cdot i^2}$ if $i \neq -1$, $K_{-1} = 1$, and we define similarly $L_i \equiv e^{-\varepsilon'\cdot a\cdot i^2}$ for $i \neq -1$ and $L_{-1} \equiv 1$. The factor L_i corresponds to the piece of power counting earned for the $n_i = 1 + a\cdot i \geq a\cdot i$ vertices produced in slice i, and the factors $e^{-\varepsilon'\cdot j\cdot k_j'}$ to the piece of power counting earned for the vertices produced at slices $j > i$.

The sum in the right hand side of (III.1.69) is bounded by a constant as $p \to \infty$, if a is taken large enough. Indeed let us choose first a so large that $|g|K < \varepsilon\cdot a/4$. In this case $K_i \cdot L_i \leq \sqrt{L_i}$. We use the binomial bound $n! \leq 2^{k'+2+2a\cdot i} k'!(2+2a\cdot i)!$ and define

$$C = 1 + \sum_{i=0}^{\infty} [(2+2a\cdot i)!(2C_1\cdot|g|)^{1+a\cdot i} e^{-\varepsilon'(a/2)i^2} \leq 1 + |g|C', \qquad \text{(III.1.70)}$$

---

†In fact here the constant ε is not small because by (I.4.7) and (II.1.14) it is simply $\log M$. However we prefer to write the bound as it would be in a generic superrenormalizable situation where the vertical decay per vertex might be small.

to get the bound:

$$|Z_\rho| \leq C \cdot \left(\sum_{k'=0}^{\infty} \frac{(4|g|C_1)^{k'}}{k'!} \sum_{k'_j<2+2a\cdot j, \sum_j k'_j = k', j=0,1,\dots} \left(\frac{k'!}{\prod_j k'_j!} \right)^3 \prod_j \left((k'_j!)^3 e^{-\varepsilon' \cdot j \cdot k'_j} \right) \right)^{1/2} . \quad \text{(III.1.71)}$$

Since $k'_j \leq 1 + 2a.j$, there exists a constant C_2 such that

$$\left((k'_j!)^3 e^{-\varepsilon' \cdot j \cdot k'_j} \right) \leq C_2^{k'_j} \cdot e^{-\varepsilon'' j \cdot k'_j}, \quad \text{(III.1.72)}$$

with $\varepsilon'' \equiv \varepsilon'/2$. Hence:

$$|Z_\rho| \leq C \cdot \left(\sum_{k'=0}^{\infty} \frac{(4|g|C_1 C_2)^{k'}}{k'!} \sum_{k'_j<2+2a\cdot j, \sum_j k'_j = k', j=0,1,\dots} \left(\frac{k'!}{\prod_j k'_j!} \right)^3 \prod_j e^{-\varepsilon'' \cdot j \cdot k'_j} \right)^{1/2} . \quad \text{(III.1.73)}$$

We define $C_3 = \sum_{j=0}^{\infty} e^{-(\varepsilon''/3)j}$, and using the multinomial identity, we have:

$$|Z_\rho| \leq C \cdot \left(\sum_{k'} \frac{(4|g|C_1 C_2 C_3^3)^{k'}}{k'!} \right)^{1/2} \leq C \cdot e^{2|g|C_1 C_2 C_3^3}, \quad \text{(III.1.74)}$$

hence we have proved a uniform upper bound on Z_ρ as $\rho \to \infty$. Furthermore, by (III.1.70) this bound is regular at $g = 0$, namely it tends to 1 as g tends to zero.

Having understood the principle of the expansion, we can now apply it to get a variety of results. First applying it to $Z_\rho - Z_{\rho-1}$ instead of Z_ρ we get an expansion identical to (III.1.62–4) except for the fact that $k_\rho = 0$ is now impossible (because this piece of Z_ρ cancels exactly with $Z_{\rho-1}$ since $d\mu^\rho$ is a normalized measure. Therefore we get a bound exactly similar to (III.1.74) but with an additional factor $e^{-\varepsilon'' \cdot \rho}$ that we can keep from the vertical decay of one vertex with highest index ρ, which has to exist since $k_\rho \neq 0$, hence $k'_\rho \geq 2$. The conclusion is that the difference $Z_\rho - Z_{\rho-1}$ is bounded by a convergent geometric series, hence the ultraviolet limit $Z(g) = \lim_{\rho \to \infty} Z_\rho(g)$ exists. Furthermore the bounds being uniform it is analytic in g. We can also analyze the

Taylor remainder in g of Z_ρ at order p in the same way. It is

$$Z_\rho^{(p)} = \int d\mu_\rho \int_0^1 dt \frac{(1-t)^p}{(p-1)!}\left(-g\int_\Delta :\varphi_\rho^4:_\rho\right)^p e^{-tg\int_\Delta :\varphi_\rho^4:_\rho} \tag{III.1.75}$$

Applying expansion (III.1.61) gives rise to expressions similar to (III.1.62–4) but with a total number of derived vertices now equal to $p + k' + n_i$. The corresponding bounds is a straightforward generalization of (III.1.74) but with an additional factor $\text{const}^p|g|^p p!$. This completes the proof that the ultraviolet limit of Z_ρ obeys the necessary Taylor remainder estimates for the validity of the Nevanlinna–Sokal Theorem I.5.1.

Altogether we have proved:

Theorem III.1.4 Ultraviolet limit of finite volume φ_2^4
For any complex coupling constant g with $\text{Re}\, g > 0$ we have

$$\lim_{\rho\to\infty} Z_\rho = Z(g) = 1 + O(g);$$

this ultraviolet limit $Z(g)$ is continuous in g at $g = 0$ and is in fact the Borel sum of its perturbative expansion.

The next step in the construction of φ_2^4 is to make a similar analysis in a finite volume Λ; one gets similar estimates but with an additional factor $K^{|\Lambda|}$ for some constant K. Starting from this result, we can then apply the cluster and Mayer expansion to construct the thermodynamic limit of the theory at weak coupling just as in the single slice case discussed above, but as previously this requires us to consider quantities like the pressure or normalized Schwinger functions, and g has to be small. The theory constructed in this way is shown easily to be still the Borel sum of its perturbative expansion (a result first proved in [EMS]). This proves that the outcome of the construction is independent of technical details such as the particular type of the cutoffs used. The corresponding theory, the weakly coupled massive φ_2^4 is presumably the simplest field theory which obeys Osterwalder–Schrader axioms, hence Wightman axioms. To check these axioms one can again rely on Borel summability, because one can make different constructions with different cutoffs preserving different subsets of the axioms and prove that they coincide because of the unicity of Borel sums. To fill the details of all these constructions is left to the reader; there are no other basic ideas needed apart from the cluster and Mayer expansion and the ultraviolet expansion (III.1.62–4).

These results also extend without much effort to the $P(\varphi)_2$ model for P any polynomial bounded below. The cluster and Mayer expansions again require small coefficients in this polynomial (which means a high temperature regime in statistical mechanics).

The conclusion we want to insist upon is that in order to control this first example of a non-trivial ultraviolet limit, one has to push perturbation theory farther and farther at higher and higher energy scales. This results in a net gain because power counting gains far exceed the losses due to the proliferation of graphs, and this in turn is true because the typical integration region of a local object like a vertex is smaller and smaller at higher energy. This idea is exploited in (III.1.62–4) and bounds like (III.1.66) in a very sloppy way, far from optimal as is obvious from the loose character of bounds like (III.1.69–70). This is possible because convergence margins in superrenormalizable theories, especially φ_2^4, are still very big. We turn now to a much better way of exploiting the same kind of ideas, and one which becomes truly necessary when ultraviolet convergence is in short supply, as is the case in renormalizable theories. This better way is the constructive multiscale or phase space expansion, in which we have to introduce a lattice of cubes adapted to each scale of momenta and test for the presence of interaction vertices of the corresponding scale in each cube of this lattice.

Chapter III.2

The Phase Space Expansion: The Convergent Case

The author feels that this technique of deliberate lying will actually make it easier for you to learn the ideas. Once you understand a simple but false rule, it will not be hard to supplement that rule with its exceptions.

—D. E. Knuth, The TEXbook

A The vertical expansion and convergent polymers

We turn to a description of the natural generalization of the cluster expansion of last section in the case of a model in which the principle of phase space chopping becomes necessary.

What are the models of this type? Consider first the massive φ^4 theory in dimension d. For $d = 2$ the renormalization problem reduces to Wick ordering. We have to do some momentum analysis but we can avoid spatial localization, hence a true phase space expansion, as shown in the preceding section. But for $d = 3$ phase space chopping starts being truly useful [GJ1]. Three dimensional theories are therefore a classical testing ground for the phase space expansionist [Ba1]. It would be interesting for the reader to test also the version of the phase space expansion defined below on φ_3^4, and compare it to the original constructions [FO][MS1]. However our formalism is really designed for marginal, just renormalizable theories, so we focus directly on this case. For φ_4^4, which is not asymptotically free, the phase space expansion simply proves that if one starts with a bare theory with a bare coupling small enough so that the first cluster expansion at the bare scale converges, then the resulting renormalized theory is a free field. This phenomenon is discussed in the next section (Theorem III.3.2).

Therefore in order to get some non-empty constructive results we have to look for other models: in the next section III.3 we discuss how to apply the phase space expansion to the infrared (critical) φ_4^4 (with fixed ultraviolet cutoff) and in section III.4 to the massive Gross–Neveu model in two dimensions, which is asymptotically free. In the concluding section III.5 we discuss the much less advanced case of the ultra violet limit of non-Abelian gauge theories, where the hope is to gain a non-perturbative understanding of asymptotic freedom.

The phase space expansion is a rather natural extension of the cluster expansion, but it leads to several additional technicalities. Sticking to our principle of dividing the difficulties into pieces easier to digest, we propose in this section an overview of the formalism and a constructive analogue of section II.1, namely an analysis of the convergent cases (with favorable power counting). For this limited purpose our standard model, massive ultraviolet φ_4^4, is still perfectly convenient.

We want to find the analogues of the phase space slicing of part II, but in a context no longer limited to Feynman graphs; we have to perform true functional integrals. Hence it is not enough to consider propagators only, and we must use the language of fields. We slice the covariance of our Gaussian measure, e.g., with α-space cutoffs as in (II.1.2–4). To this slicing of the covariance is associated the corresponding orthogonal decomposition (II.1.5) of the Gaussian measure $d\mu = \otimes d\mu^i$. The field φ becomes a sum of random variables φ^i independently distributed according to $d\mu^i$. φ^i will be called a field of frequency (or index) i, which means that it corresponds to momenta of order M^i. In each momentum slice i the Gaussian measure $d\mu^i$ is factorized and obviously one should perform one corresponding cluster expansion, this time with respect to the scaled lattice $\mathbf{D}^i$ of cubes of side size M^{-i}. These are called the horizontal cluster expansions. However this is not enough, as in the single scale model, to get the factorizations necessary for the thermodynamic limit, because now the interaction still couples the various horizontal slices. Remember that φ^4 vertices were pictured as dotted vertical links, dual to the horizontal propagators in Fig. II.1.2. One should find therefore a kind of vertical cluster expansion which is the analogue of the horizontal ones, but for these dual vertical links. Finally there is the question of the Mayer expansion, which we discuss briefly, anticipating the following.

The simplest solution is to perform first all the horizontal and vertical cluster expansions, hence to obtain a dilute gas of polymers with hard core interactions swimming in the "$d + 1$" dimensional phase space, and apply a single Mayer expansion to it. This point of view is all right as far as the bare theory is concerned (or for superrenormalizable theories like φ_2^4 or φ_3^4, where only simple mass renormalizations have to be performed). But in the case of just renormalizable theories we want to compute the analogue of the usefully renormalized expansion with effective couplings, whose advantages were detailed at length in Part II. This means that at each scale i we want to add and subtract a counterterm which in analogy with section II.4 is (in the case of the coupling constant renormalization) a sum over all polymers with 4 external variables made with the fields and propagators of higher slices. To be of the desired form $\int_\Lambda \varphi_l^4$ (with $\varphi_l = \sum_{j=0}^{l} \varphi^j$), such a counterterm has to live in all of Λ without hard core constraints. In order to combine this counterterm with the four point polymers, the hard core constraints between these polymers have to be removed. As shown in the preceding section, this is exactly what the Mayer expansion does. For this reason, we cannot wait for a single Mayer expansion only in the end, and in each new slice after the corresponding horizontal and vertical cluster expansions have been performed, a Mayer expansion has also to be performed.

The drawback is that it is difficult to visualize the result of the corresponding sequence of Mayer expansions. Indeed Mayer configurations are sequences of overlapping polymers, which can at best be pictured as superposed strata. Iterating the superposition of such strata becomes really tedious to picture. This is one of the reasons for which we decide to postpone the corresponding formalism and the problems related to renormalization to the next section, where they will be treated in the concrete case of infrared φ_4^4. The goal of this section is to familiarize ourselves with the phase space expansion in the limited context of *convergent* polymers; in this case the sandwiching of Mayer expansions between each cluster expansion is not necessary, and will not be introduced. Hence this chapter is really the constructive generalization of chapter II.1. We will discover the basic mechanism of convergence of the phase space expansion in a context free of the technicalities associated to renormalization; we hope that the experience gained in this way will be valuable to throw light on the general case.

Let us return to the "vertical" cluster expansion. What kind of interpolation formula should be applied to the vertex $\int(\sum_i \varphi^i)^4$ which imitates the interpolations of the preceding sections for the propagator? Again the key requirement is that this interpolation preserves positivity. As usual it is convenient to start from the bare theory, hence from the highest momentum slice, and proceed downwards towards the renormalized theory. Therefore the first interpolation should separate φ^p from the rest, which is $\sum_0^{p-1} \varphi^i = \varphi_{p-1}$. Also there should be a different interpolation parameter, called t_Δ (not to be confused with the s parameters of horizontal cluster expansions) for every cube Δ of $\mathbf{D}^p$, since these are the units which the pth horizontal cluster expansion tries to decouple. One of the simplest interpolations satisfying positivity is

$$\int_\Delta g(\varphi^p+\varphi_{p-1})^4 = \left[\int_\Delta g(\varphi^p + t_\Delta \varphi_{p-1})^4 + (1-t_\Delta^4)\int_\Delta g\varphi_{p-1}^4\right]\Big|_{t_\Delta=1} \quad \text{(III.2.1)}$$

with g the bare coupling constant. This is an analogue of formula (III.1.8), taking into account that we have now a quartic object (other formulas are of course possible). The pth vertical expansion consists in applying, after the pth horizontal cluster expansion the operator:

$$\prod_{\Delta\in\mathbf{D}^p}(I+R)\left[e^{-\sum_\Delta \int_\Delta g(\varphi^p+t_\Delta\varphi_{p-1})^4+(1-t_\Delta^4)\int_\Delta g\varphi_{p-1}^4}\right], \quad \text{(III.2.2)}$$

where, like in (III.1.59–60), If takes the beginning of the Taylor expansion of f in t_Δ up to some fixed order p, and Rf is the integral formula for the corresponding Taylor remainder $\int_0^1 dt_\Delta[(1-t_\Delta)^p/(p!)][d^{p+1}/(dt_\Delta^{p+1})]f$. We allow for the possibility to have $p > 1$; in fact for horizontal cluster expansions a single Taylor step always gives sufficient decay because the sliced propagator has fast decay, but this is not true for the vertical expansion. We have seen that, in the case of φ_4^4, it is only for graphs with more than 4 external low momentum legs that exponential decay in the vertical direction is guaranteed. Since each $d/(dt_\Delta)$ derivation generates at least one external low momentum leg φ_{p-1}, in dimension 4 it is wise to choose $p = 4$, so that we are sure that the remainder terms correspond to a situation in which vertical exponential decay occurs.

The inductive generalization of (III.2.2) to all scales is easy. Writing $\mathbf{D}$ for the union of all scaled lattices $\mathbf{D}^i$ covering Λ with cubes of side size M^{-i}, $0 \le i \le \rho$, we introduce a parameter t_Δ for each $\Delta \in \mathbf{D}$. Then for fixed $x \in \Lambda$ we call t_i, $0 \le i \le \rho$ the parameter of the cube $\Delta \in \mathbf{D}^i$ to which x belongs and we write:

$$\varphi_i(t) \equiv \sum_{j=0}^{i} \left[\prod_{j<k\le i} t_k \right] \varphi^j, \tag{III.2.3}$$

$$\varphi^4(x) = \left(\sum_{j=0}^{\rho} \varphi^j(x) \right)^4 = \sum_{i=0}^{\rho} (1 - t_{i+1}^4)(\varphi_i(t))^4 \bigg|_{\{t\}=1}$$
$$= \sum_{i=0}^{\rho} (1 - t_{i+1}^4) \left(\sum_{j=0}^{i} \left[\prod_{j<k\le i} t_k \right] \varphi^j \right)^4 \Bigg|_{\{t\}=1} \tag{III.2.4}$$

where by definition $t_{\rho+1} \equiv 0$.

In some cases it may be convenient to use smooth characteristic functions $\Delta(x)$ for the cubes Δ; then one defines:

$$a(x, \{t\}, i, j) = \prod_{j<k\le i} \left[\sum_{\Delta \in \mathbf{D}^k} t_\Delta \Delta(x) \right]; \quad \varphi_i(\{t\}, x) = \sum_{j \le i} a(x, \{t\}, i, j) \varphi^j(x) \tag{III.2.5}$$

and one writes

$$\int_\Lambda \left(\sum_{i=0}^{\rho} \varphi^i(x) \right)^4 dx = \int_\Lambda dx \sum_{i=0}^{\rho} \left[1 - a(x, t, i, i+1)^4 \right] \varphi_i(\{t\}, x) \bigg|_{\{t\}=1} \tag{III.2.6}$$

where again, by convention, $t_\Delta \equiv 0$ if $\Delta \in \mathbf{D}^{\rho+1}$. Formulas (III.2.5) and (III.2.6) are rather complicated, and we shall not use them in this book; but the use of sharp characteristic functions has some drawbacks too, and the reader will see below that it forces us to treat a small piece of the Gaussian measure in the form of an interaction $\exp(-\varepsilon \int \partial_\mu \varphi \partial^\mu \varphi)$, with ε a small constant.

For this reason, and for the treatment of models with counterterms, it is convenient to give also rules to interpolate fields with derivatives and quadratic interactions, like $m^2 \int \varphi^2$ and $\varepsilon \int \partial_\mu \varphi \partial^\mu \varphi$. We use the interpolations:

$$\partial_\mu \varphi_i(t) \equiv \sum_{j=0}^{i} \left[\prod_{j<k\le i} t_k \right] \partial_\mu \varphi^j, \tag{III.2.7}$$

$$\varphi^2 = \sum_{i=0}^{\rho}(1 - t_{i+1}^2)(\varphi_i(t))^2 \Big|_{\{t=1\}}, \tag{III.2.8}$$

$$\partial_\mu\varphi\partial^\mu\varphi = \sum_{i=0}^{\rho}(1 - t_{i+1}^2)[\partial_\mu\varphi_i(t)\partial^\mu\varphi_i(t)] \Big|_{\{t=1\}}. \tag{III.2.9}$$

We should think of these formulas as the inductive generalizations of natural interpolation rules analogous to (III.2.1):

$$\int_\Delta m^2(\varphi^\rho + \varphi_{\rho-1})^2$$
$$= \left[\int_\Delta (m^2)(\varphi^\rho + t_\Delta\varphi_{\rho-1})^2 + (1 - t_\Delta^2)\int_\Delta (m^2)(\varphi_{\rho-1})^2\right]\Big|_{t_\Delta=1}; \tag{III.2.10}$$

$$\int_\Delta \varepsilon(\partial_\mu\varphi^\rho + \partial_\mu\varphi_{\rho-1})(\partial^\mu\varphi^\rho + \partial^\mu\varphi^{\rho-1})$$
$$= \Bigg[\int_\Delta \varepsilon(\partial_\mu\varphi^\rho + t_\Delta\partial_\mu\varphi_{\rho-1})(\partial^\mu\varphi^\rho + t_\Delta\partial^\mu\varphi_{\rho-1})$$
$$+ (1 - t_\Delta^2)\int_\Delta \varepsilon(\partial_\mu\varphi_{\rho-1}\partial^\mu\varphi_{\rho-1}\Bigg]\Big|_{t_\Delta=1}. \tag{III.2.11}$$

Until the expansion at scale i we should put every parameter t_Δ with $\Delta \in \mathbf{D}_j$, $j \leq i$ to 1; then after the cluster expansion of scale i, the interpolating parameters t_Δ with $\Delta \in \mathbf{D}_i$ are introduced for all fields (both in the exponential of the interaction and in fields already derived by former expansions). Remark that ∂_μ-derivatives and t-dependence do not commute; indeed the t_k's are functions of x, by our convention that t_k is the parameter t_Δ such that $x \in \Delta \in \mathbf{D}^k$. To ensure that $\partial_\mu\varphi^j$ is of order $M^j\varphi^j$, as it should be, it is essential to write the t dependence of derived fields always after performing the derivative, otherwise the derivative could act on a characteristic function of a cube of much smaller size, hence be much bigger than what is expected.

The phase space expansion consists in applying at each scale, starting from ρ, a horizontal cluster expansion, for instance of the tree type as in the preceding section, then the vertical expansion, which means that we apply to the functional integral the operator

$$\prod_{\Delta\in\mathbf{D}^i}(I_\Delta^{(4)} + R_\Delta^{(4)}) \tag{III.2.12}$$

where $I_\Delta^{(4)}$ takes the beginning of the Taylor expansion t_Δ at $t_\Delta = 0$ up to 4th order and $R_\Delta^{(4)}$ is the Taylor remainder (see (III.1.59–60) for similar formulas). To compute the result of this vertical cluster expansion at scale i, we expand the product (III.2.12), and when the $I_\Delta^{(4)}$ term is chosen we draw a thick line at the bottom of the cube Δ as in Fig. III.2.1. This thick line is a sort of vertical analogue of a Dirichlet condition; it suggests the corresponding decoupling of frequencies. We call it also a "closed gate." When R_Δ is chosen the corresponding "gate" is open. The cube Δ should be thought of as connected through this open gate to the larger cube $\Delta' \in \mathbf{D}^{i-1}$, $\Delta' \supset \Delta$. Remark that dotted vertices then cross this open gate, with a total of at least 5 low momentum fields φ_{i-1} hooked to them. Such an open gate is also called a *strong* connection, and guarantees automatically a favorable power counting.

Let us make the notion of connectedness in phase space more precise. We say that two cubes $\Delta \in \mathbf{D}^i$ and $\Delta' \in \mathbf{D}^j$ are directly connected if either (we may assume $j \leq i$):

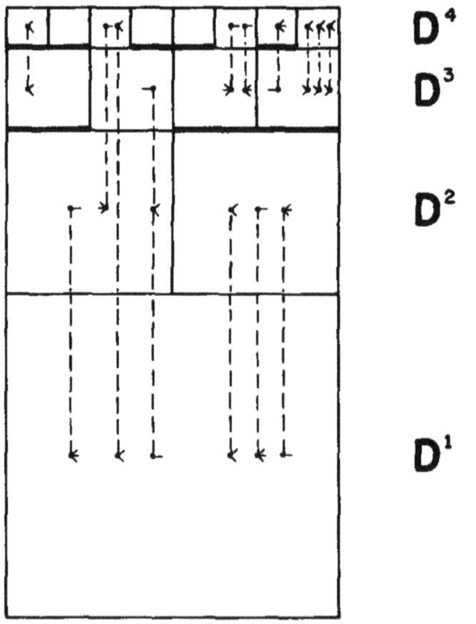

Figure III.2.1 Open and closed gates. The vertices (dotted lines) are shown. Thick lines (closed gates) are crossed by at most 4 low momentum fields. Thin lines (open gates) are crossed by at least 5.

- there is a propagator (horizontal line) between them. This is as before. It requires $i = j$ and the propagator is generated by the ith horizontal cluster expansion.
- $\Delta \subset \Delta'$ and there is a vertex (dashed vertical line) localized in Δ with two fields hooked to it, one of scale i and the other of scale j.†
- there is an open gate between them. This requires $i = j + 1$ and $\Delta \subset \Delta'$. This open gate has to be generated by the ith vertical cluster expansion. The cubes Δ and Δ' are said to be directly strongly connected.

The first and second conditions are easy generalizations of the horizontal and vertical connections of perturbation theory, but the third one corresponds to a remainder term which has no equivalent in perturbation theory. As remarked, in this third case there are at least 5 low momentum fields localized in Δ.

The notion of connectedness is then extended so that cubes Δ_i and Δ_j which can be joined by a chain of directly connected cubes are connected. Maximal sets of connected cubes in phase space are called polymers. It is also convenient to define strongly connected domains. Two cubes are said to be strongly connected if they can be linked through a chain of directly strongly connected cubes, and the corresponding maximal sets are the strongly connected domains, pictured in Fig. III.2.2. Hence they are defined exactly like the polymers, but only connections of the third type are taken into account. One should think of these strongly connected domains as elementary building blocks of the phase space expansion in which the interaction is not decoupled.

To understand better the whole expansion, it is a good exercise to visualize how it works in the first slices.

---

†In [FMRS5] another rule is used: vertical connections which connect Δ to Δ' are also required to connect together all the cubes Δ'' which are between Δ and Δ' (i.e., such that $\Delta \subset \Delta'' \subset \Delta'$; the horizontal analogue would be to require that propagators between two cubes connect together all cubes on a straight line between the two ends of the propagator. Both points of view are perfectly valid and lead to slightly different phase space expansions but which are both convergent (because by power counting the vertical links decay exponentially). But the point of view used here is certainly more natural.

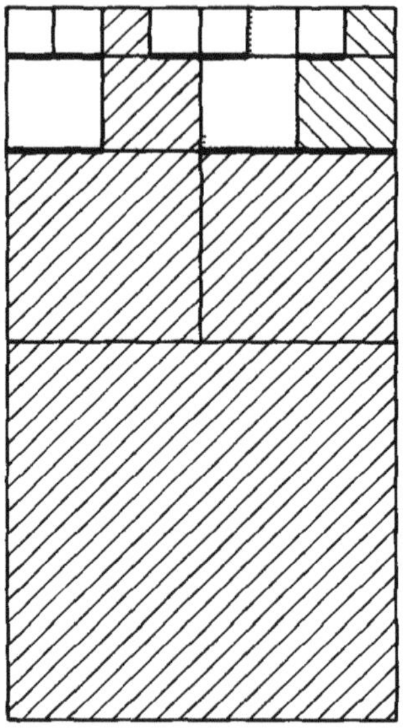

Figure III.2.2 Strongly connected domains in the previous case. There are 3 nontrivial such domains, and 6 domains reduced to a single (white) cube.

In the ρ-slice we perform the first horizontal cluster expansion with the Gaussian measure $d\mu^\rho$ corresponding to fields φ^ρ and covariance C^ρ and with respect to the cubes of $\mathbf{D}^\rho$; then we perform the first vertical expansion. We obtain a sum over a gas of disjoint ρ-polymers which is an analogue of formula (III.1.15); these polymers are defined again as maximal connected objects. By convention again, it is convenient not to consider the empty isolated cubes as polymers, which leads to a new trivial normalization of the polymers similar to (III.1.14–16). At this stage the only connections to take into account are the propagators of the ρth horizontal cluster expansion. The integration by parts analogue of (III.1.23) and the vertical expansion derive some low momentum fields $\varphi_{\rho-1}$ which (together with the true external fields, of scale -1) should be considered as external variables at this stage ($\varphi_{\rho-1}$ is usually called the "background field" at this stage).

The $\rho - 1$ cluster expansion is then performed, with respect to fields $\varphi^{\rho-1}$, the Gaussian measure $d\mu^{\rho-1}$ and the lattice of cubes of $\mathbf{D}^{\rho-1}$. Remark that this may result in propagators connecting cubes of $\mathbf{D}^{\rho-1}$ already connected (through scale ρ). Then we perform the $\rho - 1$ t-decoupling expansion (in which all the remaining former external fields $\varphi_{\rho-1}$ get decomposed into $\varphi^{\rho-1}$ or $\varphi_{\rho-2}$). We get a formula which is a sum over $(\rho-1)$-polymers which are maximal connected objects made of disjoint sets of cubes of $\mathbf{D}^{\rho-1}$ and of the ρ-polymers previously generated. The low momentum fields $\varphi_{\rho-2}$ are the new external set of variables. We iterate the process, generating i-polymers, until scale 0 is reached.

At the end of the horizontal and vertical cluster expansion the partition function is expressed directly as a sum over vacuum 0-polymers of the type (III.1.15), with hard core constraints over the whole phase space. The pressure could be computed by a global Mayer expansion of these constraints. For Schwinger functions, we would have only the true external fields (with index -1) left as external variables of the 0-polymers and we can choose the last Mayer expansion adapted to the precise quantities we want to compute (normalized or truncated Schwinger functions). This Mayer expansion generates configurations made out of these connected 0-polymers, joined through "-1 Mayer links." These Mayer links, which force the polymers to overlap, may also be considered as new connections, so that connected functions are given by a single such configuration with the prescribed set of true external variables according to (III.1.46).

It is time now to define the analogue of almost local subgraphs. Just as before with simple graphs it is useful to consider, for a given 0-polymer and a given slice i, the set of all i-subpolymers G_k^i, which are defined inductively as the maximal i-polymers (made of cubes of $\mathbf{D}^i$ and of $(i+1)$-subpolymers) which are connected through connections of indices higher or equal to i. Again the G_k^i's will be connected later together by lower scale connections and they have therefore tree structure.

A convergent 0-polymer is the obvious generalization of the convergent assignments for graphs at the beginning of Section II.3. It is a 0-polymer such that any almost local subpolymer G_k^i has at least 5 external variables (low momentum fields or true external fields) hooked to it. (As for graphs, in the last slice this condition requires that there are at least 5, and by parity in fact 6 true external fields; hence strictly speaking we have no completely convergent 0-polymers unless we consider Schwinger functions with at least 6 external arguments; but this problem may be technically circumvented if necessary by performing a single global subtraction.) It may be interesting to remark again that any G_k^i which contains a cube $\Delta \in \mathbf{D}^i$ with $t_\Delta \neq 0$ (open gate) automatically satisfies the convergence condition because the t_Δ expansion has been pushed to 4th order so that in the remainder at least 5 low momentum legs have been generated.

The reader should be aware that many technical features of the expansion are not "canonical." In particular we want to stress that the rules we have just described are simple and systematic but they are far from optimal from the point of view of "minimal expansionism." For instance, having read section III.1, the reader might remark that the brute force formulas (III.2.2)–(III.2.12) reminds him of the "pair of cubes" expansion and that presumably they do not give the minimal way to decouple slices in the vertical direction. This guess is correct; just as the "pair of cubes" expansion sometimes builds redundant horizontal loops, the vertical expansion defined above can build redundant vertical connections, i.e., link vertical regions with much more that the minimal number of fields necessary for convergent power counting. This phenomenon can be avoided and there is a more economical way in terms of expansion steps to decouple vertical regions, but as usual it is of a more inductive character. Instead of introducing one parameter t_Δ for each cube of the ρth slice, we may introduce a single parameter t for each of the regions of Λ which have been connected by the ordinary horizontal cluster expansion in the ρ slice performed before [dCdVMS]. In the previous formalism this is equivalent to equate, for all ρ-polymers P the parameters t_Δ for $\Delta \in P$ to a single parameter t_P, and to perform the Taylor expansion (III.2.12) for all t_P's rather than all t_Δ's. In the same vein one should, in the $\rho - 1$ horizontal cluster expansion, introduce interpolating s parameters which test the coupling through $C^{\rho-1}$ of the blocks of cubes connected together through the previous connections, instead of testing blindly between all cubes of $\mathbf{D}^{\rho-1}$. At the end of this $\rho - 1$ cluster expansion, we have a gas of $\rho - 1$ polymers P' and new parameters $t_{P'}$ are defined for the $\rho - 1$ vertical expansion through the collapse of the corresponding set of t_Δ parameters, $\Delta \in \mathbf{D}^{\rho-1}$, and so on. This optimized version has many advantages, in particular the superficially convergent i-polymers (i.e., those with 5 external legs φ_{i-1} or more) do coincide with the remainder terms in the vertical expansion, so that in a sense they stand out more clearly; also if phase space was used for practical computations it would be presumably optimal to expand in this minimal way. Finally it seems a general rule that expansions with minimal decoupling are always better from the point of view of preserving positivity. In models with more marginal positivity than φ^4, such as the Gross–Neveu model in three dimensions of section III.4B, the optimized expansion becomes truly necessary. Here for simplicity we choose to stick to the systematic rule (III.2.2). The convergence proofs that we provide below also apply with some modifications to the optimized expansion, for which we refer the reader to [dCdVMS].

We propose to study the convergence of the phase space expansion in the situation which corresponds to Weinberg's uniform theorem (II.1.9) or more precisely to its generalization (II.2.1). In this case every contribution has to be convergent from the point of view of power counting. Hence the analogue of Weinberg's uniform theorem in the phase space expansion is:

Theorem III.2.1 Convergence of the phase space expansion in the case of completely positive power counting

The sum over all convergent polymers of the phase space expansion is absolutely convergent, uniformly in ρ and Λ, provided g, the coupling constant, is positive and small enough; it has a limit as $\rho \to \infty$ and $\Lambda \to \infty$ which one might call the convergent piece of the φ_4^4 theory. (Of course this limit is not a field theory and depends strongly on the particular rules we have taken for slicing, etc., ..., because every polymer which violates the convergence condition has been arbitrarily set to 0).

Proof The proof does not reduce only to a tedious exercise in patching together everything we know by now about cluster expansions and tree combinatorics in the strange "$d + 1$" dimensional phase space. It involves a new difficulty which did not appear at all until now, which is the possible proliferation of low momentum fields. The solution to this problem is called the "domination" of low momentum fields, and it uses in a crucial way the positivity of the interaction. For a better understanding we provide now an informal discussion of the problem and its solution.

B Overview of the domination problem

First one should remember that although the initial functional integral has been quite chopped by the phase space expansion, the subpolymers still contain true functional integrals, hence are not simply ordinary graphs. Of course this was already true for the single scale cluster expansion of the last section. But in that case to bound the functional integral, which was of the form

$$\left(\prod \varphi_j\right) e^{\int -g\varphi^4} d\mu(\varphi), \qquad \text{(III.2.13)}$$

we applied simply a Schwartz inequality to separate $\prod \varphi_j$ from $\exp(-\int g\varphi^4)$; in other words we used positivity of $\exp(-g\int \varphi^4)$ to bound it by 1, and we bounded $(\prod \varphi_j)$ by integrating it with $d\mu(\varphi)$; this is what we call "Gaussian integration" of the fields φ_j produced by the expansions. The corresponding bound is similar to what would be obtained in perturbation theory.

This is all right for a single-scale model. However, when many scales are present, and a given vertex v_Δ in $\Delta \in \mathbf{D}^i$ produced by the horizontal and vertical expansions of scale i has both high momentum fields φ^i and low momentum fields φ^j hooked to it, $j \ll i$, we can use "Gaussian integration" for the φ^i fields, but it could be unwise to use it for the φ^j fields. Indeed, we would lose the correct power counting or generate a fraction of the factorial divergence of perturbation theory, depending on which point of view is used. We will discuss these two points of view to understand the problem well; then we will present the solution, again discussing it from both points of view. The domination problem is indeed tricky and worth being discussed thoroughly.

At scale i the computation of the cluster expansion involves functional derivations which can be expressed as $C^i(x,y)\delta/(\delta\varphi^i)(x)\delta/(\delta\varphi^i)(y)$, $x \in \Delta$, $y \in \Delta'$, $\Delta, \Delta' \in \mathbf{D}^i$. These functional derivations $\delta/(\delta\varphi^i)(x)$ can derive vertices whose three other fields are not of scale i but of a different scale, by the obvious analogue of (III.1.23):

$$\frac{\delta}{\delta\varphi^i(x)} \cdot e^{-g\int_\Delta \left(\sum_{j=1}^p \varphi^j\right)^4} = -4g\Delta(x)\left(\sum_{j=1}^p \varphi^j(x)\right)^3 e^{-g\int_\Delta \left(\sum_{j=1}^p \varphi^j\right)^4}. \quad \text{(III.2.14)}$$

As we will see below, a new problem arises when these different scales j are much lower than i.

The vertical expansion in the parameter t_Δ typically creates also similar vertices, both in the remainder term $R^{(4)}$ of (III.2.12) (at least five low momentum fields are derived) and in the term $I^{(4)}$ of (III.2.12), which is taken at $t_\Delta = 0$; remark however that in this last case the number of such fields is at most 4.

Let us consider what happens if as in the previous single scale model, we use a Schwarz inequality to evaluate the corresponding derived low momentum fields of index $j \ll i$ by Gaussian integration.

We have to remark that the Gaussian piece available in Δ for φ^j will be much weaker than for φ^i. To see this let us write intuitively the Gaussian measure like in (I.3.1) as $\exp(-\int \varphi(p^2 + m^2)\varphi).D\varphi$ in terms of an ill defined Lebesgue measure $D\varphi$. Then the slicing cut-offs tell us that typically p^2 in the slice j is of order M^{2j}, hence much smaller than p^2 in the slice i, which is typically of order M^{2i}. Therefore using $\exp(-\int_\Delta \varphi^j p^2 \varphi^j).D\varphi^j$ to integrate over φ^j is worse than using $\exp(-\int_\Delta \varphi^i p^2 \varphi^i).D\varphi^i$; by simple scaling it results in a loss of $M^{(i-j)}$ per low momentum field integrated in this way, in comparison with the estimates of the single scale model (the effect of m^2 is negligible for an ultraviolet problem where i and j are big).

The defect of this point of view is that the Lebesgue measure does not exist; nevertheless it leads to the correct conclusion, as we see now by introducing the more rigorous second point of view, in which the Gaussian integration over low momentum fields like φ^j is rigorously evaluated as a sum of graphs. In this second point of view the domination problem shows up as a piece of the divergence of perturbation theory. We know that Gaussian integration leads inescapably to local factorials of the number of fields integrated (see Lemma II.6.2). For fields φ^i the corresponding local factorials are harmless because they can be beaten by the volume argument (Lemma III.1.3). Ultimately this is because the cubes of the ith cluster expansion are of the correct size for the ith momentum slice. But this is no longer true for low momentum fields φ^j created by the ith cluster expansion; in other words, a lot of φ^j fields may be produced in a single cube Δ of $\mathbf{D}^j$, each of which coming from a different cube Δ' of $\mathbf{D}^i$, simply because there are $M^{4(i-j)}$ such different cubes Δ' in Δ. In Fig. III.2.3, a typical "worst scenario" is pictured: it is a simplified situation, in which each cube $\Delta' \in \mathbf{D}^i$ sends three low momentum fields (corresponding to a single vertex) in $\Delta \in \mathbf{D}^j$.

By our former remarks if $\Delta \in \mathbf{D}^j$ and $t_{\Delta''} = 0$ in the t decoupling expansion for every $\Delta'' \subset \Delta$, $\Delta'' \in \mathbf{D}^{j+1}$, only a finite $(4 \times M^4)$ number of fields φ^j created by higher scales expansions can enter Δ. Factorials of such a constant number are harmless. The worst-case situation of Fig. III.2.3 can occur only if the $t_{\Delta''}$ parameters for cubes Δ'' intermediate between the cubes Δ' of Fig. III.2.3 and the cube Δ are nonzero. In other words the corresponding cubes have to be connected by open gates and belong to the same strongly connected domain. These domains where the interaction is completely undecoupled are therefore truly responsible for the domination problem.

To measure precisely the local factorial effect generated by the Gaussian integration of low momentum fields, consider the situation of

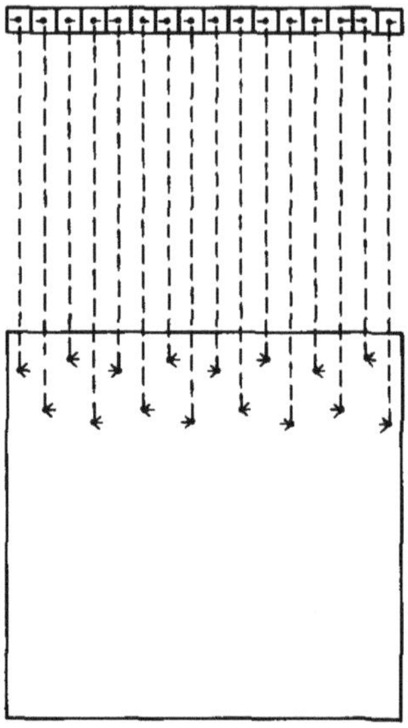

Figure III.2.3

Fig. III.2.3; let $n = M^{4(i-j)}$ be the total number of vertices, which is also the number of cubes Δ' in Δ, since we postulated one vertex per such cube. The total number of fields φ^j localized in Δ is then $3 \cdot M^{4(i-j)} = 3n$. There is no permutational symmetry among the vertices since they lie in different cubes Δ'. The Gaussian integration of the high momentum φ^i fields costs only $(\text{const})^n$ (by Lemma II.6.2), but the Gaussian integration of the $3n$ low momentum fields φ^j, by the same principle, costs $(\text{const})^n(3n/2)!$. Finally propagators C^j have a scaling factor M^{+2j} instead of M^{2i} for C^i, and this creates in the estimate a relative bonus of order $M^{-2(i-j)\cdot 3n/2} \simeq (\text{const})^n[(3n/4)!]^{-1}$. Hence the total contribution is of order $C^n(3n/4)!$, consistent with the intuitive picture that "three-fourths" of the ordinary divergence of perturbation theory has been developed, because three fourths of the n fields are packed in a single cube of the size corresponding to their frequency. It is also consistent with the heuristic prediction of the first point of view: a loss of $M^{(i-j)} = n^{1/4}$ per low momentum field means here a total loss of $n^{(1/4)\cdot 3n}$, corresponding to the factor $C^n(3n/4)!$.

A physicist might be shocked by Fig. III.2.3, and remark that "one high momentum leg and three low momentum legs" is a somewhat surprising contribution which does not obey momentum conservation, hence might be suppressed in some way or another. However momentum conservation is partly incompatible with the principle of phase space localization, so that to implement this idea is difficult. A vertex summed over a small cube corresponds in momentum space to a convolution which can bring momenta of order the inverse size of the cube. To exploit momentum conservation at the lower scale it is therefore essential to integrate the corresponding vertices over cubes of a sufficiently large size, hence to free them from the constraints of localization at the higher scale. Moreover technically it requires also cutoffs which preserve momentum rather well. In spite of these difficulties this idea is interesting and will be used in the next section for the Borel summability results. But it alone is clearly not enough to solve the domination problem. We see indeed that even in the case (certainly allowed by momentum conservation) where each vertex would have only one or two low momentum legs, a divergent factor (respectively $(\frac{n}{4})!$ or $(\frac{n}{2})!$) would still appear.

Therefore we cannot escape the conclusion that in such cases the functional integration over low momentum fields should not be done using solely the decrease of the Gaussian measure.

The only other alternative available is to make better use of the decrease of the φ^4 interaction itself, and immediately everything falls into its proper place. Returning to the first point of view, we saw that $\exp(-\int_\Delta \varphi^j p^2 \varphi^j)$ was much weaker than $\exp(-\int_\Delta \varphi^i p^2 \varphi^i)$, because in the first case $p^2 \simeq M^{2j}$ and in the second $p^2 \simeq M^{2i}$ so that in the first case the typical size of φ^j is M^{2i-j}, much larger than the typical size M^i of φ^i. But this is no longer true if we use $\exp(-g\int_\Delta (\varphi^j)^4)$, which is just as good as $\exp(-g\int_\Delta (\varphi^i)^4)$; both give a typical size $g^{-1/4}M^i$ to the field (since the volume $|\Delta|$ is M^{-4i}). In other words the φ^4 interaction is scale invariant, hence much better in this respect than the Gaussian.

In the more rigorous second point of view, where we consider the large number $(3n)$ of fields φ^j which accumulate in a single square Δ to give $(\frac{3n}{2})!$ by Wick's theorem or Lemma II.6.2, we know that a quartic integration gives only half as many factorials as a Gaussian one; this is nothing but the formula

$$\int x^n e^{-x^2} dx \sim (c_1)^n \left(\frac{n}{2}\right)!, \int x^n e^{-x^4} dx \sim (c_2)^n \left(\frac{n}{4}\right)!. \qquad \text{(III.2.15)}$$

Therefore quartic interaction will result in a $(3n/4)!$ (times $g^{-3n/4}$ to account for the coupling constant in front of φ^4). Combining this with the relative bonus $(3n!/4)^{-1}$ which is unchanged since it comes from the scaling of φ^j (now of order M^j instead of M^i) we obtain $(\text{const})^n g^{-3n/4}$, or more precisely $(\text{const})^n g^{+n/4}$ if the factor g^n in front of the n derived vertices is included. This is perfectly summable if g is small enough.

We realize that it was unreasonable to expect to earn a full small factor g for each derived vertex, when possibly only one fourth of the fields of these derived vertices were of the correct scale! We must accept to earn only a more reasonable $g^{1/4}$ factor, still sufficient of course for convergence purposes. This is done in two steps: Gaussian integration of the high momentum fields, a positive operation in which we earn $g^{1/4}$ per such field, and a bound on the unwanted low momentum fields which compares them to the exponential from which they were derived; this second operation is neutral from the point of view of estimates (no significant gain or loss except a constant per field).

This sketchy solution of the "domination" problem requires as a crucial point not only the positivity of the interaction (which was required already for the simple existence of the functional integral in a single cube) but also its decay at large φ (see (III.2.15)).

We have completed our tour of the problem and its solution, but we want to treat a particular simple example with more care to uncover some subtleties. In particular the reader might have remarked already at this stage that the exponential of the interaction contains interpolating parameters t which are not necessarily the same as those of the low momentum fields to be dominated, and might worry about that. He might also ask how to separate in practice the high momentum fields from the lower ones in order to apply the domination principle. We address these questions now.

The functional integral within a strongly connected set (set of cubes connected by open gates $t_\Delta \neq 0$) may be bounded as

$$\left| \int ABCP\, d\mu(\varphi) \right| \leq P \left[\int B^2 d\mu(\varphi) \right]^{1/2} \sup_{\varphi} |AC| \qquad \text{(III.2.16)}$$

where A is the exponential of the interaction, B is the product of all high momentum fields (i.e., those which have index equal to or greater than the index of the expansion step which created them); C is the product of the low momentum fields, and P is the product of the explicit factors produced by the cluster expansions: explicit propagators, s and t parameters and coupling constants. The interaction in the exponential is a sum of type (III.2.4), each term having a particular t dependence and a particular smearing function. We must check that the low momentum fields in C have t dependence and smearing functions compatible with the desired bounds. For simplicity let us use sharp characteristic functions. Two main technical points must be understood, namely the smearing of the low momentum fields and the reconstruction of the t dependence. Both problems are solved in the same way, by introducing some new, harmless set of "high momentum fields." Let us show in some detail how this is done for the simplest possible example, a model with only two slices, i (the higher one) and $j < i$ (the lower one). The field is $\varphi = \varphi^i + \varphi^j$ and we consider a cube $\Delta \in \mathbf{D}^i$. We want to apply a bound of type (III.2.16) to the concrete case:

$$A = e^{-g\int_\Delta (\varphi^i + t\varphi^j)^4 + (1-t^4)(\varphi^j)^4}; \tag{III.2.17}$$

$$\text{“}BC\text{»} = \int_\Delta \varphi^i(x)(\varphi^j(x))^3 d^4x. \tag{III.2.18}$$

The exponential of the interaction is smeared with the characteristic function of Δ (in simple words: integrated in Δ). The vertex is also smeared in Δ. However we cannot use this smearing function directly to compute the supremum in (III.2.16), precisely because we want to separate the high momentum field from the low momentum field. (This is an elementary point, but one that we stress because it took us some time to understand it...). We have therefore to factorize first (III.2.18), which is not yet a true product of the form BC (hence the quotes around BC). This is done by smearing each low momentum field, hence by writing:

$$\varphi^j(x) = \left(\frac{1}{|\Delta|}\right) \int_\Delta \varphi^j(y) d^4y - \delta\varphi^j(x). \tag{III.2.19}$$

The fluctuation field $\delta\varphi^j$ may be written as an integral of derived fields $\partial\varphi^j$:

$$\delta\varphi^j = \left(\frac{1}{|\Delta|}\right) \int_\Delta d^4y \int_0^1 dt (y-x)^\mu (\partial_\mu \varphi^j)(x + t(y-x)). \tag{III.2.20}$$

Such a $\partial_\mu\varphi^j$ field has an improved power counting; in this two-slice model it can be considered as a high momentum field and integrated with the Gaussian, since the factor $(y-x)^\mu\partial_\mu$ gives a bonus of M^{j-i}, exactly as in renormalization (see Sect. II.2, after II.2.8). This is because the ∂ acting on φ^j gives M^j and $|x-y| \le \sqrt{4}M^{-i}$ since both x and y belong to the same cube Δ of $\mathbf{D}^i$. This factor M^{j-i} compensates the bad factor M^{i-j} lost before, so that it is legitimate from the point of view of estimates to consider $\delta\varphi^j$ in this simple two-scale example as a high momentum field. In the general multiscale case, this is no longer true, because we need to gain more than M^{j-i}, to allow for summation over j. Hence we have to apply again formula (III.2.19), but now to $\partial_\mu\varphi$. The remainder contains $\partial\partial$ fields which can be now really considered as high momentum fields (for them we gain $M^{2(j-i)}$). We also get smoothed $\partial\varphi$ fields which can be dominated if, with some hindsight, we include in A a small piece $\exp(-\int\partial_\mu\varphi\partial^\mu\varphi)$ of the Gaussian, treated as an interaction. This rule corresponds to the fact that we are safe only when the propagator becomes summable in x-space, see Theorem III.1.1b. The corresponding formulas are given in the next section for the infrared case, so we do not elaborate further here on this subtlety.

Developing the third power of (III.2.18) we obtain a sum of terms now factorized, one of which is e.g., $B'C'$, with:

$$B' = \int_\Delta \varphi^i(x)(\delta\varphi^j(x))d^4x, \qquad C' = \left(\frac{1}{|\Delta|}\int_\Delta \varphi^j(y)d^4y\right)^2 . \qquad \text{(III.2.21)}$$

C' is not yet in a form suitable to take a supremum with A. Nevertheless we see that there cannot be any serious difficulty because the interaction in (III.2.17) interpolates between $(\varphi^j)^4$ and $(\varphi^i+\varphi^j)^4$, and because each low momentum field φ^j may be written as $\varphi^i+\varphi^j-\varphi^i$ at the cost of introducing some new high momentum fields. More precisely we write each φ^j as

$$(1-t)\varphi^j + (\varphi^i + t\varphi^j) - \varphi^i \qquad \text{(III.2.22)}$$

and expand. Again the φ^i fields are high momentum fields which lead to another redefinition B'', C'' of B', C', and in C'' we have two possible types of smeared low momentum fields left, $(1-t)\varphi^j$ and $(\varphi^i+t\varphi^j)$. We use for them a Hölder inequality, taking into account that $|\Delta| = M^{-4i}$:

$$(1-t)\left(\frac{1}{|\Delta|}\right)\int_\Delta \varphi^j(y)d^4y \le M^i\left[(1-t^4)\int_\Delta(\varphi^j(y))^4d^4y\right]^{1/4} . \qquad \text{(III.2.23)}$$

Similarly:

$$\frac{1}{|\Delta|}\int_\Delta(\varphi^i+t\varphi^j)(y)d^4y \le M^i\left[\int_\Delta(\varphi^i+t\varphi^j)^4(y)d^4y\right]^{1/4} \qquad \text{(III.2.24)}$$

and we conclude by the bound $x^{1/4}e^{-x} \leq$ const. Hence each low momentum field dominated produces (up to a constant) a factor $g^{-1/4}M^{i}$, as expected.

The general rules for domination are derived from this simple example, and shown now in more detail. Several technically different solutions exist, e.g., for the definition of the vertical decoupling expansion, for the smearing of the low momentum fields, the use of smoothed or sharp characteristic functions for the lattice cubes, the shapes of the slicing cutoffs, etc.... These technicalities are important but they may obscure the simple mechanism at work; so we suggest that the reader stop for a while to make sure that he has a good intuitive understanding of this mechanism at least in the two-scale case, before going on.

C Proof of Theorem III.2.1

We prove Theorem III.2.1 in a slightly different context, where domination is easier. We can always separate our initial Gaussian measure $d\mu$ into two pieces, writing it as $\exp(-\varepsilon \int \partial_\mu \varphi \partial^\mu \varphi) d\mu'$ with ε a small constant (we will see that $\varepsilon = g^{1/2}$, where g is the bare φ^4 constant, is a convenient choice). The slicing and horizontal cluster expansion is performed with respect to $d\mu'$, and the small quadratic piece is treated as an interaction, using formulas (III.2.7-9-11).† The slicing is such that the sliced propagator C^i has scaled exponential decrease (II.1.7) or scaled power law decrease

$$|C^i(x,y)| \leq c \cdot M^{2i}(1 + M^i|x-y|)^{-r} \qquad \text{(III.2.25)}$$

with r as large as will be necessary. We assume (III.2.25); the case of exponential decrease is easier.

In order to perform the functional integral according to the principle (III.2.16), we should give now the precise rules to determine in the general case what is a low and a high momentum field; this includes, as sketched in the above, the smearing operation and a t-dependence

†Strictly speaking this changes slightly the definition of what is the convergent piece of the phase space expansion in Theorem III.2.1. This is a minor point since this definition depends also on the arbitrary shape of the slicing cutoffs, and Theorem III.2.1 (which in fact holds in a rather general context) was only intended as a pedagogical introduction to the study of the convergence of phase space expansions.

reconstruction for the low momentum fields in order to rewrite them in a form directly suited for domination. We explain these steps now in full detail.

We define first the raw high and low momentum fields. Then we prepare the raw low momentum fields for domination, i.e., we separate them into true low momentum fields suited for domination plus some new high momentum fields. At the end of this process we have the true or final low and high momentum fields to which formula (III.2.16) is applied.

A field which is not in A, the exponential of the interaction, must have been derived from it by a cluster expansion step or by a t_Δ derivation at some scale i. In the first case it is hooked to a vertex to which an explicit propagator C^i of the ith cluster expansion is also hooked; in the second case, it is hooked to a vertex in Δ which has both fields of indices smaller than $i-1$ and bigger than i. By definition such a field derived at stage i is either a low momentum field φ_{i-1}, i.e., a sum over frequencies $i-1$, $i-2$ etc.... which we do not develop, or it is a high momentum field, which we can in contrast decompose systematically into a sum over some frequencies $j \geq i$ of sliced fields φ^j. We distinguish between the index j of such a high momentum field and the index $i \leq j$ (called its production index) of the expansion which produced it.

To be produced at scale i is the end of the story for a high momentum field, but it is not for a raw low momentum field φ_{i-1}, because these ones can be rederived later. When derived by a cluster expansion of scale $j < i$, a field φ_{i-1} simply disappears into the production of a new explicit propagator C^j and we need not worry about it any longer. But since the field φ_{i-1} becomes really a function $\varphi_{i-1}(t)$ according to (III.2.3), it can be also derived by vertical t expansions of scale $j < i$. Such a derivation selects the piece $\varphi_{j-1}(t)$ (and destroys the higher frequencies, with smaller indices). The last index j of this type is then called the production index of the low momentum field, and the cube $\Delta \in \mathbf{D}^j$ to which it belongs is called its production cube. We can essentially forget about the initial i.

Strongly connected domains are useful to describe in a precise way the range of frequencies of a high or low momentum field produced by some expansion step. More precisely a low momentum field with production cube $\Delta \in \mathbf{D}^j$ is equal either to $\sum_{k=l(\Delta)}^{j-1} \prod_{k<m<j} t_m \varphi^k$ or to $\sum_{k=l(\Delta)}^{j-1} \prod_{k<m\leq j} t_m \varphi^k$, where $l(\Delta)$ is the index of the largest cube (cor-

responding to lowest frequency) in the strongly connected domain to which Δ belongs, depending on whether the corresponding vertex was produced by a horizontal or vertical cluster expansion step and whether the low momentum field was rederived later (in which case we have always the first form). Since $t_j \leq 1$, we can bound systematically each such field by the first case (putting if necessary the additional t_j in P).

A high momentum field with production cube $\Delta \in \mathbf{D}^j$ and position x is equal to a sum

$$\sum_{k=j}^{h(x)} \prod_{k<m\leq h(x)} t_m \varphi^k \tag{III.2.26}$$

where $h(x)$ is the index of the smallest cube (hence with highest scale) containing x in the strongly connected domain to which Δ belongs (see Fig. III.2.2).

Let us prepare the low momentum fields for domination. We can immediately perform the reconstruction of the dependence in the t parameters, which generalizes (III.2.22). We write systematically such a field with production index j as:

$$\sum_{k=l(\Delta)}^{j-1} \prod_{k<m<j} t_m \varphi^k = \sum_{k=l(\Delta)}^{j} \prod_{k<m\leq j} t_m \varphi^k$$
$$+ (1-t_j)\Big[\sum_{k=l(\Delta)}^{j-1} \prod_{k<m<j} t_m \varphi^k\Big]\varphi^j. \tag{III.2.27}$$

The first two fields are considered as true low momentum fields with production index j, production cube $\Delta \in \mathbf{D}_j$ and range $[j, l]$. The last field φ^j is considered a high momentum field with production index j.

The fields with derivatives, such as those created by derivations of the interaction term $\exp(-(\varepsilon/2)\int \partial^\mu\varphi\partial_\mu\varphi)$ are treated exactly in the same way, with again the proviso that the derivatives are performed before the t-dependence (see (III.2.7)).

Finally we have to perform a smearing operation so that low momentum fields can be bounded directly by the interaction A. The formulas are analogues to (III.2.19–20), but (as announced) we push them one step further. We replace (III.2.19) by:

$$\varphi(x) = \left(\frac{1}{|\Delta|}\right)\left[\int_\Delta \varphi(y)dy + \int_\Delta (x-y)^\mu \partial_\mu \varphi(y)dy\right] + \delta\varphi(x), \tag{III.2.28}$$

$$\delta\varphi(x) \equiv \left(\frac{1}{|\Delta|}\right) \int_{\Delta} dy \int_{0}^{1} dt (1-t)(x-y)^{\mu}(x-y)^{\nu} \partial_{\mu}\partial_{\nu}\varphi(x+t(y-x)). \tag{III.2.29}$$

Similarly we write:

$$\partial_{\mu}\varphi(x) = \left(\frac{1}{|\Delta|}\right) \int_{\Delta} \partial_{\mu}\varphi(y) dy + \delta\partial_{\mu}\varphi(x), \tag{III.2.30}$$

$$\delta\partial_{\mu}\varphi(x) \equiv \left(\frac{1}{|\Delta|}\right) \int_{\Delta} dy \int_{0}^{1} dt (x-y)^{\nu} \partial_{\mu}\partial_{\nu}\varphi(x+t(y-x)). \tag{III.2.31}$$

We apply these formulas to every low momentum field with Δ the production cube of the low momentum field. In this way we replace each low momentum field by one or two corresponding fields smeared in a cube of the production scale j, plus a fluctuation field which has at least two derivatives acting on it. This fluctuation field is considered a high momentum field of production index j.

The definition of high and low momentum fields is now completed and we can apply formula (III.2.16). It remains to check that the corresponding bounds are sufficient for convergence. Let us check first the bound for smeared low momentum fields produced in $\Delta \in \mathbf{D}_j$. The fields of type φ produced by (III.2.28) are dominated using the quartic piece of A. By (III.2.27) they may be of two types:

$$\left(\frac{1}{|\Delta|}\right) \int_{\Delta} dy \sum_{k=l(\Delta)}^{j} \prod_{k<m\leq j} t_m \varphi^k \tag{III.2.32a}$$

or

$$\left(\frac{1}{|\Delta|}\right) \int_{\Delta} dy (1-t_j) \sum_{k=l(\Delta)}^{j-1} \prod_{k<m<j} t_m \varphi^k. \tag{III.2.32b}$$

We apply the Hölder inequality (III.2.23–24) to these fields. Now by definition of $l(\Delta)$, $t_{j-1}, \ldots, t_{l(\Delta)+1}$ are not put to 0 by the vertical expansion. Hence the exponential of the interaction contains the required pieces to bound these fields (this is true no matter whether t_j itself is 0 or not). Therefore we can conclude, using the bound $(x^{1/4})^m e^{-x} \leq (m/4)!$. The overall scale factor associated to such a low momentum field with production index j is $g^{-1/4}M^j$ (or $g^{-1/4}M^{j-1}$, which is the same up to a constant). In particular the power counting factor, M^j, is the same (up to a constant) as for a φ^{j-1} field in perturbation theory; and the local factorial $(m/4)!$ is better than the local factorial $(m/2)!$ that Gaussian integration of such fields φ_{j-1} would have produced.

Let us discuss now the almost similar case of smeared fields of type $\partial_\mu \varphi$ produced by (III.2.28) or (III.2.30). They are dominated using a Schwarz inequality similar to (III.2.23–24) with 4 replaced by 2; then the factor $\exp(-(\varepsilon/2)\int \partial^\mu\varphi\partial_\mu\varphi)$ in A contains the pieces required to bound these fields, using $(\sqrt{x})^m e^{-x} \leq (m/2)!$. It remains to check that the couplings are all right. Domination in such a way produces a factor $\varepsilon^{-1/2}M^{+2j}$ for such a field (up to some unimportant constant). But $\varepsilon^{-1/2} = g^{-1/4}$. When some $\partial_\mu\varphi$ legs hooked to an ordinary vertex of the φ^4 type are dominated in this way, there are at most three of them, and domination consumes $g^{-3/4}$; there remains a small factor $g^{1/4}$ for the vertex. Similarly when the vertex is of the $\partial^\mu\varphi\partial_\mu\varphi$ type, there is at most one leg dominated in this way, which consumes $\varepsilon^{-1/2}$, hence again a small factor $\varepsilon^{1/2} = g^{1/4}$ remains for the vertex (this explains our choice of $\varepsilon = g^{1/2}$).

We have to check also that the fluctuation fields produced at scale j give again the same factors as a field φ^{j-1} with Gaussian integration. This is because such a field of frequency $k < j$, apart from unimportant factors, has two derivatives and two factors $|x-y|$ of explicit size M^{-2j}. When integrated with the Gaussian (we may use a Schwarz inequality to separate these fields from the rest), they give a scale factor M^{3k} and a local factorial $(m(\Delta)/2)!$ where $m(\Delta)$ is the number of such fields per square of $\mathbf{D}^k$. The total scale factor for these fields is therefore M^{3k-2j} hence $M^{j-1}M^{-3(j-k)}$ (up to a constant). A factor $L^{-1/2} = M^{-2(j-k)}$ per field is what is needed to transform the local factorials at scale k $\sqrt{m_\Delta}!$ into the product of local factorials at scale j (or $j-1$) $\prod_{\Delta' \subset \Delta, \Delta' \in \mathbf{D}^j} \sqrt{m_{\Delta'}}!$, because $\sum_{\Sigma m_{\Delta'} = m_\Delta} m_\Delta! / \prod_{\Delta' \subset \Delta, \Delta' \in \mathbf{D}^j} m_{\Delta'}! = L^{-m_\Delta}$. The last factor $M^{-(j-k)}$ per field allows us to sum over k at fixed j. Hence again the effect on the estimates is the same as if these fields were fields φ^{j-1}.

The conclusion is that the net effect of the domination of low momentum fields φ or $\partial\varphi$ is therefore bounded (up to a constant per field) by replacing first each such field with production index j and production cube Δ by a field φ^{j-1} or $\partial\varphi^{j-1}$ localized in Δ, next replacing each coupling constant of each vertex by $g^{1/4}$ instead of g or ε and finally performing Gaussian integration over all fields.

It remains now to explain the effect of integrating the well localized high momentum fields with the Gaussian measure, according to (III.2.16).

Given a convergent 0-polymer containing the external variables, we have bounded its amplitude by a big sum:

$$\sum_T \sum_G \sum_P \sum_\mu \sum_V \sum_W A_{T,G,P,\mu,V,W}. \qquad \text{(III.2.33)}$$

This sum has to be performed over:

- the tree shapes T which describe structure of the inclusion relations of the i-subpolymers (called G_k^i, $k = 1, \ldots, l(i)$ by analogy with part II),
- the subpolymers $G = G_k^i$; each G_k^i is by definition an i polymer made of connected cubes of $\mathbf{D}^i \cup \mathbf{D}^{i+1} \cdots \cup \mathbf{D}^\rho$; remark that its support at scale i, $G_k^i \cap \mathbf{D}^i$, called also S_k^i, may be empty; this is analogous to the possibility in section II that subgraphs G_k^i could have no line at scale i,
- the procedures P which connect together these polymers; this includes whether s or t derivatives hook to already derived fields or create new vertices (hook to the exponential),
- the momentum attributions μ of scales to the high momentum fields (recall that these fields are decomposed over their allowed range into slice fields). This sum must be performed before the sum over P, otherwise we do not know exactly which vertices and fields have been created by the expansion,
- the positions V of the vertices created by the expansion,
- the Wick contractions W due to Gaussian integration applied to all fields (after the low momentum fields with production index j have been replaced by φ^{j-1} fields, as explained above).

Let us consider a square Δ in $\mathbf{D}^i$, and a vertex v. Δ is called the localization cube of v if the position of v is in Δ and the highest field hooked to v has scale i. Note that if Δ' is the production cube of v we may have $\Delta \neq \Delta'$ but Δ is necessarily in the strongly connected domain of Δ'. We should call $n(\Delta)$ the number of vertices with localization cube Δ. We call $f(\Delta)$ the number of fields of scale i whose position is in Δ.

Then we perform the Wick contractions. At each scale i, the integration over the fields φ^i with $d\mu^i(s)$ factorizes into subintegrations associated to each G_k^i (more precisely each S_k^i), hence does not modify the definition of the polymers. Using some fraction of the scaled decay of the propagator as in Lemma II.6.2 or Lemma III.1.2, we transform the corresponding sum over Wick contractions into local factorials. Finally we integrate over the positions of each vertex. We find that the contribution $A_{T,G,P,\mu} = \sum_V \sum_W A_{T,G,P,\mu,V,W}$ is bounded by:

- a constant per field (or per vertex, which is the same),
- a factor $g^{1/4}$ per vertex,
- a factor M^i per field φ^i or (by (III.2.25)) per half propagator C^i,
- a factor M^{-4i} for each vertex localized in a cube of scale i, due to the corresponding volume of integration,
- a factor $\sqrt{f(\Delta)!}$ per cube Δ by the local factorial principle,
- a scaled decrease (see (III.2.25)) for each explicit propagator C^i of the ith horizontal cluster expansion.

Let us call $N(G^i_k)$ the number of external legs of G^i_k and $l(G^i_k)$ the set of explicit propagators l of scale i in G^i_k; such a propagator joins two cubes $\Delta(l)$ and $\Delta'(l)$ of S^i_k. The bound may be summarized as:

$$|A_{T,G,P,\mu}| \leq \prod_{\Delta} \sqrt{f(\Delta)!}\,(c.g^{1/4})^{n(\Delta)} \prod_{i,k} M^{-N(G^i_k)} \cdot \prod_{l\in l(G^i_k)} [1+M^i\cdot \mathrm{dist}(\Delta(l),\Delta'(l))]^{-r}. \qquad \text{(III.2.34)}$$

Let us show now that we can perform the sum over momentum attributions and compatible procedures and get rid of the local factorials $\sqrt{f(\Delta)!}$ using a piece of the decrease in $\prod M^{-N(G^i_k)}$. For this purpose it is convenient to define $d(v)$, the range of a vertex, as its length in the vertical direction, hence the difference between its localization scale and the scale of the lowest field attached to it.

Since our polymers are convergent, $N(G^i_k) > 5$, hence $N(G^i_k) \geq 4 + N(G^i_k)/5$ and we can replace the power counting factor $\prod_{i,k} M^{-N(G^i_k)}$ in (III.2.34) by $\prod_{i,k} M^{-4}$ times $\prod_v M^{-d(v)/5}$, which gives an exponential decay in the vertical direction for each vertex. Holding the localization scale of each vertex fixed, we can perform the sum over the scale of each leg attached to this vertex, hence over momentum attributions with a third of this vertical decay (at the cost of a constant per vertex, see section II.1). Then let us control the sum over procedures. For each s or t derivative we may choose by a factor 2 whether it derives the exponential or hooks to a field already produced. In the second case we have to pay a factor $f(\Delta)$ for choosing this field. In this way we reach a bound of the same type as (III.2.34) but with $\sqrt{f(\Delta)!}$ replaced by $(f(\Delta)!)^{3/2}$ (this is a very crude estimate). It remains to beat such local factorials by the decrease of the cluster propagators.

Let us call $f^j(\Delta)$ the number of fields φ^i localized in $\Delta \in \mathbf{D}^i$ with production index j. By our rule for domination we have $j \le i+1$, hence $f(\Delta) = \sum_{j\le i+1} f^j(\Delta)$. We define also $f_k(\Delta) = \sum_{j\le k} f^j(\Delta)$ (hence $f_{i+1}(\Delta) = f(\Delta)$). Furthermore, for $\Delta' \in \mathbf{D}^j$ and $j \le i$:

$$\sum_{\Delta\subset\Delta'} f^j(\Delta) \le 15 + 3d(\Delta'), \tag{III.2.35}$$

where $d(\Delta)$ is the coordination number of Δ in the ith cluster expansion (the analogue of d_i in section III.1). This is because the t_Δ expansion produces at most 15 high momentum fields and each horizontal cluster expansion step which links Δ to an other cube creates at most 3 high momentum fields. This is also true for low momentum fields, hence we have also:

$$f^{i+1}(\Delta) \le \sum_{\Delta'\in\Delta^{i+1},\Delta'\subset\Delta} (15 + 3d(\Delta')). \tag{III.2.36}$$

Applying repeatedly the binomial argument, then applying (III.2.36) with the remark that there are M^4 cubes of Δ^{i+1} in Δ, we can write

$$f(\Delta)! \le \prod_{j=0}^{i+1} 2^{f_j(\Delta)} f^j(\Delta)! \le \left[c^{f(\Delta)}\right] 2^{\sum_{j=0}^{i}(i-j)f^j(\Delta)}$$

$$\prod_{\Delta'\in\Delta^{i+1},\Delta'\subset\Delta} [d(\Delta')!]^3 \prod_{j=0}^{i} f^j(\Delta)! \tag{III.2.37}$$

where c is some (M-dependent) large constant.

By (III.2.35) we have, for $\Delta' \in \mathbf{D}^j$ and $j \le i$:

$$\prod_{\Delta\subset\Delta'} f^j(\Delta)! \le c^{\sum_{\Delta\subset\Delta'} f^j(\Delta)} [d(\Delta')!]^3 \tag{III.2.38}$$

hence, combining (III.2.37) and (IIII.2.38), if $f \equiv \sum_{\Delta\in\mathbf{D}} f(\Delta)$ is the total number of the fields (and $c' = c^2$) we get:

$$\prod_{\Delta\in\mathbf{D}} f(\Delta)! \le (c')^f \prod_{\Delta\in\mathbf{D}} [d(\Delta)!]^6 \prod_i \prod_{\Delta\in\mathbf{D}^i} 2^{\sum_{j=0}^{i}(i-j)f^j(\Delta)}. \tag{III.2.39}$$

Provided M is chosen large enough, the factor $\prod_i \prod_{\Delta\in\mathbf{D}^i} 2^{\sum_{j=0}^{i}(i-j)f^j(\Delta)}$ is beaten by another third of our vertical decay $\prod_v M^{-d(v)/15}$; and the factor $\prod_{\Delta\in\mathbf{D}}[d(\Delta)!]^6$ is beaten by the horizontal decay of the propagators by lemma III.1.3. (We do not look for optimal bounds; in particular we may avoid to take M large by a more careful analysis, left to the reader). Finally f, the total number of fields, is at most four times n, the total number of vertices, and f is certainly larger than $|G|$, the total number of cubes in the last polymer G^0. Therefore for any large constant K, choosing g small enough we have

$$K^f g^{n/4} \leq g^{|G|/32}. \tag{III.2.40}$$

We have still at our disposal a piece of the horizontal decay of the propagators and the last third of our vertical decay. Hence at this stage we have reached the bound on $A_{T,G} = \sum_{P,\mu} A_{T,G,P,\mu}$:

$$|A_{T,G}| \leq g^{|G|/8} \prod_{i,k} M^{-4} \prod_{l \in l(G^i_k)} [1 + M^i \,\mathrm{dist}(\Delta(l), \Delta'(l))]^{-r/4} \prod_v M^{-d(v)/15}. \tag{III.2.41}$$

We have to perform finally the sum over T and G of $A_{T,G}$. At this stage it is convenient to introduce $\bar{G}$ which is obtained from G by "filling the vertical holes:" a cube Δ belongs to $\bar{G}$ if there are two cubes Δ' and Δ'' in G with $\Delta' \subset \Delta \subset \Delta''$. Similarly we define $\bar{G}^i$ as the set of cubes of $\bar{G}$ with scales $\geq i$, and $\bar{G}^i_k$ as the corresponding maximal connected components (see Fig. III.2.4). The advantage is that the set of cubes of $\bar{G}^i_k \cap \mathbf{D}^i$, called $\bar{S}^i_k$, is never empty, in contrast with S^i_k. With a factor $2^{|\bar{G}|}$ we can find G from $\bar{G}$. We may use our last piece of vertical decay $M^{-d(v)/15}$ to obtain a small factor for each cube in $\bar{G}$ but not in G; this factor will be as small as necessary provided M is chosen large enough. In particular we may use half of this small factor to perform the sum over G at fixed $\bar{G}$ and still retain a small factor per cube of $\bar{G}$.

Hence there is a constant ε, as small as we want provided M is first taken large enough and then g small enough, such that our bound contains a small constant ε per cube of $\bar{G}$. The final sum may be written $\sum_{T,\bar{G}} A_{T,\bar{G}}$. To evaluate the bound that we have at this stage for $A_{T,\bar{G}}$, there is still a small difficulty in the way one should describe the horizontal decrease still available. With the power law decrease (III.2.25) the "replica trick" (II.1.22) is no longer possible. Furthermore the cluster expansion at stage i does not exactly connect together the cubes of $\bar{S}^i_k$, but only some subsets $\bar{S}^i_{k,m}$ such that $\cup_m \bar{S}^i_{k,m} = \bar{S}^i_k$. The various $\bar{S}^i_{k,m}$ are connected together through connections of scale strictly higher than i. If we call $T^i_{k,m}$ the tree built by the ith cluster expansion between the cubes of $\bar{S}^i_{k,m}$, and define the corresponding

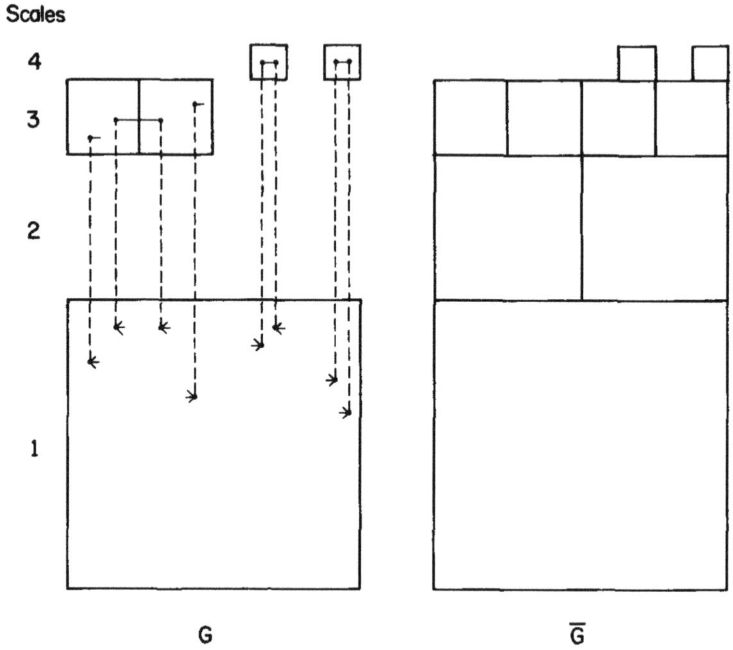

Figure III.2.4

scaled tree decay:

$$d^i_{k,m} \equiv \prod_{(\Delta,\Delta')\in T^i_{k,m}} (1 + M^i \cdot \mathrm{dist}(\Delta,\Delta'))^{-r/4}, \tag{III.2.42}$$

we have the bound (to be compared to the left hand side of (III.1.28)):

$$|A_{T,\bar{G}}| \le \varepsilon^{|\bar{G}|} \prod_{i,k} M^{-4} \sum_{\bar{S}^i_k;\mathrm{Ext}\in\bar{S}^0_1} \sum_{\{\bar{S}^i_{k,m}\};\cup_m \bar{S}^i_{k,m}=\bar{S}^i_k} \cdots$$
$$\cdots \sum_{T^i_{k,m} \text{ tree in } \bar{S}^i_{k,m}} \prod_{i,k,m} d^i_{k,m}, \tag{III.2.43}$$

where $\mathrm{Ext} \in \bar{S}^0_1$ recalls that the cubes of $\bar{G}$ at the last scale have to contain the external variables which break the last translation invariance and provide anchoring to the whole construction.

We can picture the connections of the $\bar{S}^i_{k,m}$ by the graph Γ of Fig. III.2.5. Each value of (i,k,m) is a node at height i. It is no longer a tree, but it has to be connected. From Γ we can recover T by looking at the connected components of Γ above any given height

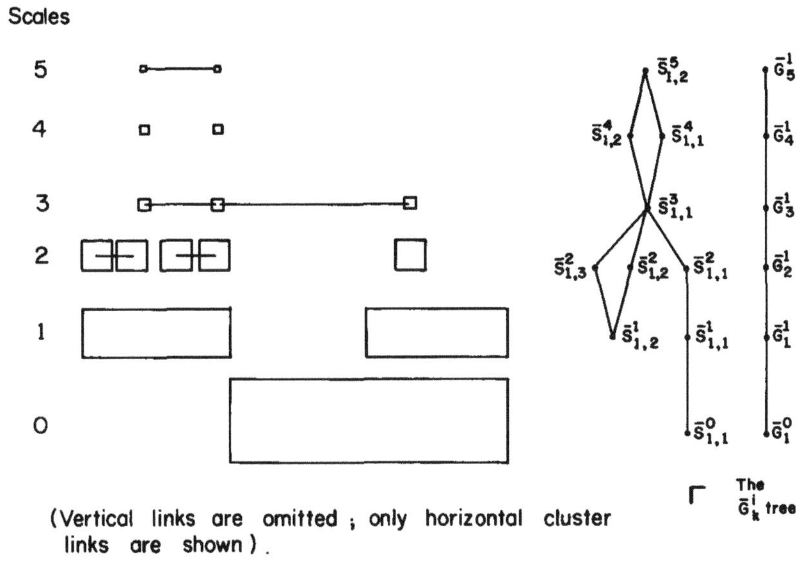

Figure III.2.5

i. In Γ there is a node at level 0 which contains external variables and which we choose as the root of the construction. Then a partial ordering relation on the nodes of Γ is defined in terms of the minimal number of lines of Γ to reach the root from the node.

We can delete some lines of Γ to form a tree T' such that for each node the number of steps in T' to reach the root is this minimal number of steps in Γ. Of course T can be recovered from T' rather than from Γ, so we divide our sum over T into smaller sums indexed by T'. Then we organize the sum over each $\bar{S}^i_{k,m}$ in the order given by the tree T', in a way which is very similar to the Mayer expansion in section III.1. We start from the nodes which are farthest from the root in T' and integrate inductively over the corresponding $\bar{S}^i_{k,m}$. For each node (i,k,m), using the corresponding factor $d^i_{k,m}$, we can perform the sum over the trees $T^i_{k,m}$ and over the position of all the cubes of $\bar{S}^i_{k,m}$, save one, $\Delta^i_{k,m}$ which is kept fixed, and which overlaps with the ancestor of (i,k,m) in T'. By the scaled analogue of Lemma III.1.4, this summation results simply in a constant to the power the number of cubes in $\bar{S}^i_{k,m}$.

We have finally to give a rule for the choice of $\Delta^i_{k,m}$. At each node we may choose by a factor 2 whether the tree T' goes up or down in phase space; this factor will be beaten provided ε in (III.2.43) is small enough. If the ancestor is some $\bar{S}^{i+1}_{k',m'}$ with scale $i+1$ (i.e., one goes up in the diagram), it is enough to know the cube of $\bar{S}^{i+1}_{k',m'}$ with which $\Delta^i_{k,m}$ overlaps, because this cube fixes $\Delta^i_{k,m}$. If the ancestor is some $\bar{S}^{i-1}_{k',m'}$ (i.e., one goes down in phase space) one has to choose the cube of $\bar{S}^{i-1}_{k',m'}$ which contains $\Delta^i_{k,m}$, but it is not enough; one has also to pay a factor M^4, the number of cubes of $\mathbf{D}^i$ in a cube of $\mathbf{D}^{i-1}$. But the corresponding factors M^4 compensate exactly the factors M^{-4} in (III.2.43), because for each value of (i,k) there is a *single* value of m such that one goes down at node (i,k,m) in T'.

Once this is realized, the choice of the cube in the ancestor is similar to the problem solved in section III.1 for the single scale Mayer expansion (equations (III.1.39–43)). If we have a coordination number d_j at node j in T', this leads to a factor $|\bar{S}^i_{k,m}|^{d_j-1}$ and there is an overcounting symmetry factor $(n!)^{-1}$ if the summations are made independent over the n nodes of the tree; by Cayley's theorem, $1/n!$ is changed into $\prod 1/(d_j-1)!$. We can bound $\sum_{d_j} |\bar{S}^i_{k,m}|^{d_j-1}/(d_j-1)!$ by $e^{|\bar{S}^i_{k,m}|}$, which itself is bounded using the small constant per cube of $\bar{G}$ in (III.2.43).

We conclude therefore that the full sum (III.2.33) (in the convergent case and for a fixed set of at least six external fields) is bounded by a convergent geometric series if M is large enough and g is small enough. This is summarized by Theorem III.2.1.

Let us also remark that we could also use the mechanism of convergence of the phase space cluster expansion Theorem III.2.1 as the starting point for a global Mayer expansion in which we could quotient out the "convergent partition function" defined as the sum over vacuum polymers which have no divergent subpolymers at any scale. But it is time now to come to more realistic situations involving renormalization.

Chapter III.3

The Effective Expansion and Infrared Φ_4^4

A Model and results

In this section we add renormalization to the phase space expansion (more precisely *useful* renormalization), to obtain finally a tool sufficiently powerful for a non-perturbative investigation of some properties of φ_4^4. In particular we give the results on the construction of the critical φ_4^4 with fixed ultra-violet cutoff, or infrared φ_4^4, based on infrared asymptotic freedom [FMRS5], and we include also some corresponding triviality results on ultraviolet φ_4^4.

In the case of infrared φ_4^4 we start with a bare theory of type (I.3.1) in a finite box Λ with a fixed ultraviolet cutoff and a mass counterterm:

$$d\nu(\varphi) = Z^{-1} \cdot e^{-\lambda \int_\Lambda \varphi^4 - (1/2) \int_\Lambda m^2 \varphi^2} d\mu_C(\varphi), \tag{III.3.1}$$

where $d\mu_C(\varphi)$ is a Gaussian measure with fixed ultraviolet cutoff. It is convenient to number the scales in this infrared problem in an order opposite to the ultraviolet one, so that the ultraviolet or bare scale has now index 0, and lower and lower momenta have higher and higher indices; it is now the last infrared scale which is called ρ, as shown in Fig. III.3.1.

Remark that there is now a “ceiling” rather than a “floor” in this picture but it would be wrong to consider that the basic picture

(Fig. II.1.2) of phase space has been turned upside down; analysis still proceeds from high to low momenta and the definition of almost-local objects is not changed. In summary shifting from an ultraviolet to an infrared problem is like changing the boundary conditions in phase space, but not the structure of the expansion.

The fixed ultra violet cutoff could be of any type, e.g., lattice, exponential or Pauli–Villars type. For a theory with real bare coupling λ any slicing may be used, in particular our favorite exponential slicing (II.1.3) (with the necessary rescaling to adapt it to an infrared problem). But in order to get the Borel summability results we need to construct the theory with λ in the complex disk of Fig. I.5.1, and it is convenient to use a slicing rule with good momentum conserving properties. This eliminates vertices with one high momentum and three low momentum legs whose "domination" is impossible for a small imaginary coupling at the border of the disk (see below). Therefore we choose to use from the beginning† a cutoff and a slicing rule which in momentum space

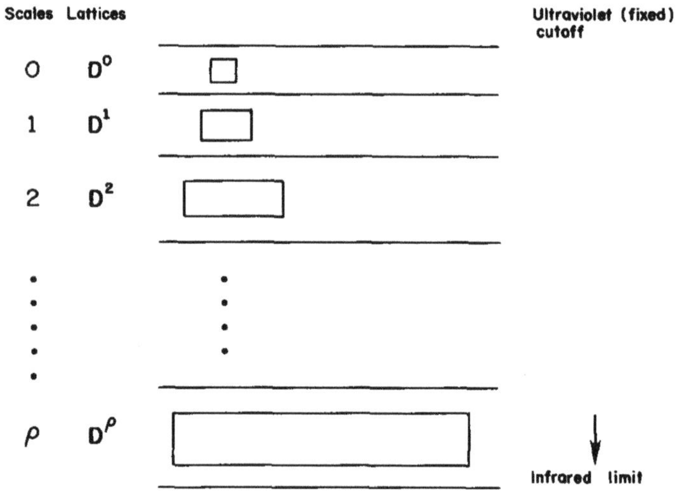

Figure III.3.1 An infrared problem (cubes are replaced by rectangles).

---

†In [FMRS5] a slicing rule suited for a cluster expansion of the "pair of cubes" type (but not good for momentum conservation) is introduced first, then for Borel summability one turns to an other slicing rule with good momentum conservation; we fear that this may be the source of some confusion.

is smooth but with compact support (this is optimal from the point of view of momentum conservation). We choose a fixed C^∞ function with compact support η with $\eta(x) = 1$ if $|x| \leq 1$ and $\eta(x) = O$ if $|x| \geq 2$, and use as cutoff:

$$C(x-y) = \frac{1}{(2\pi)^4} \int e^{ip(x-y)} \frac{\eta(p^2)}{p^2} d^4p. \tag{III.3.2}$$

This leads to the natural slicing rule with good momentum conserving properties:

$$C = \sum_{i=0}^{\infty} C^i, \qquad C^i = \int e^{ip(x-y)} \frac{\eta^i(p^2)}{p^2} d^4p, \tag{III.3.3}$$

where $\eta_0 = \eta$, $\eta^i \equiv \eta_i - \eta_{i-1}$, and $\eta_i(t) \equiv \eta(tM^{2i})$, so that C^i vanishes if $|p| \leq M^{-i}$. This last point is useful for the Borel summability results. However remark that such momentum conserving slicings can be applied to any initial cutoff, at the price of formulas less elegant than (III.3.3); so the method is really general, not limited to cutoffs like (III.3.2). Since η is smooth, the sliced covariance C^i satisfies the bound (III.2.25) for any large fixed r.

Remark that the details of the ultraviolet cutoff do not affect the long distance behavior of the massless propagator (III.3.2), which at large distances behaves as $1/|x-y|^2$. This universal behavior is called the massless Gaussian (or mean-field) behavior.

As for the study of the ultraviolet limit, it is convenient to consider first a model with finitely many momentum slices so that the index sum in (III.3.3) is bounded by ρ. The infrared limit is performed by taking *both* the infrared cutoff ρ and the volume cutoff Λ to infinity.

m^2 is the mass counterterm, which has to be fine-tuned to fix the theory at the critical point, i.e., to lead to a renormalized massless theory. For this parameter bigger than its critical value the infrared limit exists but describes an ordinary massive theory in which the two-point function decays exponentially (it is the so-called high temperature phase of the φ^4 model). At the critical value, the infrared limit still exists but the asymptotic behavior of the two point function is no longer exponential decay but power-law decay. This case corresponds to what we call the critical or infrared φ_4^4 theory. (Remark that m^2 is only meant as a generic name for the parameter in (III.3.3), and the critical value for this parameter turns out in fact to be negative in this problem, so that the associated m is purely imaginary).

The main result of [FMRS5] is:

Theorem III.3.1 Existence of infrared φ_4^4

Let R be a sufficiently large constant, and C_R be the disk of Fig. I.5.1. For λ real positive in C_R there exists a real negative critical function $m_c^2(\lambda) = O(\lambda)$ such that for m^2 in (III.3.1) greater or equal to this critical value the infrared limit of the corresponding Schwinger functions exist. Furthermore for $m^2 > m_c^2(\lambda)$ the corresponding connected functions decay exponentially at large distance (high temperature phase). For $m^2 = m_c^2(\lambda)$ (critical point), their large distance behavior is equal to the behavior of the massless Gaussian theory up to logarithmic corrections. Finally the critical mass m_c^2 and the Schwinger functions at fixed external points extend into analytic functions of λ for λ complex in the disk C_R and they are Borel summable functions of λ.

Similar results (apart from Borel summability) were obtained in [GK2–3].

Some explicit examples of what is meant by Gaussian asymptotic behavior with logarithmic corrections were proven in [FMRS5], in particular the following upper bounds, for some constants C_2 and C_3:

$$S_2(x,y) = \frac{1+O(\lambda)}{|x-y|^2}[1 + C_1(\lambda, |x-y|)]; \tag{III.3.4}$$

$$|C_1(\lambda, |x-y|)| \leq \frac{C_2}{1+\log(1+|x-y|)}; \tag{III.3.5}$$

$$|S_4(x_1,\ldots,x_4)| \leq \frac{C_3}{1+\inf_{1\leq i<j\leq 4}\log(1+|x_i-x_j|)} \int \prod_{i=1}^{4} S_2(x_i,y)d^4y. \tag{III.3.6}$$

More precise or more general results involving general N-point Schwinger functions could also be derived from the expansion by explicit computation of the leading terms in the effective phase space expansion. In particular it is easy to prove a lower bound for the two point function similar to (III.3.4–5) but with C_2 in (III.3.5) replaced by a smaller constant $C_2' < C_2$. In this way one can prove that the true critical behavior at long distances of the interacting theory is not the same as the one of the free theory, i.e., prove that the logarithmic corrections are real. The effective phase space expansion in fact provides a systematic tool to analyze these corrections.

B Overview

The construction that we give in this chapter differs in a substantial way of the one of [FMRS5]; we have tried to incorporate many improvements and hope that it will be simpler to assimilate. We thank Jacques Magnen for collaboration on these improvements. Here is a list of the main differences:

(a) A single cutoff and slicing (III.3.2–3) is used; tree cluster expansions are introduced from the beginning.
(b) Sharp characteristic functions of cubes are used, so that problems of overlap, corridors, etc ... no longer occur.
(c) The smearing of low momentum fields and the domination argument is simpler.
(d) We abandon the third order cluster expansion and modified Mayer expansion which were introduced in [FMRS5] to perform directly full mass renormalization, and return to the simpler (although less elegant) version in which the counterterm m^2 is computed by a convergent sequence of approximations, rather than by a global formula. This is discussed again below.
(e) The Mayer expansions which allow renormalization cancellations to work have been made more explicit.
(f) The proof of Borel summability is made simpler by using momentum conservation from the beginning to eliminate the corresponding troublesome piece of interaction; in [FMRS5] the rules of the expansion itself were modified to restore momentum conservation within the expansion, which is somewhat more complicated.

The improvements b) and c) are possible because we decide to keep a piece of our initial Gaussian in the form of an interaction; this will be helpful later for domination purposes. Hence rather than formula (III.3.1), our starting point is:

$$d\nu(\varphi) = Z^{-1} \cdot e^{-\lambda \int_\Lambda \varphi^4 - (\varepsilon/2) \int_\Lambda \partial_\mu \varphi \partial^\mu \varphi - (m^2/2) \int_\Lambda \varphi^2} d\mu_C(\varphi). \qquad \text{(III.3.7)}$$

Strictly speaking this would not build the theory with bare wave function constant 1, but $1 + \varepsilon$. To maintain (III.3.4) (with $O(\lambda)$ instead of $O(\varepsilon)\ldots$) we have therefore to correct slightly the propagator into:

$$C = \sum C^i, \qquad C^i = \int e^{ip(x-y)} \frac{\eta^i(p^2)}{(1-\varepsilon)p^2} d^4p. \qquad \text{(III.3.8)}$$

The constant ε has to be small, so as to lead to a convergent cluster expansion, but it should not be too small because it will be used for domination. A good choice is $\varepsilon = \lambda^{1/2}$.

The basic idea of the effective phase space expansion is to imitate the effective perturbation theory of Sect. II.4 by renormalizing in the phase space expansion of the preceding section the almost local polymers with two and four external fields. We give now an informal introduction to this idea, focusing first on the coupling constant renormalization. Renormalization, like in perturbation theory, is performed by subtracting at 0 external momenta; again in phase space this corresponds to hooking all the external fields of a four point polymer to a single point to get the corresponding counterterm. But where might this counterterm come from? A counterterm is equivalent to a modification of the value of the bare coupling constant. Is this constant not definitely fixed in (III.3.7)? These are natural questions. The correct answer is that there is of course really no change in the coupling constant, simply it is essential, in order to prove convergence, that we combine together different terms that the expansion produces in order to effectuate crucial cancellations. The rules for these cancellations lead precisely again to the definition of effective constants: hence these effective constants may be considered either as the deep solution to the problem or simply as intermediate tools to exhibit the convergence hidden in the bare formulas.

Let us give an example based on a simplified model with only two slices, two lattices $\mathbf{D}^0$ and $\mathbf{D}^1$ and two fields φ^0 (high momentum) and φ_1 (low momentum, by our infrared conventions). Remember that almost local subpolymers of $\mathbf{D}^0$ with two or four low momentum fields in slice 1 must be supported by cubes Δ of $\mathbf{D}^0$ with interpolating parameters t_Δ set to 0. But after all the interpolating formula (III.2.1) is partly arbitrary; what matters is that at $t = 1$ the theory coincides with the initial one (III.3.7) but nothing requires that at $t = 0$ both decoupled interactions have the same coupling constants. Hence the introduction of the running coupling constant is in fact extremely easy and natural: we replace the interpolation formula (III.2.1) by:

$$\int -\lambda_0(\varphi^0 + t\varphi_1)^4 - \lambda_1(1 - t^4)(\varphi_1)^4 \qquad \text{(III.3.9)}$$

where $\lambda_0 \equiv \lambda$ and $\lambda_1 = \lambda_0 + \delta\lambda$, $\delta\lambda$ being the counterterm, equal to minus the 0 momentum value of the (normalized, connected) four point function corresponding to a theory with only the 0 momentum slice and bare coupling λ. Let us sketch how interpolation (III.3.9) generates now renormalized almost local configurations. Four t derivations in the term $I^{(4)}$ in (III.2.7) may now apply to the new term $t^4\delta\lambda(\varphi_2)^4$;

hence for each cube Δ with $t_\Delta = 0$ one counterterm $\delta\lambda$ localized in Δ is generated. This counterterm should in some sense renormalize the four point polymers produced by the expansion whose "support" contains Δ. But how is this possible? There is indeed a new difficulty which has no equivalent in perturbation theory: the counterterm is a universal object (the 0 momentum value of a four point function), and there cannot be any hard core constraints in its definition. The four point polymers actually created by the expansion have in contrast hard core constraints with all other polymers produced. This problem is perhaps the single most significant technical difference between perturbative and constructive renormalization, hence in this book it should be properly emphasized. The solution was sketched in the last section and is formalized in more detail in part D below, but we pause to explain once more the problem.

We can summarize the difference between perturbative and constructive theory by the statement that perturbation theory is similar to a *free* gas of vertices (joined by propagators), while constructive theory is similar to a gas of cubes with a two body hard core interaction, each cube corresponding to some functional integration, with some derived vertices, fields and propagators which join some of the cubes together. The connection rules are defined in both cases such that there is some associated factorization. But in perturbation theory connected amplitudes really factorize; in contrast the factorization of amplitudes for the connected polymers of constructive theory is not complete because the hard core constraints between their cubes still remain. It is only after the Mayer expansion that Mayer configurations may truly factorize. This has a consequence for renormalization theory. When in perturbative renormalization we need to renormalize simultaneously several disjoint subgraphs in a given graph, the only combinatoric subtlety to check is that the counterterms for these subgraphs are generated independently in the correct way, and this is basically nothing more than the multinomial rule for packing the corresponding vertices into different boxes. But in constructive theory even the value of a particular four point subpolymer depends on the existence of other subpolymers and of their positions through the hard core constraints, so there seems to be no way to associate it with a universal counterterm. In fact the solution, like for the thermodynamic limit of section III.1, is to remove the hard core constraints via a Mayer expansion. Hence we understand the need, announced in the previous section, to sandwich the cluster and vertical expansions at each scale with a Mayer expansion. This expansion, discussed in subsection D, has to factorize the vacuum polymers (this corresponds to normalization), but it has also to free the two and four point functions from hard core interactions. There is however still a subtlety in this program; if we were to construct systematically the equivalent of connected functions by expanding blindly every hard core interactions involving two and four point almost local polymers, the result would diverge because an unbounded number of fields could accumulate at the same point. Fortunately formula (III.3.9) paves the way for the solution; we remarked that at most one counterterm can be produced per cube Δ with $t_\Delta = 0$, and this remark leads us to organize the removal of hard core interactions between two or four point polymers in such a way that also at most one of the resulting two or four point Mayer configurations is associated to a given cube. This is explained in detail in section D below.

How does the renormalization transfer convergence to improve the power counting? Of course by a mechanism very similar to the perturbative example discussed in the beginning of Sect. II.2. The main difference is that gradients created by the renormalization subtraction can no longer apply to propagators but must apply to fields: this is in a sense more natural (see the clumsy distinction between the two ends of a propagator in sections II.2–3). For example a four point polymer

$$\int A(x_1,x_2,x_3,x_4)\varphi(x_1)\varphi(x_2)\varphi(x_3)\varphi(x_4) \tag{III.3.10}$$

is renormalized by a counterterm

$$\int A(x_1,x_2,x_3,x_4)\varphi(y)^4 \tag{III.3.11}$$

by writing:

$$\int A(x_1,x_2,x_3,x_4)[\varphi(x_1)\varphi(x_2)\varphi(x_3)\varphi(x_4) - \varphi(y)^4]$$

$$= \int A(x_1,x_2,x_3,x_4)\left[\sum_{i=1}^{4}(\varphi(y))^{i-1}(\varphi(x_i) - \varphi(y)) \prod_{i<j\le 4} \varphi(x_j)\right]. \tag{III.3.12}$$

(More precise symmetrized formulas are given in (III.3.20–22)).

The difference of two fields such as $(\varphi(x_i) - \varphi(y))$, using a Taylor formula, is similar to a derived field $\partial_\mu\varphi$ at an intermediate point times some distance factor $x_i^\mu - y^\mu$ which is small when y is in the support of the polymer or close to it; when the $\partial_\mu\varphi$ field is later evaluated by Gaussian integration or domination the global effect is the same transfer of convergence discussed at length in sections II.2–3. Remark however that here also there is a subtlety; the presence of the fields at intermediate arguments creates a new type of connection between cubes, a phenomenon discussed in section D.

In perturbation theory we discovered that mass renormalization is quite different from coupling constant renormalization, in a way being simpler: it does not contain overlapping divergences, hence does not lead to sums over forests, nor does it lead to renormalon effects, provided we work at the level of one-particle irreducible objects. Hence in perturbation theory there is little incentive to pass to the effective expansion as far as mass renormalization is concerned; in fact we used directly full mass renormalization, e.g., in the construction of planar theories (section II.5). Here again perturbation theory is a good guide for what happens in constructive theory, since it turns out that direct full mass renormalization can be performed in the constructive φ_4^4 problem as well [FMRS5]. We think that it is conceptually interesting to have such a global formula for the mass counterterm. Nevertheless the technical price to pay is heavy, so that in fact this global formula is not very transparent. One has to perform the analysis in phase space at the level of one particle irreducible objects. This requires "third order" cluster expansions and a rather complicated particular Mayer expansion on the chains of one-particle irreducible two point subgraphs, as mentioned in section III.1. In a sense one can argue that the limiting process has been simply hidden in this complicated Mayer expansion.

Therefore here we decided to use a fixed-point argument, less elegant but technically simpler. We construct a sequence of theories and of mass counterterms such that the renormalized mass (corresponding to the coefficient of exponential decay of the two point function) is closer and closer to 0, and pass to the limit.

In practice by this fixed point argument we can construct as well the theory with renormalized mass m_r for any $m_r \geq 0$; the case $m_r > 0$ corresponds to $m^2 > m_c^2$ in Theorem III.3.1 and it is a rather trivial massive theory with an ultraviolet cutoff such as the one slice models considered in section III.1. What is not trivial is that the case $m_r = 0$ (the infrared theory) can be treated and that in fact all convergence estimates are uniform in m_r near $m_r = 0$ (hence the approach to the critical point from the high temperature phase is under control and can be studied by this method, which leads to a rigorous confirmation of the corresponding renormalization group predictions for the logarithmic corrections to mean field critical exponents [KW]).

The construction at $m_r = 0$ relies in an essential way on the fact that the recursion relation $\lambda_2 = \lambda_1 + (\delta\lambda)_1$ iterates into $\lambda_{i+1} = \lambda_i + (\delta\lambda)_i$, where $(\delta\lambda)_i = -\beta_i\lambda_i^2 +$ higher order terms, with $\beta_i > 0$ and $\lim_{i\to\infty}\beta_i = \beta > 0$ (as in section II.5, β is the scale invariant part of the bubble graph). Hence the behavior of λ_i is $\lambda_i \simeq \lambda/(1+i\beta\lambda)$; the theory is infrared asymptotically free. The wave function constant flow is bounded and small for small λ, as in the planar theory studied in section II.5, because $\sum \lambda_i^2$ is finite (and $O(\lambda)$).

Finally let us sketch the difficulties associated with the proof of Borel summability. We want our expansion to work also in the case of a complex bare coupling constant λ in the disk $C_R = \{\lambda, \mathrm{Re}\,\lambda^{-1} > R^{-1}\}$ of Fig. I.5.1, as should be the case to apply Theorem I.5.1. A new difficulty arises with domination,† because we can use only the decrease of the real part of the interaction. On the border of the disk C_R near $\lambda = 0$ we have $\mathrm{Re}\,\lambda \simeq (\mathrm{Im}\,\lambda)^2/R$ (with in fact $|\mathrm{Im}\,\lambda| \le \sqrt{R\cdot\mathrm{Re}\,\lambda}$ for $\lambda \in C_R$). Hence we can dominate at most two low momentum fields hooked to an imaginary vertex and still keep a small factor. We may for instance use the bound:

$$\mathrm{Im}\,\lambda\Big(\int \varphi^4\Big)^{1/2} e^{-\mathrm{Re}\,\lambda\int\varphi^4} \le c.\sqrt{R} \qquad \text{(III.3.13)}$$

for some constant c. But we can no longer dominate three low momentum fields hooked to an imaginary vertex, as in the worst case situation of Fig. III.2.1. Fortunately this type of vertex violates momentum conservation, as remarked already. With our slicing rule (III.3.3) we can even claim that a vertex $\int_{\mathbb{R}^4}\varphi^{i_1}\varphi^{i_2}\varphi^{i_3}\varphi^{i_4}$, with $i_1 \le i_2 \le i_3 \le i_4$ is 0 with probability one with respect to the Gaussian measure $d\mu(\varphi) = \prod_i d\mu(\varphi^i)$ as soon as $i_2 > i_1 + 1$. Indeed with probability one the fields φ^i have their support in Fourier space out of the region where C^i is 0, hence they are supported by momenta p with $M^{-2i} \le p^2 \le 2M^{-2(i-1)}$, and

†This difficulty is specific to φ_4^4; Borel summability for superrenormalizable theories ([EMS][GGS][MS2]) is somewhat simpler, because the coupling constant is dimensional so that one can trade analyticity in the coupling constant against analyticity in the mass, which is easier to obtain. This is no longer possible for φ_4^4. Also in contrast with the superrenormalizable case, it does not seem easy to improve over the Nevanlinna–Sokal domain of analyticity of Fig. I.5.1.

there is no way that three momenta lower than $\sqrt{2}M^{-i_1+1}$ add up to a momentum larger than M^{-i_1} (provided $M \geq 3\sqrt{2}$, which we assume now). It would seem that we can therefore identify the theory with these vertices to a theory in which they have been suppressed. This is not exactly so because we cannot define functional integrals directly in the infinite volume limit. There is no obvious solution to this subtlety. After comparing various possibilities it seems to us that a reasonably simple solution is to construct the theory in which the imaginary part of the corresponding bare vertices is suppressed, but not the real part (to suppress also the real part does not seem possible, because the corresponding interaction would no longer be positive). The domination problem therefore never occurs, but the corresponding sequences of approximations to the infrared limit are not analytic in λ, so that at the end of the construction one has to give a separate argument to conclude that the limit is analytic in λ. This is however almost obvious: one can check the Cauchy–Riemann equation $\partial f / \partial \bar{\lambda} = 0$. Indeed the Cauchy Riemann operator creates at least one vertex with coefficient ($\mathrm{Re}\,\lambda$) of the non-momentum conserving type, and such objects vanish in the thermodynamic limit $\Lambda \to \mathbb{R}^4$. Other alternative solutions would be to keep the unwanted vertices but to use a smoothed characteristic function for our volume cutoff Λ. The unwanted vertices do not completely disappear but they become tiny boundary effects linked to $\partial\Lambda$. However one still has to check that they are sufficiently tiny to compensate for the factorials generated by the failure, in their case, of the ordinary domination argument. This is of course true, but painful. Still more complicated perhaps is the solution of [FMRS5]: to modify the expansion rules so as to partly restore translation invariance when such vertices are produced.

This completes our sketch of the construction of infrared φ_4^4. This lengthy discussion leads us to use as starting point for λ complex a bare theory which is not exactly (III.3.7), but:

$$d\nu(\varphi) = Z^{-1} \cdot e^{-\mathrm{Re}\,\lambda \int_\Lambda \varphi^4 - i\,\mathrm{Im}\,\lambda \int_\Lambda \varphi^4_{MC} - (\varepsilon/2)\int_\Lambda \partial_\mu\varphi\partial^\mu\varphi - (m^2/2)\int_\Lambda \varphi^2}\, d\mu_C(\varphi) \tag{III.3.14}$$

where by definition φ^4_{MC} (Momentum Conserving) does not contain the momentum violating pieces:

$$\varphi^4_{MC} = \varphi^4 - \varphi^4_{MV}; \qquad \varphi^4_{MV} \equiv 4\sum_i \varphi^i \varphi^3_{i+2}. \tag{III.3.15}$$

The next sections give additional details on the various aspects of the construction.

C The cluster and vertical expansion

Let us turn now to the details of the expansion. We introduce the orthogonal decomposition of fields $\varphi = \sum \varphi^i$ associated to the slicing (III.3.8), and the corresponding lattice of cubes $\mathbf{D}^i$ of side size M^i. Recall that now $\varphi_i \equiv \sum_{j=i}^{\rho} \varphi^j$.

The expansion is performed scale after scale. At scale i the first part of the expansion is an ordinary cluster expansion among the cubes of $\mathbf{D}^i$ as described in Section III.1, with respect to the covariance C^i.

The second piece of the expansion is the vertical decoupling expansion, designed to decouple the fields of frequency i from the fields with lower frequencies. As in (III.2.7) an operator is applied which computes a fifth order Taylor formula within each cube of $\mathbf{D}^i$ in an interpolating parameter t_Δ. Since we have introduced a different interaction for the imaginary and real part of λ, we should give the corresponding interpolation rules in these vertical decoupling parameters t_Δ. We introduce the natural generalization of (III.3.15):

$$(\varphi_i(t))^4_{MC} \equiv (\varphi_i(t))^4 - (\varphi_i(t))^4_{MV}$$

$$(\varphi_i(t))^4_{MV} \equiv 4 \sum_{j=i}^{\rho-2} \left[\prod_{i \leq k < j} t_k \right] \varphi^j (\varphi_{j+2}(t))^3 \qquad \text{(III.3.16)}$$

where $\varphi_i(t)$ was defined in (III.2.3); we use again the convention that each t_k stands for t_Δ, Δ being the cube of $\mathbf{D}^k$ to which x belongs.

The formula which generalizes (III.2.4) and (III.3.9) is:

$$\sum_{i=0}^{\rho} (1 - t_{i-1}^4) \left[\operatorname{Re} \lambda_i (\varphi_i(t))^4 + i \operatorname{Im} \lambda_i (\varphi_i(t))^4_{MC} + i \operatorname{Im}(\delta\lambda_{i-1})(\varphi_i(t))^4_{MV} \right] (x) \qquad \text{(III.3.17)}$$

with $\delta\lambda_{i-1} = \lambda_i - \lambda_{i-1}$, and $t_{-1}, \lambda_{-1} \equiv 0$. This may seem complicated, but remark that for real λ it reduces to formula (III.2.4). Remark also that we tolerate momentum violating interactions for the counterterms $\delta\lambda_{i-1}$ because those will be dominated easily (they are of order λ^2 rather than λ) and because otherwise some configurations in the expansion would not be properly renormalized; to show that they are in fact harmless (because they violate momentum conservation) would require a separate argument.

When all parameters t are set equal to 1, (III.3.17) coincides with the quartic interaction in (III.3.14). The effective parameters $\lambda_i = \lambda_{i-1} + \delta\lambda_{i-1}$ are defined inductively; $\delta\lambda_k$ will be defined below as the 0 momentum value of the almost local 4 point configurations at scale k; by definition $\lambda_0 \equiv \lambda$.

Finally we have to give a rule for the interpolation of the quadratic pieces which are considered as interaction in (III.3.14) namely $m^2 \int \varphi^2$ and $\varepsilon \int \partial_\mu \varphi \partial^\mu \varphi$. We may or may not perform the wave function renormalization (which turns out to be finite). In order to have always exponentially convergent sums, we choose to perform it. Hence we use the interpolations generalizing (III.2.8–9):

$$\sum_{i=0}^{p} (1 - t_{i-1}^2)(m^2)_i (\varphi_i(t))^2(x), \tag{III.3.18}$$

$$\sum_{i=0}^{p} (1 - t_{i-1}^2)(\varepsilon + a_i)[\partial_\mu \varphi_i(t) \partial^\mu \varphi_i(t)](x), \tag{III.3.19}$$

with the convention $a_0 = 0$, and $(m^2)_i = (m^2)_{i-1} + (\delta m^2)_{i-1}$, $a_i = a_{i-1} + \delta a_{i-1}$.

It remains to give, at each scale, the precise definition of the counterterms δm^2, δa and $\delta\lambda$.

D The Mayer expansion and the definition of counterterms

At each scale, after each corresponding cluster and vertical expansion has been performed, we have to define and perform a Mayer expansion. As previously stressed, the main conceptual difficulty is that cluster expansions generate polymers (sets of cubes) but Mayer expansions generate configurations, i.e., sequences of polymers joined by Mayer links. Iterating this process leads to sequences of sequences of sequences ... and it is not easy to picture the result. This seems unfortunately an intrinsic difficulty, which can be avoided only by hiding it in an induction: we prefer to underline it right from the beginning.

The Mayer expansion that we want to apply is designed to free the vacuum, two and four point functions from hard core constraints. However there is an obstacle to apply this program fully: since there can be an arbitrarily large number of polymers with four external low momentum fields, if we were to suppress all hard core constraints between them, there would be the possibility that an arbitrarily large number of low momentum fields accumulate at the same place, and by the local factorial principle, this would lead later to divergences when these fields are estimated.

We should remark also that the vertical expansion can produce only *one* counterterm per cube Δ with $t_\Delta = 0$. Therefore it would be nice if the two and four point functions had all their external legs in a single external cube, and if such external cubes remained disjoint (with hard core constraints). After playing sometime with this idea we are led to the conclusion that the Mayer expansion should apply to the vacuum, two and four point functions, but not to their external cubes, i.e., to the cubes which contain their external fields, and that to prepare the function for renormalization we should separate it into a local and a renormalized piece, in such a way that the local piece has only one external cube, hence will match exactly with the counterterm.

Let us give more precisely the corresponding rules. Suppose we are working with a given slice i. The previous cluster and Mayer expansions of slice $0, 1, \ldots, i-1$ and the cluster expansions (horizontal and vertical) of slice i have produced a set of i-polymers with various numbers of external low momentum fields φ_{i+1}. These polymers are really made of $i-1$ Mayer configurations plus true polymers made of disjoint cubes of $\mathbf{D}_i$ (because the ith Mayer expansion is not performed yet; we are going to define it). We temporarily use simply φ as the notation for low momentum fields φ_{i+1} which play the rôle of external fields. The vacuum polymers are called $V_1, \ldots, V_n$. The two and four point polymers are then separated into a local part and a renormalized part according to formulas analogous to (III.3.12) but slightly more sophisticated. Let us start with the four point polymers. Such a given polymer W has a support made of cubes of $\mathbf{D}_i$ and of the previous slices. Each of the four external fields is located at some point x_n integrated in some cube E_n of $\mathbf{D}_k$, $k \leq i$, $n = 1, \ldots, 4$. E_n has to lie in the support of W; k is the production index of the corresponding vertex. Without loss of generality we will assume that $k = i$ and that the external cubes E_n are distinct for $n = 1, \ldots, 4$, the other cases being treated in the same way.

The amplitude for the polymer W is

$$\int_{E_1} \varphi(x_1) \int_{E_2} \varphi(x_2) \int_{E_3} \varphi(x_3) \int_{E_4} \varphi(x_4) A(x_1, x_2, x_3, x_4, W) \qquad \text{(III.3.20)}$$

where A is a function of the positions x_j and of the support of the polymer (which is also denoted W with a slight abuse of notations). A is a complicated sum of the type (III.2.33) over the various procedures in the previous expansions which connect together the support W, but the important point is that in A the low momentum field φ no longer occurs since all corresponding parameters t_Δ have been set to 0.

We can replace $\varphi(x_1)\varphi(x_2)\varphi(x_3)\varphi(x_4)$ by

$$(1/4)\left[\sum_{n=1}^{4}\frac{1}{|E_n|}\int_{E_n}\varphi(y)^4 dy\right]+[\varphi(x_1)\varphi(x_2)\varphi(x_3)\varphi(x_4)]_{\text{ren}} \qquad \text{(III.3.21)}$$

where $[\varphi(x_1)\varphi(x_2)\varphi(x_3)\varphi(x_4)]_{\text{ren}}$ is an expression with at least one derivative acting on φ. It is better (although not necessary) to use fully symmetric formulas. For instance we can write:

$$[\varphi(x_1)\varphi(x_2)\varphi(x_3)\varphi(x_4)]_{\text{ren}} = \frac{1}{4}\sum_{n=1}^{4}\frac{1}{|E_n|}\int_{E_n} dy \cdot$$
$$\frac{1}{4}\sum_{k=1}^{4}\left(\int_0^1 dt (x_k - y)^\mu \partial_\mu \varphi(y + t(x_k - y))\right)\cdot$$
$$\left[\prod_{l\neq k}\varphi(x_l) + \varphi(y)\cdot\sum_{l\neq k}\prod_{m\neq k, m\neq l}\varphi(x_m) + \varphi(y)^2\cdot\sum_{l\neq k}\varphi(x_l) + \varphi(y)^3\right]. \qquad \text{(III.3.22)}$$

The formulas for the two point polymers are similar. Using parity considerations as in section II.3 (see (II.3.43–47), we can simplify these formulas and replace $\varphi(x_1)\varphi(x_2)$ by

$$\frac{1}{2}\left[\sum_{n=1}^{2}\frac{1}{|E_n|}\int_{E_n} dy\left(\varphi(y)^2 + \frac{1}{2}[(x_1 - y)^\mu\partial_\mu\varphi(y)]^2 \cdots\right.\right.$$
$$\left.\left. + \frac{1}{2}[(x_2 - y)^\mu\partial_\mu\varphi(y)]^2\right)\right] + [\varphi(x_1)\varphi(x_2)]_{\text{ren}}, \qquad \text{(III.3.23)}$$

where $[\varphi(x_1)\varphi(x_2)]_{\text{ren}}$ is an expression symmetric in x_1 and x_2 with at least three derivatives acting on φ.

Using this decomposition for all polymers, we obtain what we call local and renormalized two and four point polymers. (We may treat in the same way the polymers with one or three legs but they vanish by parity considerations so let us neglect them).

We remark that the renormalized polymers have improved power counting. Indeed the distance factors $y - x_k$ are controlled by the corresponding horizontal decay at scale i in the same way than in perturbative renormalization, hence they lead to a factor M^i in the estimates. But when integrated with the Gaussian measure, the derivatives applied to fields of lower momentum (here of slices $j > i$) deliver in contrast the better scaling factor M^{-j}. In this way renormalization transfers convergence, and results in a net gain of $M^{-(j-i)}$ as usual.

Remark also however that renormalized polymers contain derived fields at interpolated points such as $y + t(x_k - y)$ and since the rules for connectivity do not imply that polymers are convex, we must consider in this case that the cube Δ containing the interpolated point is connected to the polymer (otherwise there would not be correct factorization). This means that there is a new type of link (with decrease obviously similar to the regular decrease of an ordinary propagator) to add to the list of the previous section. We must decompose the t integral in (III.3.22) into a sum over Δ, and redefine connected polymers taking this new link into account. In this way renormalized polymers may connect with other polymers already occupying the cube Δ. However no new two or point polymers can be created in this way so that we have not a cascade of corresponding redefinitions but just a single one.

After this preparation, let us come to the Mayer expansion. This is a device to remove hard core constraints between polymers: therefore it can only be applied to polymers which have factorized amplitudes depending only on their geometrical shape. The vacuum polymers V_k at a given level have such factorized amplitudes $A(V_k)$ (their vertical coupling parameters being set to 0), but the polymers with more than 4 external legs or the renormalized polymers have not really factorized amplitudes, because they contain low momentum fields, still not integrated, which may therefore connect them in some still unknown way at a later stage. For the local two and four point polymers, labeled as W_j, $j = 1, \ldots, m$, the situation is slightly more complicated. After the preparation steps (III.3.21–23), they contain a single given external vertex of the type $\varphi^2(y)$, $(\partial\varphi)^2(y)$ or $\varphi^4(y)$; this vertex is integrated in a single well defined cube E_j, called the external cube of W_j. It plays the rôle of an external source for the polymer. They contain also other internal fields or propagators. Combining (III.3.20) and (III.3.21–23) the amplitude for such a local (say 4 point) polymer W_j is equal to $\int_{E_j} dy\varphi^4(y)dy$ times an amplitude $A(W_j)$ which is now independent of the particular position y within E_j of the external vertex. This amplitude factorizes from everything else, and is therefore solely a function of the support of W_j, also called the internal domain of the polymer. Remark that this internal domain has to contain the particular cube

E_j. This is a translation breaking constraint somewhat similar to the condition $0 \in Y$ in (III.1.20).

Polymers with more than 4 external legs or renormalized polymers with two and four external legs will be treated similarly from now on because they have good power counting. They are labeled as Y_l, $l = 1, \ldots, p$. By some slight abuse of notation, let us also call E_j the factorized factor corresponding to the external vertex of the polymer W_j integrated in E_j (e.g., the factor $\int_{E_j} dy\varphi^4(y)dy$ for a four point polymer). Then we gather every contribution containing low momentum fields and the corresponding remaining functional integral over these low momentum fields in a single big amplitude $A(Y_1, \cdots, Y_p, E_1, \cdots, E_m)$ which is *not* factorized at this stage (since the corresponding cluster and Mayer expansions for scales $\geq i$ are not yet performed).

We define the Mayer expansion as removing all hard core constraints involving one vacuum polymer V_r or one internal domain W_j. Hence in the end hard core constraints between the Y's and the external cubes E_j do remain. More precisely we have a big sum:

$$\sum_{\{V_k\},\{W_j \supset E_j\},\{Y_l\}} A(Y_1, \cdots Y_p, E_1, \cdots E_m) \left[\prod_k A(V_k) \prod_j A(W_j)\right] P_1 P_2 \quad \text{(III.3.24)}$$

where the hard core interactions are divided into P_1 and P_2:

$$P_1 \equiv \prod_{k \neq k'} e^{-V(V_k, V_{k'})} \prod_{k,j} e^{-V(V_k, W_j)} \prod_{k,l} e^{-V(V_k, Y_l)} \prod_{j \neq j'} e^{-V(W_j, W_{j'})} \prod_{j,l} e^{-V(W_j, Y_l)}, \quad \text{(III.3.25)}$$

$$P_2 \equiv \prod_{j \neq j'} e^{-V(E_j, E_{j'})} \prod_{j,l} e^{-V(E_j, Y_l)} \prod_{l \neq l'} e^{-V(Y_l, Y_{l'})} \quad \text{(III.3.26)}$$

where $V(X, Y)$, as in section III.1 is the hard core interaction, with value 0 if X and Y are disjoint and $+\infty$ otherwise. Of course the reader may remark that the hard core constraints $e^{-V(E_j, E_{j'})}$ are redundant being implied by the constraints $e^{-V(W_j, W_{j'})}$, but we need to emphasize them. Indeed the Mayer expansion consists in expanding P_1 according to the algebraic formula (III.1.33), in the inductive manner shown in (III.1.37–38); P_2 remains unexpanded, hence in particular the cubes $E_1, \ldots, E_m$ remain all distinct. The result is expressed as a sum over configurations, i.e., sequences of polymers. The normalization (sum over vacuum configurations) is fully factorized as in (III.1.44). However the computation does not correspond exactly to truncated functions, but rather to a partial truncation at the level of two and four point functions, which furthermore is restricted by the fact that the truncations relative to the "external cubes" are not performed. Of course the convergence theorems of section III.1 still apply to such situations. The important point is that with this restricted rule, as in the convergent case, the number of external fields φ in any cube of $\mathbf{D}_i$ cannot get large (except when many corresponding propagators hook to distant other cubes, so that the corresponding local factorials remain controlled by the volume effect (lemma III.1.3)†).

The definition of the W's and Y's objects and of P_1 and P_2 is dynamical, i.e., it changes within the inductive Mayer expansion. For instance as soon as a four point polymer, say W_1, is joined by a Mayer link to something else than a vacuum polymer (say an other four point polymer W_2) it becomes an object with more than four point, i.e., it is put in the Y category and in particular the corresponding constraints of W_1 and W_2 with other W objects are no longer considered of type P_1 but become of type P_2 and are therefore not expanded any more.

The result of this expansion is a sum over i-configurations, i.e., sequences of i-polymers joined by Mayer links. Remark that a "local two point configuration" is a sequence made of a *single* local two point polymer and an arbitrary number of vacuum polymers, but a "local four point configuration" can be made of one local four point polymer plus any number of vacuum polymers or made of two local two point polymers plus any number of vacuum polymers.

This remark leads us to modify the expansion in a way that according to D. Knuth's pedagogy we postponed until now. Each time a Mayer configuration is generated by the overlap of two local two point functions both corresponding to mass terms (i.e., both with external fields $\varphi^2(y)$), a four point function is generated which is neither local nor renormalized. So we have to apply again to these objects the decomposition (III.3.21–22). In the new local four point functions

---

†It would be possible to remove all the hard core constraints involving a *two-point* polymer, because when n such objects accumulate in a given cube, we recover a $1/n!$ symmetry factor which compensates for the $n!$ obtained by Gaussian integration of $2n$ fields. This possibility does not extend to higher point functions, so we will not actually use it.

defined in this way there is again an external cube and an internal domain, which contains the previous external cube of one of the two point functions whose overlap formed the four point function. Therefore the global definition of external cube and internal domain at the beginning is not exactly correct.

One correct rule (not unique) is to decompose first the two point functions into local and renormalized parts and define the corresponding internal domain and external cubes of the two point functions; then expand all hard core constraints involving the internal domain of local two point functions. Only when this is finished the four point functions are defined, decomposed into local and renormalized parts (with the corresponding external cube and internal domain for the local parts). Then the remaining constraints, involving vacuum polymers or internal domains, are expanded.

Performed in this way the Mayer expansion factorizes all vacuum polymers and generates configurations with two or four external fields which are either of the local type (with all external fields hooked at the same point) or of the convergent or renormalized type (hence have favorable power counting). Moreover the two and four point local configurations are completely free of any hard core constraints with the other objects of the expansion, apart from the fact that they contain a *single* well defined external cube with a corresponding factorized external source φ^4, φ^2 or $(\partial^\mu\varphi)^2$ integrated in this cube.

Iterating this process, for each scale i we generate i-polymers and i-configurations. It remains to show that with the correct definition of the counterterms δm^2, δa and $\delta\lambda$, every i-subconfiguration in the expansion is renormalized, i.e., every local two and four point subconfiguration disappears exactly (this corresponds to the perturbative rule for the effective expansion that each almost local subgraph is renormalized).

We have yet to give the precise definition of the counterterms and perform renormalization. For clarity this is decomposed in three steps. The ith cluster, vertical and Mayer expansions are performed with $(m^2)_{i+1} = (m^2)_i$, $a_{i+1} = a_i$ and $\lambda_{i+1} = \lambda_i$; they are called the first level expansions (for slice i). Then we define first the mass and wave function counterterms (respectively $(\delta m^2)_i$ and δa_i) as minus the sum of all local two-point i-configurations with fixed external cube Δ generated at this first level (respectively with external fields $\varphi^2(y)$ and with external fields $\partial_\mu\varphi\partial^\mu\varphi(y)$). To transform $(1/2)[(x-y)^\mu\partial_\mu\varphi(x)]^2$ into

$(1/8)|y-x|^2\partial_\mu\varphi\partial^\mu\varphi(x)$ is similar to (II.3.44–46) and we do not repeat it here.

In this way we obtain correctly normalized values for these counterterms, which clearly do not contain hard core constraints with other configurations and are independent of Δ (up to boundary effects†).

The full expansion for slice i is then recomputed with the new values $(m^2)_{i+1} = (m^2)_i + (\delta m^2)_i$ and $a_{i+1} = a_i + \delta a_i$ fed into formulas (III.3.18) and (III.3.19), but still with $\lambda_{i+1} = \lambda_i$. The result is called the second level expansion, in which every 2 point local i-configuration is *exactly* cancelled.

Indeed the case in which a local two point configuration with external cube Δ is generated by the former first level expansion cancels exactly against the new case in which two t_Δ derivatives have generated exactly one counterterm of the type $(\delta m^2)_i$ or δa_i, and there is no other possibility for local two point i-configurations.

To check that the cancellation is exact, one has to remark that the counterterm, $(\delta m^2)_i \int \varphi_{i+1}^2(x)dx$ is completely independent of the fields of slices i and above. Therefore after the second level cluster and t expansions but before the second level Mayer expansion the value of a two point i-polymer with one such counterterm produced in Δ reduces to this counterterm times the value of the vacuum polymer to which Δ belongs. After the second level Mayer expansion, the vacuum graphs are factorized, and the value of the i-configuration becomes therefore 1 times the counterterm, hence cancels exactly with the sum of the former (first level) local i-configurations, as announced. This statement is somewhat tautological. Indeed with the cluster and Mayer expansions of section III.1, we may compute normalized Schwinger functions even in the case where the external fields decouple. When there is a single such external source δm^2 in the cube Δ, the unnormalized Schwinger function is obviously $\delta m^2.Z$, where Z is the normalization. Computed by formula (III.1.44), it is also equal to $A^T(M).Z$, in which M is the Mayer configuration containing the source. Therefore $A^T(M) = \delta m^2$, which is the desired result.

---

†Because of the boundary effects due to our finite box Λ, δm^2, δa and $\delta\lambda$ have still weak dependence on Δ. We decide to neglect this dependence since it has no effect on the results. A way to avoid it completely is to pass to the thermodynamic limit $\Lambda \to \mathbb{R}^4$ in each slice successively rather than at the end of the expansion. We leave to the reader to develop this nice possibility.

Finally the counterterm $\delta\lambda_i$ is computed in the same way as minus the sum of all local four point i-configurations generated at the second level. The new value $\lambda_{i+1} = \lambda_i + \delta\lambda_i$ is fed into (III.3.17), and the expansion is recomputed for the third and last time. This leads to a third level expansion in which all two and four point i-configurations are of the renormalized type. Remark that $\delta\lambda_i$ depends on $(\delta m^2)_i$, since two local two point functions (at least one of which is a counterterm) may still combine at second level to give new four point functions (which are automatically decomposed into local and renormalized pieces, hence taken into account for the value of $\delta\lambda_i$). For instance the vertical t_Δ expansion, which is a Taylor expansion to fourth order, may create two mass counterterms in a single cube Δ with $t_\Delta = 0$ which is a new kind of four point function. It is for this reason that we introduced these "levels"; in the following we forget about them, and the expansion we speak about is always the final one (third level).

In the case of complex λ there is one additional small technical detail linked to the special form of (III.3.17) which has to be mentioned here: if we study what φ^4_{MC} means in formulas (III.3.16–17) we realize that not every four point i-configuration generated by the expansion has necessarily four external fields of the type φ_{i+1}, hence is suitably renormalized by a counterterm of the form φ^4_{i+1}. There is indeed the possibility of a (single!) border vertex of a 4 point configuration having only one field in slice i and three fields below, in which case the three fields become of type $(\varphi_{i+1})^3$ only if we add the missing momentum violating piece $4\,\mathrm{Im}\,\lambda(\varphi^i(\varphi_{i+2})^3$. If we write

$$4\varphi^i(\varphi_{i+2})^3 = 4\varphi^i[(\varphi_{i+1})^3 - 3(\varphi_{i+1})^2\varphi^{i+1} + 3\varphi_{i+1}(\varphi^{i+1})^2 - (\varphi^{i+1})^3] \quad \text{(III.3.27)}$$

we obtain two classes of four point i-configurations, the normal ones with 4 external fields φ_{i+1} and the exceptional ones with some of their external fields being φ^{i+1}. The obvious rule is then not to take into account these exceptional objects in the computation of $\delta\lambda_i$. They should not be considered as true four point i-configurations. As a consequence they remain unrenormalized at scale i; this has no consequence at all, because there is no logarithmic divergence associated to such objects; recall that the need for renormalization of four point almost local objects comes solely from the lack of vertical decrease between their internal and first external scale; but for these exceptional objects, the first external scale is $i + 1$, next to the last internal scale i, so there is obviously no divergence associated to them. We will not return again to this small technical detail in the following.

E The functional bounds

When the expansion is completed, we write it as a sum of terms

$$\sum \int ABCP\,d\mu(\varphi), \tag{III.3.28}$$

where, as in (III.2.16), A is the exponential of the interaction (including the quadratic pieces), B is a product of high momentum or well localized fields, and C is the product of the low momentum or badly localized fields. As before, P is the product of all explicit factors (not depending on fields) created by the expansion: this includes the former factors such as propagators derived by the cluster expansions, coupling constants derived by the cluster and vertical expansions and the new $e^{-V(Y,Y')} - 1$ factors (Mayer links) derived by the Mayer expansions. The principle of the functional bounds is then the same as (III.2.11), namely high momentum fields are integrated with the help of the Gaussian measure, and low momentum fields are "dominated," i.e., bounded with the help of the interaction.

The precise rules to determine what is a low and a high momentum field are the same as in the previous section III.2.C. Beware that in this infrared problem indices run in the opposite way, so that many formulas of the previous section cannot be used literally but summations or inequalities involving indices usually have to be reversed. There is also a minor modification, concerning the $\mathrm{Im}\,\lambda$ vertices, in which φ_{i+1} must be replaced by $\varphi_{i+2} + \varphi^{i+1}$ and φ^{i+1} must be treated as a high momentum field. We might even do this systematically for all vertices, so that by convention the low momentum fields would be always of the form φ_{i+2}; the corresponding small technical changes would not be relevant for the main issue of convergence of the expansion.

The t dependence reconstruction and the decomposition of low momentum fields into smeared pieces plus fluctuation fields is exactly similar to the previous formulas (III.2.26–30). Domination of the smeared low momentum fields proceeds as before, using the φ^4 interaction for fields and the $\partial_\mu\varphi\partial^\mu\varphi$ interaction for derived fields $\partial\varphi$.

One delicate point however has to be discussed: the coupling constant which equips the vertex to which such a low momentum field φ_{j+1} is hooked is λ_i where i is the *initial* production scale. We will soon check that by asymptotic freedom, $\lambda_j \simeq \lambda/(1+\lambda\beta.j)$ may be much smaller than $\lambda_i \simeq \lambda/(1+\lambda\beta.i)$ (see (III.3.29)). The total factor associated to a vertex (apart from power counting factors) is therefore at worst

$\lambda_i\lambda_j^{-3/4} \leq c.\lambda^{1/4}(j-i)$ (since $(j/i)^{3/4} \leq j-i$), for some constant c (the high momentum fields, integrated with the Gaussian, do not modify this estimate). We have therefore to find a factor to compensate for the potentially harmful difference $j-i$ (when this difference gets large). Fortunately power counting in a usefully renormalized expansion is always favorable, so there is exponential decay in the vertical direction; moreover each pair of lines hooked to a vertex has a separate exponential decrease (which is in fact at least $M^{-|i-j|/18}$, see (II.1.19)). We can pick a fraction of this decay $M^{-\varepsilon(j-i)}$ and use it to bound the corresponding factors $(j-i)$, up to a constant. Remark that different factors $j-i$ correspond to different pairs of fields hooked to a vertex, hence there is no risk of factorials of the renormalon type generated in this way.

Similarly when the vertex is of the $\partial^\mu\varphi\partial_\mu\varphi$ type, there is at most one leg dominated. This consumes $(\varepsilon+a_j)^{-1/2}$, so that at worst $(\varepsilon-|a_i|)/\sqrt{\varepsilon+|a_j|}$ remains for each such vertex. But in the next section we find that $a_i=\sum_{j=0}^{i}\delta a_j$ is finite and of order λ for small λ, independently of i (see (III.3.33)). By our choice of $\varepsilon=\sqrt{\lambda}$, again a small factor $\varepsilon^{1/2}=\lambda^{1/4}$ (up to a constant) remains for such a vertex.

It remains to discuss the case of mass vertices $\delta m^2\varphi^2$. It will be shown below (in (III.3.30)) that the correct choice of counterterms is such that $|(\delta m^2)_i| \leq c\cdot\lambda_i M^{-2i}$, so that such a vertex, produced in a cube of side M^i and volume M^{4i} is neutral just as a φ^4 vertex; in this case at most one leg is a low momentum leg; its domination with the φ^4 interaction costs at most $\lambda_i^{-1/4}$, hence again a small factor (here at least $\lambda_i^{3/4}$) remains for convergence purposes.

The conclusion is that in every case a small factor per vertex does remain for the convergence of the expansion.

Finally let us examine what happens in the case of a complex coupling λ. As sketched above, domination is all right for all the vertices with at most two badly localized legs, provided the radius of the disk C_R is chosen small enough. (This includes in particular the case of the quadratic vertices $\delta m^2\varphi^2$). The only dangerous case is the "momentum violating piece" φ^4_{MV}, with imaginary coefficient. We chose formula (III.3.17) precisely so as to eliminate the term $\mathrm{Im}\,\lambda\varphi^4_{MV}$. The formula (III.3.17) does contain imaginary vertices of the momentum violating type, but they are counterterms. We will show below that $\delta\lambda_i \simeq \beta\lambda_i^2$. Therefore, since $\lambda^2\,\mathrm{Re}\,\lambda^{-3/4}$ is small for $\lambda\in C_R$, these imaginary pieces do not create any problem when bounded by the corresponding $\mathrm{Re}\,\lambda\varphi^4$ interaction.

The rest of the argument is essentially similar to section III.2.C, and we do not repeat it. Renormalized two and four point functions have indeed the same power counting as a six point function. Let us remark however that after domination and evaluation of the Gaussian integrations over high momentum fields, summation is no longer over subpolymers G^i_k but over subconfigurations, so that the full results of section III.1 (not only the results on the cluster expansion but also theorem III.1.3 on convergence of the Mayer expansion) have to be used in order to control such summations.

A final comment is in order about the external variables $\varphi(x_1)$, ..., $\varphi(x_N)$. Their treatment depends in some respects on the kind of results we are after. For a simple proof of existence of the correlation functions in the infrared limit, these external fields are smeared against some test functions of the scale 0 (or taken at fixed lattice sites if we work on a lattice). Hence, as in the ultraviolet problem it is convenient to consider by convention these external fields as having index -1, independently of their true index in the decomposition of each external field φ as a sum of φ^i fields. But this rule has the opposite effect as in the ultraviolet problem as far as renormalization is concerned: the polymers with e.g., four external fields (and no low momentum fields) have not to be renormalized in the effective expansion. Indeed their power counting is already favorable, because translation invariance is broken at the highest possible scale by the presence of the external fields, which are smeared over well-defined cubes of the smallest size in the problem.

F The behavior of the effective constants

The behavior of the effective couplings is studied in the manner of section II.5, i.e., inductively. We may do this in two steps. For the purpose of domination, hence of convergence of the partially renormalized phase space expansion we may prove first inductively a crude bound, then we can study in more detail the recursion relations which give the flow of effective constants, and prove asymptotic formulas like (III.3.4–6). Some additional work would lead also to the rigorous proof of the more detailed results about logarithmic deviations from mean field theory in this model which have been derived by the use of the renormalization group. Here we will limit ourselves to check some rather crude bounds, in the form of:

Lemma III.3.1

There is some constant β and some large constant c such that:

$$\mathrm{Re}(\lambda_i^{-1}) \geq \mathrm{Re}(\lambda^{-1}) + \beta \cdot i - O(\log i); \tag{III.3.29}$$

$$|\delta(m^2)_i| \leq c \cdot \lambda_i M^{-2i}; \tag{III.3.30}$$

$$|\delta a_i| \leq c \cdot (\lambda_i)^2. \tag{III.3.31}$$

These three bounds are derived by looking at the first non-trivial term in the definitions of the counterterms. For the coupling constant, this is our friend the bubble graph, which means that the recursion relation is $\lambda_i = \lambda_{i-1} - \beta\lambda_{i-1}^2 + O(\lambda_{i-1}^3, \lambda_i^2/i)$, and this leads to the recursion relation $\lambda_i^{-1} = \lambda_{i-1}^{-1} + \beta + O(\lambda_{i-1})$. Starting from some place in C_R we are driven to the real axis and to the origin, hence (III.3.29) holds.

Similarly the leading graph for the mass counterterm is the single vertex "tadpole," which is proportional to λ_i; for the wave function constant the leading graph is B_0 in Fig. II.5.1 and this leads to the quadratic bound (III.3.31). Combining (III.3.30) and (III.3.31) with (III.3.29) we obtain some bounds for the total flow of the mass and wave function terms:

$$\sum_{i=0}^{\infty} |\delta(m^2)_i| \leq O(\lambda); \tag{III.3.32}$$

$$\sum_{i=0}^{\infty} |\delta a_i| \leq O(\lambda). \tag{III.3.33}$$

The reader may wonder why the leading orders are the natural radiative corrections in λ since in this problem there are interactions $\varepsilon\partial_\mu\varphi\partial^\mu\varphi$ and $m^2\varphi^2$. In particular the coefficient ε is $\sqrt{\lambda}$ so subgraphs with such vertices seem to dominate. However this is only apparent. We know that if we push the expansion up to a given order n in λ (hence e.g., up to $2n$ in ε), the graphs with ε vertices will recombine exactly to change the factor $1-\varepsilon$ of p^2 in the denominator of each propagator into 1 up to $o(\lambda^n)$. Similar remarks apply to the mass vertices, which are shown below to be $O(\lambda)$. Hence the renormalization group flows are indeed given by the ordinary radiative corrections.

Remark that the flow of effective constants is simpler in this infrared problem than in the ultraviolet problem of section II.5, because at least for the coupling constant and wave function constant, there are really no renormalization conditions; we start with given values λ and 1, and let them evolve as they want (in fact they go to the Gaussian free field theory with some wave function constant close to 1). In particular there is no need of introducing the intermediate values λ_i^ρ as was the case for the g_i^ρ of section II.5. The only non-trivial renormalization condition is that the renormalized mass should be set to 0. We address this issue now.

G The inductive choice of $m_r = 0$

The expansion converges, but what does it construct so far? Not exactly the massless theory, unless the initial mass term m^2 in (III.3.1) is fine-tuned to the correct value. To obtain this, we have to compute

$$m_r^2 = m^2 + \sum_{i=0}^{\infty}(\delta m^2)_i \tag{III.3.34}$$

and to prove that there exists m^2 such that $m_r^2 = 0$. Of course the counterterms $(\delta m^2)_i$ are themselves functions of m^2, as every other quantity in the phase space expansion. But from (III.3.30) and (III.3.32) we know that the series in (III.3.34) converge, and that

$$f_\lambda(m^2) = -\sum_{i=0}^{\infty}(\delta m^2)_i(m^2, \lambda)$$

is $O(\lambda)$. Therefore for ε small and λ small enough f maps the interval $[-\varepsilon, \varepsilon]$ into itself. To show that f has a fixed point, it remains to prove that it is contracting, hence to compute df_λ/dm^2. The $d/(dm^2)$ operator creates a mass term $m^2 \int \varphi^2$ from the exponent of the initial functional integral (III.3.1), (III.3.7) or (III.3.14); it remains to evaluate the effect of such a term on the counterterms $(\delta m^2)_i$. This is done with the phase space expansion. The result is simply bounded by $m^2 \cdot O(\lambda)$, hence for ε and λ small the mapping f_λ is a contraction and has a single fixed point m^2; this fixed point is $O(\lambda)$ and at the fixed point we have $m_r = 0$, as desired. To derive the asymptotic decay (III.3.4–5) of the propagator at the fixed point is then a simple exercise.

There is one issue which we have postponed until now. We remark that the tadpole graph is negative (for positive λ), hence the leading counterterm $(\delta m^2)_i$ is positive. Therefore m^2 itself is negative.† The careful reader may worry about the functional integration with such a

†We recall that m^2 is simply a notation for a parameter which was not assumed to be the square of a real number. We may consider m to be in fact purely imaginary in this problem.

term. It is true that one can no longer bound the exponential of the interaction by 1 simply as in (III.1.25). But $m^2 = O(\lambda)$, and

$$m_j^2 \equiv m^2 + \sum_{j=0}^{i-1} (\delta m^2)_j = M^{-2i} O(\lambda_i) \tag{III.3.35}$$

since $\sum_{j \geq i} (\delta m^2)_j = M^{-2i} O(\lambda_i)$ by an analogue of the analysis above. Hence when we have to bound a functional integral in a cube Δ of $\mathbf{D}^i$ we can bound the dangerous positive mass term in the exponential by half of the negative φ^4 term, uniformly in φ. A prototype of the corresponding inequalities is

$$e^{-\lambda_i \int_\Delta \varphi_i^4 + (m^2)_i \int_\Delta \varphi_i^2} \leq K \cdot e^{-(\lambda_i/2) \int_\Delta \varphi_i^4} \tag{III.3.36}$$

for some constant K, since by a Schwarz inequality

$$\int_\Delta \varphi^2 \leq M^{2i} \left(\int_\Delta \varphi^4 \right)^{1/2} . \tag{III.3.37}$$

The large constant K in the empty cube disappears in the normalization and in non-empty cubes is compensated by taking λ still smaller. Half of the large field decrease of the φ^4 interaction remains available for domination tasks; again some big constants in the estimates get still bigger, but of course convergence itself is not in danger for λ sufficiently small.

H Analyticity and Taylor remainders

To complete the proof of Borel summability, it remains to prove that the correlation functions built by the expansion become analytic in λ when $\rho, \Lambda \to \infty$ (this is not true in finite volume, because (III.3.14) is not analytic). Then we have to compute the Taylor remainder at nth order in the bare coupling constant λ of these functions and verify a factorial bound (I.5.2).

To verify analyticity, we apply the $\partial/(\partial\bar{\lambda})$ operator to such a correlation function, as defined by the effective phase space expansion. This operator must act on the only piece which is not analytic, hence it creates at least one momentum-violating vertex φ_{MV}^4 swimming somewhere in the phase space expansion. It is therefore similar to a new kind of external source, except that this one is smeared over the whole of Λ, not over a cube of slice 0. We treat this vertex in the same way as other external sources, namely the corresponding fields have scale -1 by convention, hence the divergent polymers which contain them are not renormalized. The limit $\rho, \Lambda \to \infty$ is performed as before, and this limit vanishes because such a vertex vanishes when $\Lambda \to \infty$. Of course a true proof requires to check that this vertex does not destroy the convergence of the expansion. If we perform the thermodynamic limit with a smooth C_0^∞ volume cutoff $\Lambda(x)$ and link the limit $\rho \to \infty$ and $\Lambda \to \infty$ by the natural rule $\int \Lambda(x)dx = M^{4\rho}$, we find that a momentum violating vertex, integrated in such a volume Λ, gives a very small factor $M^{-k\rho}$, where k, which depends on the exact shape of the smooth function Λ, can be adjusted to any large fixed integer we want. In this way we can compensate for the large volume factor $M^{4\rho}$ associated to the fact that this new external source is summed over the whole of Λ, instead of being smeared over a small cube of size 0. This proves that in the limit $\rho \to \infty$ analyticity is recovered. If we insist on Λ being not smooth but being the characteristic function of some cube of $\mathbf{D}^\rho$, we have to take boundary effects into account and the proof is slightly more complicated, but of course the result remains true.

Finally the Taylor remainder bounds are obtained by developing explicitly n vertices from the exponential before applying the whole phase space expansion. The result converges in the same way as before, but with a prefactor $n!$ which is simply the large order bound on perturbation theory detailed at length in part II. Therefore we can apply the Nevanlinna–Sokal theorem and complete the proof of Borel summability of the correlation functions in λ.

I Weak triviality

The construction of infrared φ_4^4 is interesting because it allows to put on a mathematically rigorous basis the famous analysis of this model by the renormalization group [KW]. The logarithmic asymptotic freedom of the model leads to the possibility of a non-trivial infrared fixed point of the renormalization group below four dimensions; in its most naive form, this is because the coupling constant is dimensional for $d < 4$. The corresponding evolution equation for the effective constant λ_i is therefore $\lambda_{i+1} - \lambda_i \simeq \varepsilon\lambda_i - \beta\lambda_i^2$, with $\varepsilon = d - 4$, and it has a non-trivial fixed point at $\lambda \simeq \varepsilon/\beta$. This fixed point is small for small ε. These observations are at the basis of the analysis of non-trivial infrared fixed points in powers of ε (the Fisher–Wilson expansion). This expansion leads to reasonably good numerical computations of critical exponents in $d = 3$ ($\varepsilon = 1$).

But the construction of infrared φ_4^4 sheds also some light on the ultraviolet problem for φ_4^4. Theorem III.3.1 can be immediately rephrased as a triviality theorem for ultraviolet φ_4^4 at weak bare coupling. More precisely:

Theorem III.3.2: triviality at weak bare coupling

Let us consider the φ_4^4 model with fixed (non-zero) renormalized mass and wave function constants (for instance computed in the BPHZ scheme), a cutoff ρ and a bare coupling λ_ρ. There exists some $\varepsilon > 0$ such that the corresponding ultraviolet limit is a free field provided the bare coupling λ_ρ is smaller than ε for all ρ.

Of course the weak coupling condition is very restrictive, but it cannot be lifted if we want the phase space expansion to converge. There are more general triviality statements [Aiz][Fr] which apply for any bare coupling but they are limited to a lattice regularization and they operate fully only for $d > 4$, with some problems remaining at $d = 4$. They rely on rigorous inequalities, the only alternative tools we know so far when convergent expansions fail.

On this issue we have to conclude rather sadly that for the moment neither the construction of ultraviolet φ_4^4 nor a fully general triviality theorem does seem accessible with phase space expansions. This problem remains of more than purely mathematical interest since φ_4^4 is expected to govern the dynamics of the Higgs particle, and it is not clear how the triviality issue changes when coupled to non-Abelian gauge fields and how it exactly affects the corresponding physics.

As a final, more optimistic note, let us mention that the relationship between the bare and the renormalized couplings is not yet fully understood and that some interesting or surprising results may await us in this domain. In particular it should be interesting to study the relationship between the infrared and ultraviolet β functions of φ_4^4. The infrared β function is defined as the limit, when $x \to \infty$ of the derivative of the bare coupling with respect to $x = \log(\kappa/m_r)$, where κ is the ultraviolet cutoff scale and m_r the renormalized mass. This limit has to be computed by holding the renormalized coupling fixed, and has to be expressed as a power series in the bare coupling. In our notations x is roughly $\rho \log M$. This infrared beta function is the natural infrared analogue of the ultraviolet β function, which is computed by holding the bare coupling fixed and looking at the variation with x of the renormalized coupling, reexpressed in terms of the renormalized coupling (see the end of section II.5). In the phase space expansion, recall that up to terms in $(\log M)^p$, $p > 1$, which reflect the discrete nature of our flow equations, the ultraviolet β function is what we obtain when we compute the discrete change $\delta\lambda$ in λ. This is certainly a well defined quantity, but phase space expansion expresses it as a power series in all the previous running couplings, and the ultraviolet power series for the β function is obtained only by expressing all these intermediate couplings in terms of the last one, a potentially dangerous operation since it generates useless counterterms usually associated with renormalons. In contrast the infrared beta function is related to the precise behavior of the renormalized coupling, beyond the obvious x and $\log x$ terms, expressed in terms of the bare coupling. In [Kop] it is shown that there is a renormalization scheme in which the ultraviolet β function as a power series in the renormalized coupling is equal to the infrared one as a power series in the bare coupling, and that this infrared β function, being related to the infrared Schwinger functions can also be defined by Borel summation. This scheme is a minimal subtraction scheme in a Pauli–Villars regularization. It resembles therefore the minimal subtraction scheme in dimensional regularization. This gives some weight to the belief that in the dimensional minimal subtraction scheme there are also no renormalons in the β function, a belief which would partly justify its use in the computation of the ε expansion [BDZ].

Even in the BPHZ scheme the presence or absence of renormalons for the β function is unclear. Let us recall that no renormalon singularity has been proved to exist in any nontrivial BPHZ-renormalized function (see section II.6). But for many functions (and in particular the γ function [IZ]) renormalons are found easily in the first order of the $1/N$ expansion, which is a non-trivial resummation of perturbative contributions. This fact is taken at least by theoretical physicists as a clear sign of their existence [Pa1][BeDa1]. However even this test does not give up to now a clear-cut answer for the β function because its first non-trivial order in the $1/N$ expansion is simply a second order monomial $\beta_2^{N=\infty} g^2$. The next order involves already a complicated sum of graphs and we found that some cancellations do affect the first expected renormalon; subleading effects do not seem to cancel completely but at the moment we have not been able to prove this completely because some contributions are very hard to evaluate [KopRi].

If renormalons turn out to exist in the BPHZ β function we are apparently locked into a logical loop on the renormalon problem because a rigorous non-perturbative definition of the ultraviolet β function would be very useful for a rigorous study of the renormalons themselves, as explained in section II.6. Presumably to break this loop requires more hard work on the dependence of the Borel transform of the series involved in both the ultraviolet cutoff and the Borel parameter.

The triviality problem itself remains a challenging issue in dynamical systems, which might perhaps have to be tackled numerically rather than analytically. The equations of transformations of the parameters of a cutoff-φ_4^4 theory under a change of this cutoff are well defined non-perturbatively, and well defined differential equations do have local solutions. Here the problem comes from the fact that the coupling constant becomes big in a finite time, and we lose information on what it does after. If for a given fixed renormalized coupling the bare coupling were to go to infinity only as the cutoff is removed that might not prevent the construction of the corresponding theory, since after all the bare coupling is not physically observable and might tend to infinity with the cutoff; this is precisely what was expected in the old fashioned renormalized perturbation theory. A more serious problem would occur if this bare coupling were to explode to infinity in finite time as the Landau ghost coming from the iteration of the simple function $x+\beta_2 x^2$ suggests. Such dynamical flows are said to be not complete over $\mathbb{R}$. In this case mathematicians often cure this defect by reformulating the flow equations in another space. For instance on a compact space all smooth flows are obviously complete. It is perhaps in this direction that one might reach in the future some positive results concerning φ_4^4, although for the moment we have no geometrical insight on how to change the ordinary space on which the φ_4^4 parameters such as the coupling constant live. Let us conclude simply by a philosophical remark: in mathematics non-existence theorem, although quite common, rarely remain the last word on a subject. Often a problem with no solution is simply badly formulated and has to wait until the proper formalism in which it does have a solution is found.

Chapter III.4

The Gross–Neveu Model

A Two dimensions

The Gross–Neveu model in two dimensions [MW][GrNe] is a model of fermions with a color index $N > 1$ and a quartic interaction. The model was introduced in [MW] and shown to be asymptotically free in the ultraviolet direction. In [GN] it was further studied and advocated as a toy model for non-Abelian gauge theories. In particular it was argued in [GN] that physical phenomena like dynamical symmetry breaking and the non-perturbative generation of mass (which is an analogue of the non-perturbative confinement of quarks) could be studied in this model in a much simpler way than in non-Abelian gauge theories. This is certainly true; however to illustrate the unfortunate distance between theoretical and mathematical physics, let us remark that on the rigorous level there is yet no published proof that the non-perturbative generation of mass for the massless Gross–Neveu model really occurs or that a lattice N components σ model in two dimensions has exponentially decaying correlations at any temperature, even at large N. However for large number of components N we expect that such proofs are within reach of phase-space expansion techniques in the style of this book and of [dCdVMS]. Here in what follows we concern ourselves with the more modest task of building the ultraviolet limit of the massive models and establishing rigorously the property of asymptotic freedom.

The case of a single fermionic field with quartic interaction ($N = 1$) is somewhat special and goes under the name of the massive Thirring model. This model can be mapped exactly on the bosonic sine-Gordon model and is the subject of an extensive literature, but it is not asymptotically free in the usual sense because its β function is identically 0. It is therefore what one may call an asymptotically safe model. It would be interesting to apply the phase space analysis below directly to this model; the coupling constant flow would be trivial, but there would be presumably a wave function flow which is no longer bounded, hence which has to be treated with some care. We leave this problem to the reader and in what follows we restrict ourselves to the case $N > 1$.

The simplest massive two dimensional Gross–Neveu model has formal Euclidean action:

$$S(\bar{\psi}, \psi) = \bar{\psi}(ia \not{\partial} + m)\psi - \lambda(\bar{\psi} \cdot \psi)^2 \qquad \text{(III.4.1)}$$

where $\bar{\psi} \cdot \psi \equiv \sum_{a,\alpha} \bar{\psi}^a_\alpha \psi^a_\alpha$. The letters a, b, ... are used for color indices, hence take values from 1 to N and the letters α, β, ... are used for spinor indices (which in two dimensions take two values). Pairs of a color and a spinor index such as (a, α), (b, β) are noted A, B, ... Except in the explicit computations of the leading graphs which drive the discrete evolution of our effective constants, the spinor and color indices may be forgotten. The model has a perturbation expansion similar to the one of φ^4 (more precisely to the one of N-vector φ^4, for which vertices should show the circulation of indices). But it is just renormalizable in two dimensions, just as φ^4 is in four, because a fermionic propagator decreases at large momenta like p^{-1}, not p^{-2}. There are also similar models with a pseudo-scalar rather than scalar interaction which could be treated along the same lines.

We introduce the following conventions for two dimensional γ matrices:

$$\gamma_0 = \begin{pmatrix} i & 0 \\ 0 & -i \end{pmatrix} \qquad \gamma_1 = \begin{pmatrix} 0 & 1 \\ -1 & 0 \end{pmatrix} \qquad \text{(III.4.2)}$$

and the usual relations $\not{\partial} = \gamma_0 \partial_0 + \gamma_1 \partial_1$ etc ....

The $2p$ point functions are formally defined as:

$$S_{2p}^{A_1 \dots A_p B_1 \dots B_p}(y_1, \dots, y_p, z_1, \dots, z_p)$$

$$= \frac{1}{Z} \int \psi_{a_1}^{a_1}(y_1) \dots \psi_{a_p}^{a_p}(y_p) \bar{\psi}_{b_1}^{\beta_1}(z_1) \dots \bar{\psi}_{b_p}^{\beta_p}(z_p) \dots$$

$$\dots e^{\int \lambda(\bar{\psi}\psi)^2 - \bar{\psi}(i a \not{\partial} + m)\psi} \prod_x d\bar{\psi}(x) d\psi(x) \qquad \text{(III.4.3)}$$

with the usual rules of fermionic (Berezin) integration. We do not know how to define fermionic functional integrals rigorously except by performing them explicitly, which is possible here. Indeed the perturbative expansion of an unnormalized Schwinger function S_{2p}^{u} is

$$S_{2p}^{u,A_1\dots A_p B_1\dots B_p}(y_1,\dots,y_p,z_1,\dots,z_p) = \sum_{n=0}^{\infty} \sum_{C_i,D_i} \frac{\lambda^n}{n!} \int_\Lambda d^2x_1 \dots d^2x_n$$

$$\begin{pmatrix} y_1 \dots y_p x_1 x_1 x_2 x_2 \dots x_n x_n \\ z_1 \dots z_p x_1 x_1 x_2 x_2 \dots x_n x_n \end{pmatrix}_{B_1\dots B_p C_1 D_1 \dots C_n D_n}^{A_1\dots A_p C_1 D_1 \dots C_n D_n}, \qquad \text{(III.4.4)}$$

where the upper variables correspond to the ψ fields, the lower variables correspond to the $\bar{\psi}$ fields, and we use the notation:

$$\begin{pmatrix} u_1 \dots u_n \\ v_1 \dots v_n \end{pmatrix}_{B_1\dots B_n}^{A_1\dots A_n} \equiv \det(C_{B_j}^{A_i}(u_i, v_j)). \qquad \text{(III.4.5)}$$

The propagator C_B^A is diagonal in color space (vanishes unless $a_i = b_i$) and for $a_i = b_i$ it is equal to $(-\not{p} + m)/(p^2 + m^2)$. Let us replace it by a propagator C_ρ with the same index dependence and our favorite ultraviolet cutoff (of course other ones may be accommodated):

$$C_\rho(p) = C(p) e^{-M^{-2\rho}(p^2+m^2)} = \sum_{i=0}^{\rho} C^i(p), \qquad \text{(III.4.6)}$$

$$C^i(p) = C(p)\left(e^{-M^{-2i}(p^2+m^2)} - e^{-M^{-2(i-1)}(p^2+m^2)}\right) \qquad \text{if } \; i \geq 1,$$

$$C^0(p) = C(p)(e^{-(p^2+m^2)} - 1). \qquad \text{(III.4.7)}$$

The indices now run again in the regular way for an ultraviolet problem, and the lattice of cubes $\mathbf{D}^i$ is again the lattice of cubes with side size M^{-i}. The sliced propagators are very similar to the bosonic ones; in particular they satisfy for any given fixed r the bound:

$$|\partial^m C^i(x,y)| \leq K \cdot M^{i.(m+1)} e^{-c.M^i|x-y|} \qquad \forall m \leq r, \qquad \text{(III.4.8)}$$

where K and c are constants (depending only on r), and ∂^m is any partial derivation of order m.

With such an ultraviolet cutoff and a finite volume Λ, the series (III.4.4) have in fact an infinite radius of convergence, so that they can be taken as a well defined starting point for the bare theory. Indeed, as we know now, in the cutoff φ^4 theory the divergence of perturbation theory is a local phenomenon due to the fact that vertices can accumulate in a small spatial region where they become undistinguishable for the propagator. But fermions cannot behave in this way thanks to the Pauli principle.

Mathematically this corresponds to the fact that if we develop the determinant (III.4.4) we recover the usual φ^4 graphs, hence $n!$ contributions at order n. However we know that in a determinant there are many changes of signs, so that typically tremendous cancellations can occur. Hence let us start our discussion with a simple bound [IM2], which proves the convergence of (III.4.4) and is also useful for the phase space analysis of the model.

Lemma III.4.1

For any positive integer r there exists a constant $K(r)$ such that, if $C^i(x_j, y_k)$ is the n by n matrix with (j,k) entries $C^i(x_j, y_k)$ and for any $\Delta \in \mathbf{D}^i$, n_Δ (respectively $\bar{n}_\Delta$) is the number of x_j variables in Δ (respectively the number of y_k variables in Δ), we have:

$$|\det C^i(x_j, x_k)| \leq [K(r)M^i]^n \cdot \prod_\Delta \frac{1}{(n_\Delta!)^r(\bar{n}_\Delta!)^r}. \tag{III.4.9}$$

Proof Let us give the proof for any dimension d of space time. The total number of color and spinor pairs is $2N$. Let us define $p = (4r+2)d$ and divide each cube $\Delta \in \mathbf{D}^i$ into $n_\Delta/4Nd^p$ cubes, each of side size $M^{-i}.(4Nd^p/n_\Delta)^{1/d}$. Such a new cube is noted Δ_a, its center is noted z_a, and the number of variables x_j (respectively y_k) that it contains is noted n_a (respectively $\bar{n}_a$). If z_a is the center of the cube Δ_a containing x_j we apply a Taylor expansion to the propagator $C(x_j, y_k)$, writing (with the convention that repeated indices are summed from 1 to d):

$$\begin{aligned} C(x_j, y_k) = &\sum_{m=0}^{p-1} \frac{1}{m!} \prod_{l=1}^{m} (x_j - z_a)^{\mu_l} \partial_{\mu_l} C(z_a, y_k) \\ &+ \frac{1}{(p-1)!} \int_0^1 dt (1-t)^{p-1} \\ &\prod_{l=1}^{p} (x_j - z_a)^{\mu_l} \partial_{\mu_l} C(x_j + (1-t)(z_a - x_j), y_k). \end{aligned} \tag{III.4.10}$$

In this way each row of the initial determinant is a sum of at most d^{p+1} vectors. Expanding the determinant, we get a sum of at most $d^{n\cdot(p+1)}$ determinants. In any of these non-zero new determinants, at least $n_a - 2Nd^p$ of its n_a rows with arguments x_i localized in Δ_a must be remainder terms in (III.4.10). For each row containing a term with a qth order derivative localized in Δ, we have using the bound (III.4.8) a net gain in the estimate of the propagator of $(4Nd^p/n_\Delta)^{q/d}$. If we expand any such determinant by brute force we obtain therefore the bound

$$|\det C^i(x_j, y_k)| \leq K_1^n \prod_{\Delta} \prod_{\Delta_a \subset \Delta} (1/n_\Delta)^{(n_a - 2Nd^p)p/d} \cdot B \qquad \text{(III.4.11)}$$

where B is evaluated as a sum of Feynman graphs without any cancellations taken into account, hence by the local factorial lemma (Lemma II.6.2):

$$B \leq K_2^n M^{ni} \prod_{\Delta} \sqrt{(n_\Delta!)(\bar{n}_\Delta!)}. \qquad \text{(III.4.12)}$$

Since

$$\sum_{\Delta_a \subset \Delta} (n_a - 2Nd^p) = n_\Delta/2 \qquad \text{(III.4.13)}$$

we obtain

$$|\det C^i(x_j, y_k)| \leq M^{i.n}.K_1^n K_2^n \prod_{\Delta} (1/n_\Delta)^{n_\Delta(p-d)/2d} \sqrt{(\bar{n}_\Delta!)} \qquad \text{(III.4.14)}$$

and repeating the argument with the columns and taking the geometric mean of the two bounds achieves the proof of the lemma, since $p = (4r+2)d$.

Furthermore we can apply the lemma to C_ρ instead of C^i; if ρ is fixed to a constant, C_ρ itself satisfies a bound such as (III.4.8) (e.g., with $i = 0$), with K a large ρ-dependent constant. Therefore the announced result that the bare perturbation expansion for the Schwinger functions with both ultraviolet and finite volume cutoff Λ has an infinite radius of convergence follows. We can integrate the x and y positions over the whole of Λ, which gives a factor of at most q^{2n} in the bound, if $q = |\Lambda|$ is the number of unit cubes in Λ. The product over these cubes of e.g., $n_\Delta! \bar{n}_\Delta!$ is at least $(2n)!/(2q)^{2n}$, hence applying the lemma with respect to the cubes of $\mathbf{D}^0$ with $r = 1$ we have a series bounded by a series in λ^n with an infinite radius of convergence. Of course the corresponding bounds are very crude, and would not allow a study of the ultraviolet or infinite volume limits. The phase space analysis that we introduce now will precisely allow this by vastly improving this bound.

Starting with the well defined bare theory (III.4.4–5) and $C = C_\rho$ we call the bare quantities for λ, m and a respectively λ_ρ, m_ρ and a_ρ as usual.

We apply a phase space expansion similar to the previous case, the determinant in equation (III.4.4) playing the rôle of the exponential of the interaction in φ^4; it is also an expression which cannot be developed completely without losing the structure of cancellations responsible for convergence. Again the expansion applied to it must preserve most of this structure but develop it sufficiently to show a minimal set of connections in each slice, in order to perform correct power counting analysis and the necessary useful renormalizations (which here again correspond to coupling constant, mass and wave function renormalization).

If we compare further to φ^4 we realize that the model here is in many respects simpler. The local bound (III.4.9) replaces the domination in an advantageous way because Lemma III.4.1 means that any power of local factorials can be beaten by the Pauli principle, which is therefore in a sense much more powerful than the decrease of $\exp(-\int \varphi^4)$ at large φ. In fact it is a good intuitive picture to consider Berezinian anticommuting variables as *bounded* variables. Also there is no analogue of the positivity of the φ^4 interaction, hence no need to preserve it in our interpolation schemes.

The initial constructions of the Gross–Neveu model [FMRS4] [GK4] provided the first renormalizable field theory. It was shown to obey the Osterwalder–Schrader axioms (for fermions), and the correlation functions were also shown to be the Borel sum of their renormalized perturbation expansion [FMRS4]. Following results on this model include the definition of general irreducible kernels [IM2], a study of the Bethe–Salpeter equation and of two-particle asymptotic completeness [IM4] and the analysis of large momentum properties and the Wilson–Zimmermann short distance expansion [IM3].

In this section we sketch the proof of the main result on the existence of the ultraviolet limit of the model, summarized in:

Theorem III.4.1 Existence of Gross–Neveu$_2$

Let C be some large constant. With the bare ansatz:

$$\lambda_\rho = [(-\beta_2 \log M)\rho + (\beta_3/\beta_2)\log\rho + C]^{-1}; \qquad \text{(III.4.15)}$$

$$m_\rho = m.\rho^{-\gamma}, \qquad a_\rho = 1; \tag{III.4.16}$$

$$\beta_2 \equiv -2(N-1)/\pi \qquad \beta_3 = 2(N-1)/\pi^2,$$

$$\gamma = (N - 1/2)/(N-1), \tag{III.4.17}$$

the normalized Schwinger functions have a limit as $\rho, \Lambda \to \infty$, which corresponds to finite values of λ_{ren}, m_{ren} and a_{ren} in the BPHZ scheme. m_{ren} is close to m, a_{ren} is close to 1, and λ_{ren} is small and non-zero, so that the theory is not trivial. Furthermore the corresponding theory is the Borel sum of its renormalized series (and therefore can be analytically continued to a disk C_R; this requires to take C in (III.4.15) complex with a large real part). It is therefore independent of the particular limiting process used to construct it, and in particular it obeys the O.S axioms for fermions.

We give only a sketch of the proof since it overlaps strongly with the previous constructions of section II.5 and III.3, and also because the reader has already the choice between several rather detailed published constructions [FMRS4][GK4][IM2]. Again we could avoid to consider running masses by performing full (useful and useless) mass renormalization. As discussed above, this requires an analysis at the level of one particle irreducible objects, rather than simply connected objects. We refer to [FMRS4] for such an approach.

We have to check first that the only counterterms of the theory are of the expected type, namely $(\bar\psi \cdot \psi)^2$, $\bar\psi \cdot \psi$ and $\bar\psi \cdot (i \not\partial)\psi$. This is an analogue of Furry's theorem in electrodynamics, involving some algebraic manipulation of gamma matrices. For its proof we refer to [FMRS4, Lemma 2.1]. Remark that a four point function not of the form $(\bar\psi \cdot \psi)^2$ is not identically 0, but it is convergent and does not need any renormalization.

As in the previous section, a phase space expansion is applied to (III.4.4). For each scale i, starting with $i = \rho$ and ending at $i = 0$ we perform first a horizontal cluster expansion with respect to C^i and the lattice $\mathbf{D}^i$ of cubes of side size M^{-i}, then a fifth order vertical decoupling expansion, and finally a Mayer expansion with respect to the vacuum graphs and the internal domains of the two and four point functions, keeping their external cubes fixed, as explained in section III.3D above.

At the end of the expansion we have explicit propagators or vertices (which realize the explicit connections of the expansion) and the remaining fields still have the structure of a determinant. This determinant is bounded using lemma III.4.1. Hence as announced there is no longer any distinct treatment of the "high" and "low momentum" fields.

We compute at each scale counterterms $\delta\lambda_i$, δm_i and δa_i exactly like in the previous section. These counterterms are introduced in the vertical interpolations so as to renormalize all the local two and four point configurations generated by the expansion.

In the last slice we obtain a sum of usefully renormalized contributions with ordinary vertices and counterterm vertices; in particular there are 2-leg vertices with coefficients $\sum_{j=i+1}^{\rho} \delta m_j$ and $\sum_{j=i+1}^{\rho} \delta a_j$ respectively for a mass or wave function vertex with highest leg in slice i. To relate oneself to the BPHZ normalization conditions one has to resum all two point insertions into an effective propagator. This is possible (since the corresponding series is geometric). This effective propagator is of the form

$$e^{-V}\left[\not p\left(1-\sum_{i=1}^{\rho}\delta a_i e^{-V}\right)+m_\rho-\sum_{i=1}^{\rho}\delta m_i e^{-V}+O(p^2)\right]^{-1} \tag{III.4.18}$$

with $V=(p^2+m_\rho^2)$.

The discrete flow of λ, m and a is governed by the leading graphs, and to land on the desired renormalized coupling we must, as in section II.5, push the analysis up to third order graphs for the β function. These discrete equations for λ and a are:

$$\begin{aligned}\delta\lambda_i &= -\beta_2\log M\lambda_i^2+\beta_2\log M(\beta_2\log M-\beta_3)\\ &\quad +O\left(\frac{\log i}{i^4}\right)+\lambda_i^2[O(e^{-(\rho-i)})+O(e^{-i})],\end{aligned} \tag{III.4.19}$$

$$\delta a_i = \gamma_2\log M\lambda_i^2\left(1+O(\lambda_i)+O(e^{-(\rho-i)})\right)+O(e^{-i}) \tag{III.4.20}$$

where $\gamma_2=(2N-1)/(2\pi)^2$ is a numerical constant corresponding to the graph B_0 in Fig. II.5.1; β_3 is again (as in the planar φ_4^4 theory, see lemma II.5.2) the sum of two contributions $\gamma_3=(N-3/2)/\pi^2$ which corresponds to the analogue of the graph Q_3 in Fig. II.5.1, and $\delta_3=2\gamma_2$, which corresponds to a reaction of the wave function renormalization on the coupling constant flow similar to the one due to the graph Q_5 in Fig. II.5.1.

Only the discrete flow of the mass is significantly different from the "wrong sign" planar φ_4^4 flows. Indeed by parity considerations the mass renormalization is only logarithmically divergent. The leading

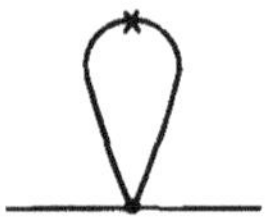

Figure III.4.1

contribution is given by a tadpole with mass counterterm insertion (Fig. III.4.1).

The corresponding numerical factor is $\gamma_1 = -(2N-1)/\pi$. If we used a resummed propagator with running mass m_i, the corresponding equation would be $\delta m_i \simeq -\gamma_1 m_i \lambda_i \log M + O(\lambda_i^2)$; but since we leave the mass counterterms as interaction vertices we find an equivalent but slightly more complicated discrete flow equation, which is [IM2]:

$$\delta m_i = \left(\lambda_i \sum_{i+1}^{p} \delta m_j C(j-i)\right) [1 + O(\lambda_i) + O(e^{-(p-i)})] + O(e^{-i}) \quad \text{(III.4.21)}$$

where

$$C(k) = \int \frac{d^2p}{p^2} \left(e^{-(p^2+m_p^2)} - 1\right) \left(e^{-M^{-2k}(p^2+m_p^2)} - 1\right). \quad \text{(III.4.22)}$$

In both points of view, if we define:

$$\gamma = -\gamma_1/\gamma_2 = \lim_{i\to\infty} \lim_{p\to\infty} \frac{\delta m_i}{m_i \lambda_i \beta_2} = (N-1/2)/(N-1) \quad \text{(III.4.23)}$$

we have the approximate behavior $\delta m_i/m_i \simeq -\gamma/i$, which is consistent with the ansatz (III.4.16). More precisely we can show that equations (III.4.15–16) lead by an easy induction to the behavior:

$$\lambda_i = [(-\beta_2 \log M)i + (\beta_3/\beta_2)\log i + C + f(i)]^{-1}, \quad \text{(III.4.24)}$$

$$m_i = m.i^{-\gamma}(1+g(i)), \qquad i \geq \text{const} \cdot C/(-\beta_2 \log M), \text{(III.4.25a)}$$

$$K^{-1} \leq m_i/m \leq K, \qquad i \leq \text{const} \cdot C/(-\beta_2 \log M), \quad \text{(III.4.25b)}$$

$$\left|\sum_{j=i+1}^{p} \delta a_j\right| \leq K \sup\{\lambda_i, M^{-2i}\}; \qquad |a_j - 1| \leq 1/2 \ \forall j, \quad \text{(III.4.26)}$$

for some large constant K, with $|f(i)| \leq 1/2$ and $|g(i)| \leq 1/2$. All these behaviors assume that C is chosen sufficiently large in (III.4.15), and they are sufficient to verify the convergence of the usefully renormalized phase space expansion.

It remains to relate precisely the ultraviolet limit obtained in this way to the BPHZ conditions, which are:

$$\lambda_{\text{ren}} = S_4^A(C, m)(0,0,0,0); \tag{III.4.27}$$

$$(m_{\text{ren}})^{-1} = S_2(C, m)(0,0); \tag{III.4.28}$$

$$a_{\text{ren}} = (m_{\text{ren}})^2 \left[-i \frac{d}{dp_0} S_2(C, m)(p) \right] |_{p=0}, \tag{III.4.29}$$

where S_4^A is the amputated four point function, in which the Fourier transform of the propagator for the four external legs has been taken out.

Then one can check that the parameters C and m in (III.4.15–16) correspond to another renormalization scheme (which in terms of the Schwinger functions is implicit rather than explicit like the BPHZ scheme). More precisely, as in section II.5, we can study the map from (C, m) to $(\lambda_{\text{ren}}, m_{\text{ren}})$, and prove that for m close to m_{ren} and $\operatorname{Re} C$ large enough, we can obtain any prescribed value of $(\lambda_{\text{ren}}, m_{\text{ren}})$ with λ_{ren} in a disk C_R. The resulting theory is analytic in this disk and again Taylor remainders at large order may be evaluated by the combination of the large order estimate of part II and the convergence of the phase space expansion, so that the theory is the Borel sum of its perturbative expansion.

This result is particularly welcome for this model, which is a full fledged field theory. Indeed it can be called upon to check that the ultraviolet limit constructed in this way is universal (independent of the technical details of the construction) and it allows a quick proof of the O.S. axioms: since different regularization schemes preserve different subsets of the axioms, it is easy to check the corresponding axioms. But by the Borel summability result, the theory constructed using these different regularizations is the same, hence verifies all axioms! For instance we can use the cutoffs (III.4.6) to check every axiom except reflection positivity (OS3 in section I.2), and a lattice cutoff to check this last axiom. Indeed lattice regularization preserves reflection positivity with respect to the symmetry hyperplanes of the lattice, which become denser and denser as the lattice spacing tends to zero; in the limit reflection positivity is recovered with respect to every hyperplane (here in dimension two these hyperplanes are simply straight lines).

B Three dimensions

In the previous sections the technique of phase space expansion appears as a compromise between the perturbative expansion, which allows rather precise information on the model but diverges, and the functional integral which is a beautiful resummation of this expansion, but does not allow detailed estimates. However this main idea can be used of course in a more general context, and in particular phase space expansion can be used as a compromise between functional integration and another expansion scheme than the ordinary perturbative one; what seems however necessary up to now for this method to work is the existence of a small parameter which, when tending to zero, drives the theory towards a Gaussian one.

The Gross–Neveu model in three dimensions is an example of such a situation: it is no longer renormalizable in the ordinary perturbative sense but it is renormalizable in the sense of the $1/N$ expansion, as is explained below. Furthermore its heuristic renormalization group analysis shows a nontrivial fixed point which is close to a Gaussian one if N, the number of colors, is big enough. This fixed point may be considered the remnant of the two dimensional asymptotic freedom of the model. Rigorous construction of the model using the phase space analysis is in progress [dCdVMS], and we will discuss briefly here the content of the model, the expected results and the technical difficulties to overcome.

The starting point is the same Lagrangian density as for the two dimensional model (III.4.1):

$$S(\bar\psi,\psi) = \bar\psi C^{-1}\psi - (\lambda/2N)(:\bar\psi\psi:)^2 \qquad \text{(III.4.30)}$$

with $C^{-1} = (ia\,\not{\partial} + m)$. The main change is that we are in a three dimensional space, so we can choose the Fermi fields to be 4 component spinors, and use the first three of the usual 4 by 4 gamma matrices of four dimensional space time. The usual γ_5 matrix may be used also to introduce chirality in the model. Another change is that the parameter N, the number of color species, will be taken large, so that it is convenient to normalize the coupling constant as in (III.4.30).

One can no longer treat the model at a purely perturbative level like in the two dimensional case, because perturbation theory is not renormalizable in this case. Indeed the convergence degree of a graph in the quartic fermionic theories is $\omega(G) = (2-d)n(G) + (d-1)(N(G)/2) - d$ and for $d = 3$ there are divergent graphs with an arbitrary number of external legs. It is therefore very surprising that there is an other expansion scheme, the $1/N$ expansion scheme, in which the theory becomes renormalizable. This scheme corresponds to a resummation of the chains of bubble graphs, which form an alternating geometric series. In the initial perturbation expansion each bubble diverges linearly. In the resummed expansion the corresponding divergences appear in the denominator of the resummed propagator, and they improve the power counting from non-renormalizable to renormalizable!

To perform bubble resummation or $1/N$ expansion in N-component vector theories it is convenient to use the well known method of Matthews–Salam, in which the Fermi functional integral is exchanged for a bosonic integration over an auxiliary field σ. Formally one writes at each point:

$$e^{\frac{\lambda}{2N}(\bar{\psi}\psi)^2} = \int e^{-(1/2)\sigma^2 + \sqrt{\lambda/N}\sigma\bar{\psi}\psi} d\sigma \tag{III.4.31}$$

and integrating over the Fermi fields, the partition function of the system becomes:

$$\int e^{-\bar{\psi}(\imath a\partial\!\!\!/ + m)\psi - (\lambda/2N)(:\bar{\psi}\psi:)^2} \prod_x d\bar{\psi}(x) d\psi(x)$$

$$= \int e^{-(1/2)\sigma^2 - \sqrt{\lambda N}\,\mathrm{Tr}\, C\sigma} \det\left(C^{-1} + \sqrt{\frac{\lambda}{N}}\sigma\right)^N \prod_x d\sigma(x) \tag{III.4.32}$$

(the partial Wick ordering is responsible for the extra linear term in σ; without it the measure would blow up as $N \to \infty$). This formula has now a natural Gaussian limit as $N \to \infty$. Indeed we can write:

$$\det(C^{-1} + \sqrt{\frac{\lambda}{N}}\sigma)^N = (\det C)^{-N} \det(1 + \sqrt{\frac{\lambda}{N}}C\sigma)^N$$
$$\simeq e^{\sqrt{\lambda N}\,\mathrm{Tr}\, C\sigma - (\lambda/2)\mathrm{Tr}\, C\sigma C\sigma + O(1/\sqrt{N})}. \tag{III.4.33}$$

Therefore up to a normalization the limit $N \to \infty$ leads to the natural Gaussian measure in σ:

$$\int e^{-(1/2)\sigma^2 - (\lambda/2)\,\mathrm{Tr}\, C\sigma C\sigma} \prod_x d\sigma(x). \tag{III.4.34}$$

It is important to remark that this measure is no longer a simple local mass term in σ^2 (a δ function in x space); the non-local correction is given by the bubble graph $\mathrm{Tr}\, C(x-y)C(y-x)$. To define more rigorously this $N \to \infty$ limit we have to worry about ultraviolet finiteness (because the bubble graph is linearly divergent); in order to replace an ill defined

Lebesgue measure by a well defined Gaussian measure as in section I.3, we have to apply Minlos theorem; therefore we need to check the positivity of the corresponding Gaussian measure, which is not obvious because the bubble graph is negative (the reader is urged to check this: in field theory it is a well known fact that fermionic loops have a minus sign when compared to bosonic loops).

An ultraviolet cutoff is first applied to the fermionic propagator C which is therefore changed into C_ρ as in (III.4.6). The covariance of the measure (III.4.34) for the σ field (corresponding to the cutoff bubble graph) is:

$$\Gamma^0_\rho(p) = (1 + \lambda_\rho \pi_\rho(p))^{-1} \tag{III.4.35}$$

with a well defined value

$$\pi_\rho(p) = \frac{1}{(2\pi)^3} \int \operatorname{Tr} C_\rho(k) C_\rho(p-k) d^3k, \tag{III.4.36}$$

which is negative and minimal for $p = 0$, so that $\pi_\rho(0) \leq \pi_\rho(p) \leq 0$ and $\pi_{\text{ren}}(p) \equiv \pi_\rho(p) - \pi_\rho(0) \geq 0$ (hint: use the α representation to check this). Since $\pi_\rho(p) \to -\infty$ as $\rho \to \infty$, we expect the propagator (III.4.35) to have an unphysical pole, similar to the Landau ghost of electrodynamics (I.1.6) if λ_ρ is kept fixed as $\rho \to \infty$. In fact we will choose to send λ_ρ to zero as $\rho \to \infty$ so fast that we never meet such a pole.

For such a choice of λ_ρ the covariance (III.4.35) will be positive (the factor 1 in (III.4.35) dominates) and the other hypotheses for Minlos theorem are easy to check so that the Gaussian measure $d\mu_{\Gamma^0_\rho}$ corresponding to the covariance Γ^0_ρ is well defined on Schwarz space of distributions. In order to see this more precisely, it is convenient to rescale first the σ field by $\sqrt{\lambda_\rho}\sigma \to \sigma$ so that λ_ρ disappears from the interaction. The σ covariance becomes $1/(\lambda_\rho^{-1} + \pi_\rho(p))$, and the constant term λ_ρ^{-1} is chosen so as to control the negative zero momentum value of π_ρ. More precisely if we define $\lambda^{\text{eff}}_\rho = \lambda_\rho/(1 + \lambda_\rho \pi_\rho(0))$, and ask $\lambda^{\text{eff}}_\rho$ to have a positive limit as $\rho \to \infty$, the covariance becomes

$$\frac{1}{(\lambda^{\text{eff}}_\rho)^{-1} + \pi_{\text{ren}}(p)}. \tag{III.4.37}$$

The advantage is that $\pi_{\text{ren}}(p)$ is positive, hence for positive $\lambda^{\text{eff}}_\rho$ the covariance is of positive type. It is easy to check that the corresponding asymptotic behavior of λ_ρ as $\rho \to \infty$ is:

$$\lambda_\rho \simeq \left(c^{-1} + M^\rho \frac{e}{1 - M^{-1}}\right)^{-1}. \tag{III.4.38}$$

In this way the limit $N = \infty$ is a well defined Gaussian theory in terms of the σ field.

This Gaussian limit and the formula (III.4.33) is the correct starting point for the construction of the theory, and the expansion scheme will be based on the $1/N$ rather than the perturbative expansion. To factorize the Gaussian piece one uses the standard notation:

$$\det_{n+1}(1+K) = \det(1+K)e^{\mathrm{Tr}(-K+\frac{K^2}{2}+\cdots+(-1)^n\frac{K^n}{n})} \qquad \text{(III.4.39)}$$

and one introduces a coupling constant in the form $\lambda_\rho g_\rho^2$ where λ_ρ behaves as in (III.4.38) up to $1/N$ corrections and g_ρ should converge to 1 as $N \to \infty$. Then a normalized Schwinger function, with test functions f_i, g_j is given by:

$$\begin{aligned} S_{\Lambda,\rho}&(f_1,\ldots,f_p;g_1,\ldots,g_p) \\ &= \frac{1}{Z}\int \bar\psi(f_1)\ldots\bar\psi(f_p)\psi(g_1)\ldots\psi(g_p)e^{\lambda_\rho g_\rho^2\int_\Lambda(:\bar\psi\psi:)^2}d\mu_\rho(\bar\psi,\psi) \\ &= \frac{1}{Z'}\int \det\left(f_i\frac{1}{1+K_\rho}g_j\right)\det{}_3(1+K_\rho)^N d\mu_{\Gamma_\rho}(\sigma), \end{aligned} \qquad \text{(III.4.40)}$$

where K_ρ is the regularized kernel:

$$K_\rho(x,y) = \frac{g_\rho}{\sqrt{N}}C_\rho(x-y)\Lambda(y)\sigma(y) \qquad \text{(III.4.41)}$$

and

$$\Gamma_\rho(p) = (\lambda_\rho^{-1} + g_\rho^2\pi_\rho(p))^{-1}. \qquad \text{(III.4.42)}$$

The power counting using the covariance of the σ field is now better than for the initial perturbative expansion. Apart from vacuum diagrams, the only divergent diagrams are those with one σ external field (quadratically divergent), two Fermi or σ fields (linearly divergent) and two Fermi fields plus one σ field (logarithmically divergent). Diagrams with three external σ fields, although apparently divergent are in fact finite because of the vanishing of the trace of an odd number of γ matrices. These renormalizations basically correspond to the renormalization of the initial parameters (masses and coupling constant) of the model.

Therefore rewriting the theory in terms of the σ field and absorbing the first term in the $\det_2$ in the definition of a Gaussian measure for this field has changed the theory from non-renormalizable to just renormalizable. In terms of the initial perturbative series, this operation is simply the resummation of all chains of bubble diagrams (see Fig. II.5.4).

To proceed further towards the rigorous construction of the theory one has to introduce a momentum slicing, a phase space expansion and effective parameters for the theory (III.4.40), and to analyze the corresponding renormalization group flows. These flows are controlled as before by a few leading order graphs. The corresponding fixed point is not exactly Gaussian but close to a Gaussian for N sufficiently large, and the definition of the corresponding renormalized parameters is not particularly difficult; we refer the interested reader to [Sen]. Nevertheless the construction of the model is more difficult than for the previous examples because there is less positivity in terms of the bosonic σ field. This point is of great constructive importance and has no counterpart in the ordinary standard (perturbative) analysis, so let us try to sketch it briefly.

The interaction of the theory in the σ field point of view is the $\det_3$. We remember that for bosonic theory there is a stability problem at large fields, which show up in the definition of the functional integral and (in a phase space expansion) in the domination process. The interaction $\det_3(1+K_\rho)$ is much worse than a positive φ^4 interaction in this respect. Indeed the standard bound on a $\det_3$ is:

$$\log(\det_3(1+K)) \le \frac{1}{2}\,\mathrm{Tr}(K^2+KK^*). \tag{III.4.43}$$

$\mathrm{Tr}\,K^2$ corresponds to the bubble graph (with two external σ fields) and is negative.

But $\mathrm{Tr}\,KK^*$ is nothing but the bubble graph with both external fields at the same point (see Fig. III.4.2). Therefore we can rewrite (III.4.43) as

$$|\det_3(1+K)| \le e^{\pi_{\mathrm{ren}}/2}, \tag{III.4.44}$$

where π_{ren} stands for the renormalized bubble graph. On the other hand the functional bound on the interaction is of the same order as

Figure III.4.2

the positivity of the Gaussian measure (remember that in (III.4.42) λ_p is adjusted so as to cancel the zero momentum value of $\pi(p)$ so that apart from a remaining constant, the positivity of the Gaussian measure is exactly that of $\pi_{\text{ren}}(p)$). Therefore the ultimate positivity of the theory is much more marginal than in previous models, and this leads to many technical complications.

In particular a new feature of the expansion is the decomposition of the functional integral according to whether the mean of the σ field is large or small. In the small field region, the $\det_3$ can be expanded; in the large field regions the bound (III.4.44) is applied and compensated by the positivity of the Gaussian; furthermore a small factor per cube is earned from the remaining piece of the Gaussian (corresponding to the term λ_p^{eff} in (III.4.38)). The cluster expansions are not performed inside the large field regions, which can be treated as a single block.

However in addition there are also some important technical changes at the level of the interpolation parameters corresponding to the vertical and horizontal cluster expansions. In the φ^4 problem of section III.3 there was separate positivity of the Gaussian measure and of the interaction, so it was enough to write independent interpolations on both of them which preserved separately the positivity of each of them. Now by (III.4.34) it is only the combination of the $\det_3$ interaction and of the Gaussian measure which is positive hence the horizontal and vertical expansions cannot be fully independent, but must be defined in such a way that for any value of the interpolating parameters the combined interaction and Gaussian measure remains positive. This is possible, but requires an inductive definition of these parameters, which takes into account already derived connections. We remark that just as for the horizontal tree like expansion, it seems a general property that inductive interpolations which are minimal in the sense that they never build redundant connections are also optimal from the point of view of preserving positivity requirements.

To conclude the discussion let us return to the question of non-renormalizability versus renormalizability. Surely the three dimensional Gross–Neveu model will be the first field theory with a non-renormalizable ordinary perturbation theory to be constructed rigorously; the details of the construction should be soon available [dCdVMS]. It is also true that the corresponding ultraviolet renormalization group fixed point is non-Gaussian. Indeed at N large but finite, the ultraviolet behavior of the theory is not Gaussian like it would be for example for φ_2^4 or φ_3^4. However in our opinion the important point to be emphasized is that the construction is again possible because of the existence of a renormalizable perturbative expansion (in the $1/N$ parameter) around a Gaussian limit (the $N \to \infty$ limit). In this sense, although it contains additional difficulties from the technical constructive point of view, the model is really similar to infrared φ_4^4 or Gross–Neveu$_2$.

This ends our brief description of the Gross–Neveu model in three dimensions. Unfortunately in four dimensions, bubble chain resummations are not sufficient to make the $1/N$ expansion of the Gross–Neveu model renormalizable. Hence it seems that non-Abelian gauge theories, in spite of all their complexities, are the only candidates for full-fledged four dimensional field theories. Our next chapter is devoted to them.

Chapter III.5

The Ultraviolet Problem in Non-Abelian Gauge Theories

I see nothing wrong with it because any nontrivial idea is in a certain sense correct.

—A. M. Polyakov, Gauge Fields and Strings

A Introduction

The efforts to understand renormalization theory better should culminate in a rigorous solution of the ultraviolet problem for non-Abelian gauge theories. Most physicists are convinced that the problem is well understood and void of any surprises, because of its asymptotically free character. However there is only one rigorous program of study of this problem completed so far, the one of Balaban [Ba2–9]. This program defines a sequence of block-spin transformations for the pure Yang–Mills theory in a finite volume on the lattice and shows that as the lattice spacing tends to 0 and these transformations are iterated many times, the resulting effective action on the unit lattice remains bounded. From this result the existence of an ultraviolet limit for *gauge invariant* observables such as "smoothed Wilson loops" should follow, at least through a compactness argument using a subsequence of approximations; but the limit is not necessarily unique. Clearly this is a point which requires further work.

Although very impressive, Balaban's work reaches the limits of human communicability, partly because the use of the lattice regularization is the source of many technical complications and partly because the results are scattered over many publications. Also it does not address the problem of constructing the expectation values of products of the field operators in a particular gauge (the Schwinger functions), because these are not gauge invariant observables. These remarks also apply to the related program of Federbush [Fed2–7] (an other impressive task which, like European common market or the Channel tunnel, is scheduled for completion by 1993). It is true that physical quantities should be gauge invariant. Nevertheless the gauge fixed framework is obviously the most convenient for perturbative computations, and one can consider in fact that the ultraviolet problem for the Yang–Mills *field* theory is not yet understood until this point is clarified.

In collaboration with J. Feldman, J. Magnen and R. Sénéor, we tried also our own study of this problem with the phase space method described in this book. Our ambitious and perhaps naive goal was to construct the Schwinger functions of the field e.g., in the Feynman or Landau gauge, with an infrared cutoff. In spite of hard work, at least in term of the hours spent, we did not succeed. The functional integrals obtained always lacked sufficient positivity for control. For long we hoped that at some point in the construction the phase space analysis would "kill" the related Gribov problem, or "chop" it into manageable subpieces, but in the author's present opinion this hope was unfortunately ill-founded.

We do not have at the moment many intermediate results of such obvious value that we owe an explanation of them to the community. Nevertheless at the perturbative level we are convinced that our approach, which is based on a gauge breaking regularization, can be used to recover the results on renormalizability of the Yang–Mills perturbation theory, without the use of dimensional regularization (which is not a useful constructive device up to now), and with explicit uniform bounds at large order such as those of part II. Furthermore with this method, one can rewrite the bare or renormalized series as an effective series with useful renormalization at the level of power series in the renormalized coupling like in section II.4. At this level of formal power series the running coupling constant seems clearly to display asymptotic freedom.

At the constructive level, the bare theory with cutoff that we consider is well defined (and this is somewhat nontrivial, since it depends on the shape of the ultraviolet cutoff that we use). But whether the ultraviolet limit exists remains a wide open question. Also if it exists it is not clear that it obeys the standard (perturbative) renormalization group computations: in [DeZw1] it is argued that this may not be the case, at least for the two point function, because of the non-perturbative restrictions on the classical configuration space which correspond to fixing the "Gribov ambiguities." We discuss briefly this surprising result below. If it turns out to be correct, one might perhaps still reconcile it with Balaban's work, and the standard "Monte-Carlo" wisdom if for gauge invariant observables the standard renormalization group happens to be correct.

In conclusion it is the goal of this section to explain our tentative approach to the Yang–Mills ultraviolet problem and to summarize the little we know about the Gribov problem, from our point of view.

B The model

We consider the pure Yang–Mills theory with an infrared cutoff, which we never try to lift. This cutoff may be imposed on the propagator, or we could consider the theory on a finite volume with some boundary conditions, or on the sphere S^4, the torus $\Lambda = \mathbb{R}^4/\mathbb{Z}^4$ or an other compact Riemannian four-dimensional manifold. Naive infrared regularization breaks gauge invariance, but compactification of space and the choice of a particular principal bundle with fiber G defines an unbroken group of gauge transformations. For instance in the case of the torus with the trivial SU(2) bundle, the gauge transformations are simply the functions $x \to g(x)$ from $\mathbb{R}^4$ to G which are periodic with period lattice $\mathbb{Z}^4$. The momentum space corresponds to discrete Fourier analysis on the dual lattice $\Lambda^* = \mathbb{Z}^4$. Moreover the constant fields or the zero mode in Fourier space is deleted in all our functional integrals, hence there is no infrared problem.

For the pure SU(2) Yang–Mills theory the vector potential is a field A^a_μ, $\mu = 1, \ldots, 4$, $a = 1, 2, 3$ with Lorentz (greek) indices and Lie algebra (roman) indices (the group is noted SU(2) and the algebra su(2)). Geometrically A is a connection on the considered principal bundle; again in the case of the trivial SU(2) bundle one can consider that each A_μ is simply a function with values in su(2). Our conventions are those of [IZ], which we recall briefly; later to simplify the notations we will forget indices most of the time. We write $A = \sum_{a=1}^{3} A^a t_a$, with $t_a = (i\sigma_a/2)$ where the σ's are the three usual hermitian Pauli matrices. With this convention the covariant derivative is $D_\mu = \partial_\mu - \lambda[A_\mu, .]$. We have $\mathrm{Tr}\, t_a t_b = -\delta_{ab}/2$. The field curvature is:

$$F_{\mu\nu} = (\partial_\mu A_\nu - \partial_\nu A_\mu) - \lambda[A_\mu, A_\nu] = (\partial \wedge A - \lambda[A,A]), \tag{III.5.1}$$

λ being the coupling constant; the second notation is a condensed one in which indices are omitted (and $\partial\wedge$ is the exterior derivative). Remark that in the three dimensional su(2) space, the commutator is a wedge product: $[A^a_\mu, A^b_\nu] = \varepsilon^c_{ab} A^a_\mu A^b_\nu$, ε being the usual antisymmetric symbol. The pure Yang–Mills action is (for Euclidean canonical metric on the flat torus the raising of "Lorentz" indices is trivial so that $F_{\mu\nu} = F^{\mu\nu}$):

$$-\frac{1}{2}\int_\Lambda d^4x \,\mathrm{Tr}\, F_{\mu\nu}F^{\mu\nu} = \frac{1}{4}\int_\Lambda d^4x \sum_a F^a_{\mu\nu}F^{\mu\nu a}. \tag{III.5.2}$$

To simplify, we define a scalar product $\langle A, B\rangle$ on space time tensors of the same type with values in the Lie algebra A and B, by the convention that a trace is taken over all corresponding space time indices and minus a trace over group indices, so that it is positive definite with a factor 1/2 in component notation. We also write simply A^2 for $\langle A, A\rangle$, and with this convention we can write the action as $\frac{1}{2}\int_\Lambda F^2$. We distinguish between the quadratic, trilinear and quartic pieces of F^2, writing:

$$F^2 = F_2 + \lambda F_3 + \lambda^2 F_4. \tag{III.5.3}$$

This action is invariant under the gauge transformations:

$$A \to A^g; \; (A^g)_\mu = gA_\mu g^{-1} + \partial_\mu g \cdot g^{-1}. \tag{III.5.4}$$

In what follows these gauge transformations are limited to a particular topological sector, for instance the functions from the compact space to G. It is often useful to consider the infinitesimal gauge transformations ε with values in the Lie algebra, which are tangent to the gauge transformations; the corresponding formula is

$$A \to A^\varepsilon; \; (A^\varepsilon)_\mu = A_\mu + D_\mu \varepsilon, \tag{III.5.5}$$

where $D = \partial - \lambda[A, .]$ is the covariant derivative. Our starting point is the Yang–Mills theory in the Feynman gauge. In this gauge there is an additional factor $\exp(-\int\langle\partial.A,\partial.A\rangle)$ which is the gauge fixing term, and a Faddeev–Popov determinant which is $\det\partial.D$. This determinant can formally be written in terms of ghost fields η and $\bar\eta$ as

$$e^{\int\partial_\mu\bar\eta_a(\partial_\mu\eta^a-\lambda\varepsilon^a_{bc}A^b_\mu\eta^c)}. \tag{III.5.6}$$

The propagators for the gauge and ghost fields are, respectively,

$$\frac{\delta_{\mu\nu}\delta_{ab}}{p^2}, \qquad \frac{\delta_{ab}}{p^2}, \tag{III.5.7}$$

and the Feynman rules are exactly the standard ones that can be found in [IZ] (after rotation to Euclidean space).

The class of ultraviolet cutoffs we consider is defined as follows. κ is a fixed function which is 1 near 0 and decreases at infinity. For instance it could be an exponentially decreasing function or a C_0^∞ function, which is 0 for $|p|\geq 2$ and is 1 for $|p|\leq 1$ (the C_0^∞ character is perhaps not essential but it should be such that the slices built out of it using (III.5.10) have good spatial decay; it might be interesting to have also good momentum conserving properties).

Then we define our scaled momentum cutoff κ_ρ to be:

$$\kappa_\rho(p)=(1/2)[\kappa(apM^{-\rho})+\kappa(pM^{-\rho})] \tag{III.5.8}$$

where a is a small constant (this unusual form leads to a stabilizing A^4 counterterm, as shown below). We can consider the well defined normalized Gaussian measure in A, $d\mu_\rho$, whose propagator is:

$$C_\rho=\delta_{\mu\nu}\delta_{ab}\frac{\kappa_\rho}{p^2} \tag{III.5.9}$$

and create the first momentum slice by writing, as before:

$$C^\rho=C_\rho-C_{\rho-1}. \tag{III.5.10}$$

The same cutoff and slicing is used for the ghost propagator. We will use the notation

$$\det_\rho(A)\equiv e^{-\lambda\int\partial_\mu\bar\eta_a\varepsilon^a_{bc}A^b_\mu\eta^c}d\nu_\rho(\bar\eta,\eta) \tag{III.5.11}$$

where $d\nu_\rho$ is the formal free measure for the anticommuting fields $\bar\eta$ and η with propagator $\delta_{ab}\kappa_\rho/p^2$. This notation is useful, but the true rigorous definition of the corresponding regularized determinant $\det_\rho$ is in fact as in the previous section on the Gross–Neveu model, through convergent power series. There is indeed both an ultraviolet and an infrared cutoff, so by an analogue of Lemma III.4.1, for any smooth sample field A the power series for the determinant $\det_\rho(A)$ indeed has infinite radius of convergence.

Corresponding to the slicing (III.5.10) there is an orthogonal decomposition of the field A^ρ which is the random variable associated to $d\mu_\rho$ as $A_\rho = A^\rho + A_{\rho-1}$ and of the ghost fields $\bar\eta_\rho = \bar\eta^\rho + \bar\eta_{\rho-1}$, $\eta_\rho = \eta^\rho + \eta_{\rho-1}$, but this is again only a convenient notation for the manipulation of determinants.

From standard renormalization group analysis we learn that in order to get a finite nontrivial renormalized theory at the unit scale of our finite box, we should use a bare coupling constant which has the usual asymptotic behavior with ρ implied by asymptotic freedom. Hence a good ansatz for the bare coupling λ_ρ should be:

$$\lambda_\rho = \frac{1}{-\beta_2(\mathrm{Log}\, M)\rho + \beta_3/\beta_2 \log\rho + C} \tag{III.5.12}$$

where C is a large constant, and β_2 and β_3 are the usual first non-vanishing coefficients of the β function, whose numerical value is given in standard textbooks like [IZ]; as is well known, β_2 is negative for such a pure non-Abelian gauge theory. Then one hopes that the renormalized coupling constant λ_{ren}, which should be defined as the last one in a sequence of effective constants, is finite and arbitrarily small as C becomes arbitrarily large by the same mechanism as in section II.5 (if perturbative renormalization group analysis turns out to be correct).

One could believe that the bare theory to start with is simply obtained by multiplying the well defined measure $\det_\rho d\mu_\rho$ by the interaction terms $e^{\lambda F_3 + \lambda^2 F_4}$. However first it would not be obviously well defined since even with a cutoff the cubic term F_3 is not positive and to bound it would require extracting the full positivity of the Gaussian measure. Second, the true bare theory must be more complicated in order to respect gauge invariance in the limit $\rho \to \infty$. Indeed our ultraviolet cutoff breaks gauge invariance. This gauge breaking must be compensated by appropriate gauge-variant counter terms. In fact only the relevant and marginal counterterms must be included because they are the only ones whose effect on finite scales does not vanish as $\rho \to \infty$. We show now how to compute these effects.

C Computation of the counterterms due to the ultraviolet cutoff

Our ultra violet cutoff does not break global SU(2) or Euclidean invariance (small Euclidean breaking effects nevertheless occur due to the infrared cutoff; for instance in the case of a torus there exist such effects due to the lattice structure of Λ^*, but they are tied to the unit scale and do not need counterterms). Therefore the only new relevant or marginal operators that we should consider are $-\operatorname{Tr} A_\mu A_\mu$, $(-\operatorname{Tr} A_\mu A_\mu)^2$, $(-\operatorname{Tr} A_\mu(-\Delta)A_\mu)$ and $-\operatorname{Tr}(\partial_\mu A_\mu)^2$ which we abbreviate respectively as A^2, A^4, $A(-\Delta)A$ and $(\partial A)^2$ (recall the convention that traces are negative definite). This is only true for SU(2) theory; for an SU(N) theory there would be a longer list of operators to consider and the analysis would be more complicated.

In fact our gauge breaking cutoff also disturbs the magic relation $Z_2 Z_4 = Z_3^2$ which relates the multiplicative renormalization of F_2, F_3 and F_4 in F^2 and expresses the fact that up to a rescaling of A only the coupling constant λ is renormalized [IZ]. To correct this problem, using the possibility of rescaling A, we need only to introduce a single counterterm, for instance of the type F_4.

Therefore the counterterms that we introduce are:

$$e^{CT} = e^{-a_\rho \int_\Lambda A^4 - b_\rho \int_\Lambda A^- c_\rho \int_\Lambda A(-\Delta)A - d_\rho \int_\Lambda (\partial A)^2 - e_\rho \int_\Lambda F_4}. \qquad \text{(III.5.13)}$$

The relevant counterterm $b_\rho \int_\Lambda A^2$ must be fine tuned exactly to have a renormalized mass which is zero. This is the same problem as fixing the critical bare mass in infrared φ_4^4 and should be solved either by a fixed point argument as in section III.3 above or using a full renormalization of the two point function (and a one particle irreducible analysis) as in [FMRS5]. For the marginal counterterms, an analysis to lowest order in perturbation theory is in fact enough for our purpose (because of asymptotic freedom, higher orders again should give no contributions to finite scales in the limit $\rho \to \infty$). We obtain:

Lemma III.5.1

$$a_\rho \simeq a\lambda_\rho^4, \qquad b_\rho \simeq bM^{2\rho}\lambda_\rho^2, \qquad c_\rho \simeq c\lambda_\rho^2,$$
$$d_\rho \simeq d\lambda_\rho^2, \qquad e_\rho \simeq e\lambda_\rho^4. \qquad \text{(III.5.14)}$$

Furthermore by choosing the cutoff of the form (III.5.8) with a large enough (depending on the shape of κ), the coefficient a is strictly positive.†

---

†It is not clear whether a cutoff for which a would be negative (or zero) can be used in a constructive way. The answer may depend on adding irrelevant

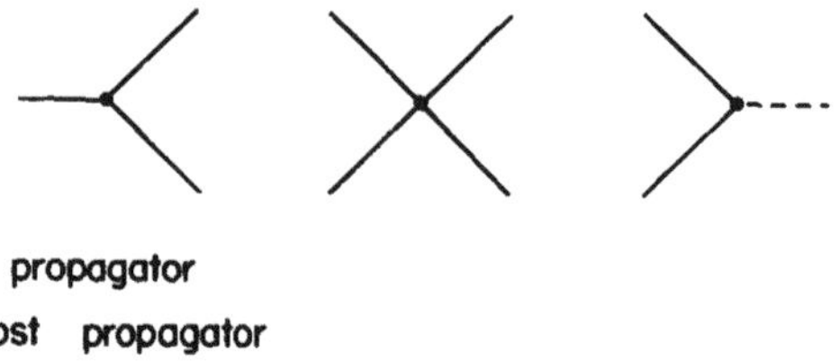

Figure III.5.1 The vertices of pure Yang–Mills.

Proof We recall the Feynman rules for the pure SU(2) gauge theory in the Feynman gauge [IZ]. The propagators are given in (III.5.7). The interaction vertices are of three kinds. For simplicity we always forget to write the overall multiplication factor (of 2π) and the δ function which expresses momentum conservation which equips them. These three kinds of vertices are then pictured in Fig. III.5.1.

We concentrate on the computation of the A^4 counterterm, which is the most interesting, and include also the computation of the A^2 counterterm. The other ones are less interesting and left to the reader.

At one loop, which also means at order g^4 in perturbation theory, there are 4 graphs which may contribute to the A^4 term. They are pictured in Fig III.5.2 and called G_1, G_2, G_3 and G_4. To compute their contribution, we may assume by symmetry that in all four external legs, both the space time and group indices are equal to 1.

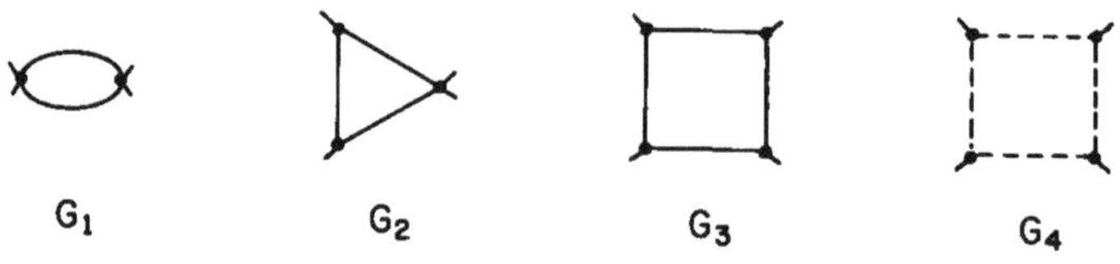

Figure III.5.2

counterterms of higher order generated by the cutoff, which may stabilize the theory. The analysis of globally invariant such terms becomes more and more complicated as the order increases and we will therefore not try to explore this possibility here.

a Computation of G_1

The graph is obtained by applying 4 derivatives $\partial/(\partial A_1^1)$ on $(1/2!)(-F^2/4)^2$. The result is $3(\partial^2 F^2/4)^2$ where ∂ in what follows is a short notation for $\partial/(\partial A_1^1)$. The only non-vanishing pieces come from the derivatives acting on the commutator in F, hence $\partial^2 F^2/4$ gives $(1/2)(\partial F)^2$. Moreover we have $\partial F_{\alpha\beta}^c = \varepsilon_b^{c1}[A_\beta^b \delta_{\alpha 1} - A_\alpha^b \delta_{\beta 1}]$. But remark that if $\alpha = \beta = 1$ the term vanishes. Hence when developing the square $(1/2)(\partial F)^2$ the cross terms vanish. Therefore this square gives $(\varepsilon_b^{c1})^2 (A_\beta^b)^2 \delta_{\alpha 1}$, $\beta \neq 1$. There are now two possible Wick contractions, a sum over three values (2, 3 and 4) for β and a sum over 2 values (2 and 3) for b. Collecting all factors we obtain a positive coefficient $3 \cdot 2 \cdot 3 \cdot 2 = 36$ in front of the integration over the loop momentum of the two propagators of G_1.

b Computation of G_2

We apply 4 derivatives on $(1/3!)(-F^2/4)^3$. The result is $-6(\partial^2 F^2/4)$ $(\partial F^2/4)^2$ where derivatives are again with respect to A_1^1. The term in $\partial^2 F^2/4$ is the same as before, hence gives $(\varepsilon_b^{c1})^2 (A_\beta^b)^2 \delta_{\alpha 1}$, $\beta \neq 1$. But we have now two trilinear vertices in $\partial F^2/4$ hence terms with derivative couplings; remark that a partial derivative ∂_μ can be replaced by $-ik_\mu$. The computation of this term leads to two identical vertices, one which gives $\varepsilon_{mn}^1 A_\mu^n [\partial_1 A_\mu^m - \partial_\mu A_1^m]$, and the other with m, n, μ respectively replaced by p, q, λ. In the Wick contraction schemes we can first contract to form the line between these two trilinear vertices. Since the two half legs of the remaining vertex bear the same index $\beta \neq 1$, a tedious computation gives that the only term compatible with future contractions is $(\varepsilon_{mn}^1)^2 (A_\mu^m)^2 [4k_1^2 + k_\mu^2]$. Using Euclidean symmetry, this is equivalent to $(\varepsilon_{mn}^1)^2 (A_\mu^m)^2 [5k_1^2]$. Contracting with the remaining vertex, we have now as before two possible Wick contractions, a sum over three values (2, 3 and 4) for β and a sum over 2 values (2 and 3) for b. Collecting all factors we obtain a negative coefficient $-6 \cdot 2 \cdot 3 \cdot 2 \cdot 5 \cdot k_1^2 = -90 \cdot 4k_1^2$, again equivalent by Euclidean symmetry to $-90k^2$ in front of the integration over the loop momentum of G_2

c Computation of G_3

We apply 4 derivatives on $(1/4!)(-F^2/4)^4$. The result is $+(\partial F^2/4)^4$ where derivatives are again with respect to A_1^1. The term in $\partial F^2/4$ gives the same trilinear vertex as before, hence gives $\varepsilon_{mn}^1 A_\mu^n [\partial_1 A_\mu^m - \partial_\mu A_1^m]$, In the Wick contraction schemes we can first choose one particular leg of vertex 1 to form a first line between two trilinear vertices. To choose the vertex (2, 3 or 4) to which this leg contracts gives a factor 3. After this contraction has been performed, the line equipped with two not yet contracted fields gives a term $(\varepsilon^1_{mn})^2[2k_1^2(A^m_\mu)^2 + k_\mu^2(A^m_1)^2 - 3k_1k_\mu A^m_1 A^m_\mu]$. Here we can assume $\mu \neq 1$. We can now contract once more to create one line between the two remaining vertices, and this can be done in all possible ways, hence gives a different term, which is $(\varepsilon^1_{mn})^2[4k_1^2(A^m_\mu)^2 + k_\mu^2(A^m_1)^2 - 6k_1k_\mu A^m_1 A^m_\mu + k_\mu k_\lambda A^m_\mu A^m_\lambda]$. We can assume that $\mu \neq 1$ in the first three terms and that $\mu = \lambda = 1$ is excluded in the last one. It remains to contract together both expressions. We have as before two possible Wick contractions, a sum over three values (2, 3 and 4) for μ and a sum over 2 values (2 and 3) for m. After collecting all factors, taking into account Euclidean symmetry we obtain a positive contribution $9 \cdot 12 \cdot k_1^4 + 10 \cdot 12 k_1^2 k_2^2$ in front of the integration over the loop momentum of G_3. Converting it into units of $(k^2)^2$, we find a final combinatoric factor $3(9 \cdot 12/8 + 10 \cdot 12/24) = 55.5$.

d Computation of G_4

We apply 4 derivatives on $(1/4!)(F.P.)^4$, where $F.P.$ means the Faddeev–Popov term $\partial_\mu \bar\eta_a (D_\mu \eta)^a$, with D the covariant derivative. The result is $(\partial_1 \bar\eta_a \varepsilon^{1a}_b \eta^b)^4$. The combinatoric is easier. We obtain a factor 6 for the Wick contractions, a factor 2 for summations over roman indices and a minus sign corresponding to the fermionic loop, which comes from reordering correctly the anticommuting fields η and $\bar\eta$. Hence the contribution is $-12 \cdot k_1^4$ in front of the integration over the loop momentum of G_4. Applying the same conversion rate, we obtain in units of $(k^2)^2$ a final combinatoric factor of -1.5.

Remark that $36 + 55.5 - 90 - 1.5 = 0$, hence all 4 coefficients add up to 0. This is a particular case of the famous miracle of renormalizability (at one loop ...) of four dimensional gauge theories.

Let us perform now a similar analysis for the A^2 counterterm. There are three graphs contributing at order g^2, pictured in Fig III.5.3.

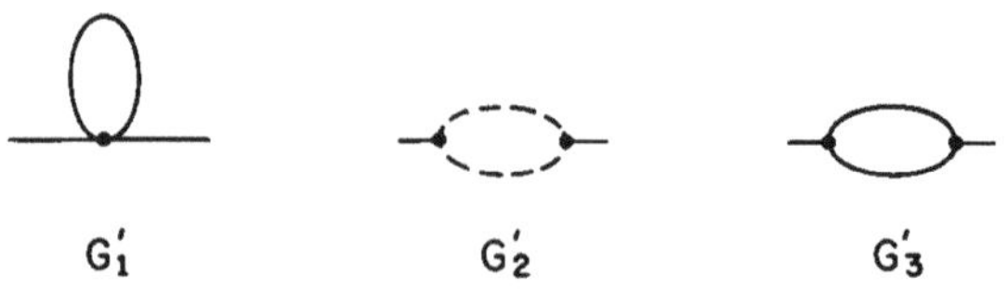

Figure III.5.3

With the same notations than above, the first graph, G_1', gives $\partial\partial(-F^2/4)$, hence $-(1/2)(\partial F)^2$, hence $-(1/2)(\varepsilon_b^{c1})^2(A_\beta^b)^2\delta_{a1}$, $\beta \neq 1$, like in the computation of G_1. There is therefore a factor-6 in front of the integration over the loop momentum (3 for the sum over β and 2 for the sum over b). The second graph, G_2', gives $\partial\partial(1/2)(F.P.)^2 = (\partial F.P.)^2 = (\partial_1 \bar{\eta}_a \varepsilon_b^{1a} \eta^b)^2$. There is a minus sign due to the fermion loop, but there is an additional minus sign from the rule $\partial_\mu \to -ik_\mu$. The contribution is therefore $2k_1^2 = (1/2)k^2$ in front of the integration over the loop momentum. The last graph, G_3', is given by $\partial\partial(1/2)(F^2/4)^2 = (\varepsilon_{mn}^1 A_\mu^n[\partial_1 A_\mu^m - \partial_\mu A_1^m])^2$ (which is non-zero only for $\mu \neq 1$). The computation is similar to the case of the graph G_3 and the contribution is $18k_1^2 = (9/2)k^2$ in front of the integration over the loop momentum.

The result for the A^2 term is then $(-6 + 1/2 + 9/2) = -1$ times the loop integration.

To complete the Lemma, we want to study the sign of the A^4 counterterm. Let us explain why it is important to us. Our strategy is to cancel explicitly the A^4 and A^2 contributions due to the gauge breaking character of our ultraviolet cutoff by appropriate counterterms. Remark that strictly speaking, only the A^2 contribution diverges as $\rho \to \infty$ and requires a counterterm (for the A^4 term the coefficient of the divergent piece is 0, as computed above). However this A^2 counterterm is positive (since the contribution is negative, see the -1 above). This is dangerous for stability estimates. We will use the (finite) A^4 counterterm to control this dangerous A^2 term and stabilize the theory. But this requires that we use an ultraviolet cutoff such that the A^4 counterterm is negative, hence such that the total A^4 contribution induced by the cutoff is positive. As a consequence of our expansion the leading contribution is the one-loop contribution; we want its sign to be positive. We show now that this is possible provided the covariance defining our ultraviolet cutoff is taken to be the arithmetic mean of two covariances, one with cutoff $K \cdot M^{2\rho}$ and the other with cutoff $(1/K) \cdot M^{2\rho}$, K being a large constant. The case (III.5.8) is then similar, with a playing the rôle of K^2, up to a redefinition of the unit scale.

Let $\kappa(p)$ be the ultraviolet cutoff function in momentum space. The phenomenon that we will discuss here is universal, and does not depend of the particular form of κ. Using the coefficients computed in the preceding section, the one loop contribution to the A^4 term is, for a single cutoff $\kappa_\rho(p) = \kappa(pM^{-\rho})$ (all our integrals are infrared regularized and "finite" means finite as $\rho \to \infty$):

$$\int \frac{d^4p}{p^4}[36\eta^2(pM^{-\rho}) - 90\eta^3(pM^{-\rho}) + 54\eta^4(pM^{-\rho})] = 0 \cdot \rho + \text{finite terms.} \tag{III.5.15}$$

For the arithmetic mean of two cutoffs, the corresponding contribution is:

$$\begin{aligned}\int d^4p \Big[&\frac{36}{4}\left(\kappa(KpM^{-\rho}) + \kappa(K^{-1}pM^{-\rho})\right)^2 \\ &- \frac{90}{8}\left(\kappa(KpM^{-\rho}) + \kappa(K^{-1}pM^{-\rho})\right)^3 \\ &+ \frac{54}{16}\left(\kappa(KpM^{-\rho}) + \kappa(K^{-1}pM^{-\rho})\right)^4 \Big].\end{aligned} \tag{III.5.16}$$

But for any given integers q and r and any reasonable cutoff (and certainly for the ones we use) we have the following rules:

$$\int \frac{d^4p}{p^4}\kappa^q(K \cdot pM^{-\rho}) = -\log KM^{-\rho} + \text{finite terms if } q \geq 1 \tag{III.5.17}$$

and, if $K \gg 1$, $M > 1$:

$$\begin{aligned}\int \frac{d^4p}{p^4}&\kappa^q(K \cdot pM^{-\rho})\kappa^r(K^{-1} \cdot pM^{-\rho}) \\ &= -\log KM^{-\rho} + \text{finite terms if } q \geq 1; \\ \int \frac{d^4p}{p^4}&\kappa^q(K \cdot pM^{-\rho})\kappa^r(K^{-1} \cdot pM^{-\rho}) \\ &= -\log K^{-1}M^{-\rho} + \text{finite terms if } q = 0, r \geq 1.\end{aligned} \tag{III.5.18}$$

As a consequence the one loop A^4 contribution behaves as

$$\begin{aligned}\Big(-\frac{36 \times 2}{4} + \frac{90 \times 6}{8} - \frac{54 \times 14}{16}\Big)&\log K + \text{finite terms} \\ &= 2.25 \log K + \text{finite terms,}\end{aligned} \tag{III.5.19}$$

where "finite terms" now means terms which are uniformly bounded both as ρ and K tend to $+\infty$. Therefore taking K large enough (depending on the details of our cutoff, which are responsible for the particular value of the finite terms) we can always achieve our goal of a positive total A^4 contribution, hence of a negative stabilizing counterterm.

In conclusion our starting ansatz for the bare theory is:

$$e^{J_\rho} d\mu_\rho(A) d\nu_\rho(\bar{\eta}, \eta) \tag{III.5.20}$$

with

$$J_\rho = J(A_\rho, \bar{\eta}_\rho, \eta_\rho) = -\frac{1}{2}\left(\lambda_\rho \int_\Lambda F_3 + \lambda_\rho^2 \int_\Lambda F_4\right) - \lambda_\rho \partial_\mu \bar{\eta}_a \varepsilon_{abc} A_{\mu,b} \eta_c + CT \tag{III.5.21}$$

and λ_ρ as in (III.5.12). This starting point is now clearly well defined because we have both finite volume and ultraviolet cutoff. Hence the sample fields are smooth. At large field the exponential of the interaction is clearly bounded for any sample field, in every direction of the configuration space. This is true because the leading terms at large field are the A^4 and F_4 terms which, thanks to the sign of the A^4 term, are respectively positive definite and positive (the determinant $\det_\rho(A)$ may be bounded at large A for instance by an analogue of (III.4.43), hence by a quadratic term). Remark however that it is only for fields of order λ^{-1} that the A^4 term provides convergence, so this term does not confine the field to the true perturbative region ($A \ll \lambda^{-1}$).

D Perturbative results

For perturbative results, we can expand the e^{J_ρ} term in (III.5.20) as a formal (bare) power series, which integrated with the Gaussian measure $d\mu_\rho d\nu_\rho$ gives a bare perturbative expansion. Then we must rewrite this bare expansion as an effective expansion following the method of section II.4. However perturbative results are limited to statements about formal power series, and we cannot in this context justify asymptotic freedom. In particular we must abandon the ansatz (III.5.12) and consider the coupling λ_ρ simply as the parameter of the series. The violations of gauge invariance due to the ultraviolet cutoff are still of the form (III.5.13) but we can no longer use asymptotic freedom (III.5.12) and Lemma III.5.1 to argue that for instance only the one loop contribution has to be included in the A^4 counterterm. For perturbative results we need to express a_ρ, b_ρ, c_ρ, d_ρ and e_ρ as full formal power series in λ_ρ, computing each order as is done explicitly above for the one loop order. However there is an advantage: we have no longer positivity requirements for the coefficient a in (III.5.14) hence we are no longer limited to cutoffs of the type (III.5.8). We can use any shape we want for the ultraviolet cutoff function.

If we compare to the case of φ_4^4 treated in section II.4, there is one main additional problem. We want to show that the resulting usefully renormalized theory has not as many running parameters as the naive number of relevant or marginal operators (which, including non-Euclidean invariant ones, is in the order of the hundreds even for pure SU(2), since the field has twelve components). First by Euclidean and global SU(2) invariance of the slice cutoffs we show that there is no divergence associated to subgraphs whose external leg indices correspond to non-Euclidean or non-global SU(2) invariant operators. Therefore the corresponding subgraphs do not require renormalization at all, and in particular no useful renormalization; no running parameters are therefore associated to the corresponding operators.

Still among the coefficients of the remaining operators in the field A, which are A^2, A^4, $(\partial A)^2$, F_2, F_3, and F_4 ($A(-\Delta)A$ being a combination of F_2 and $(\partial A)^2$) we need some relations to hold in order to show that the effective parameters at scale i reconstruct a theory similar to (III.5.20) but with running parameters. There are also relations to be checked involving the divergent operators with ghosts; the ghost wave function renormalization, which we abbreviate as $\bar{\eta}\partial^2\eta$ and the ghost-field coupling in (III.5.11), which we abbreviate as $\partial\bar{\eta}A\eta$ (other ghost terms, in particular the relevant ghost mass renormalization are zero by unbroken symmetries).

Since a wave function renormalization is allowed both for the field and the ghosts, there are two main relations to be checked, which up to a rescaling of the fields express that only λ, the coupling constant, is renormalized; these are the two relations (12–126a) of [IZ], which we prefer to rewrite (with hopefully slightly more logical notations) as:

$$\frac{Z_4}{Z_3} = \frac{Z_3}{Z_2} = \frac{\tilde{Z}_3}{\tilde{Z}_2}, \tag{III.5.22}$$

where Z_2, Z_3 and Z_4 are the multiplicative coefficients of F_2 F_3 and F_4, and $\tilde{Z}_2$ and $\tilde{Z}_3$ are the corresponding multiplicative coefficients of the ghost-ghost wave function term $\bar{\eta}\partial^2\eta$ and ghost-ghost-field coupling $\partial\bar{\eta}A\eta$. In the initial expression the power of the coupling constant is equal to the degree of the term minus two; relations (III.5.22) automatically express this rule and prove that up to a rescaling of the fields there is only a single coupling constant renormalization.

Relations (III.5.22) were initially proved "by hand" (see e.g., [LZ]) using so called Slavnov–Taylor identities [Sla][Tay]. The modern method [IZ] uses BRS symmetry [BRS] and dimensional regularization, which does not break the symmetry. A correct use of dimensional regularization remains a delicate task, however [BeDa2], and more important, it has no functional counterpart up to now. In our case, the regularization is not dimensional, and we prefer to return to the old fashioned analysis of [LZ]. In this point of view the relations (III.5.22) are deduced from some identities on Schwinger functions. At the formal level one introduces the generating functional for the theory with gauge condition $a(\partial A)^2$ (the case $a = 1$ corresponds to the Feynman gauge):

$$W(J) = \langle e^{-(1/2)F^2 - a(\partial A)^2} \det K e^{J \cdot A} \rangle \tag{III.5.23}$$

where $K = \partial \cdot D$ is the Faddeev–Popov operator, and the expectation value is with respect to the formal Lebesgue measure. As in [LZ][IZ], we write $G = K^{-1}$, where it is understood that in K and K^{-1} the A field should be replaced by the corresponding $\delta/\delta J$ derivation. Then in (III.5.23) we perform a change of variables $A \to A + D\gamma$ with $\gamma = K^{-1}\omega$; by (infinitesimal) gauge invariance, there is no first order dependence in ω, which gives the Ward–Takahashi–Slavnov–Taylor–$\cdots$ equation (omitting group indices):

$$\left(a \frac{\partial}{\partial x^\mu} \frac{\delta}{\delta J_\mu(x)} + \int d^4y J_\lambda(y) D_\lambda(y) K^{-1}_{y,x} \right) W(J) = 0. \tag{III.5.24}$$

We apply another $\frac{\delta}{\delta J}$, take a divergence and put $J = 0$ to get an identity on the two point function:

$$\left\langle \frac{\partial}{\partial x^\mu} A^a_\mu(x) \frac{\partial}{\partial y^\nu} A^b_\nu(y) \right\rangle = \frac{\delta^{ab}}{a} \delta(x - y). \tag{III.5.25}$$

Since in a general gauge with parameter a the propagator is diagonal in su(2) space and is equal, in Fourier space, to

$$(k^2 \delta_{\mu\nu} - (1 - a) k_\mu k_\nu)^{-1} = \frac{\delta_{\mu\nu} k^2 - k_\mu k_\nu}{(k^2)^2} + \frac{k_\mu k_\nu}{a (k^2)^2}, \tag{III.5.26}$$

the first term is killed by $-\partial_\mu \partial_\nu = k_\mu k_\nu$, hence (III.5.25) expresses both that there is no mass renormalization and that the parameter a (the gauge condition) does not renormalize.

Differentiating two more times and taking divergences before putting $J = 0$ leads to a particularly simple identity analogous to (III.5.25) but involving the four point function, which (omitting all indices) takes the form:

$$\langle \partial A \partial A \partial A \partial A \rangle^T = 0, \tag{III.5.27}$$

where the truncation corresponds to the subtraction of three terms of the type $\langle \partial A \partial A \rangle \langle \partial A \partial A \rangle$. It is possible to grasp intuitively that relation (III.5.27) leads to the first of the magic relations (III.5.22), namely to $Z_2Z_4 = Z_3^2$ [LZ]; indeed in terms of one particle irreducible components (vertices) the truncated four point function piece in (III.5.27) may be pictured as in Fig III.5.4, with the first term proportional to the four point vertex Z_4 and the other ones proportional to $Z_3^2Z_2^{-1}$ (because they are made of two three point vertices and one propagator).

Finally one needs to derive from (III.5.24) a last relation, involving the three point function (unfortunately slightly more complicated), to obtain the second magic relation in (III.5.22), which ensures that the renormalization of the ghost-ghost-field vertex remains synchronized with the rest. It is ([LZ]):

$$\left\langle \left(\frac{\partial}{\partial x^\mu} A_\mu(x) \frac{\partial}{\partial y^\nu} A_\nu(y) A_\lambda(z) \right) + \frac{1}{a} D_{\lambda,z} K_{zx}^{-1} \frac{\partial}{\partial y_\nu} A_\nu(y) \right\rangle = 0. \quad \text{(III.5.28)}$$

The strategy for computing the effective expansion for the well defined theory (III.5.20–21) is to find the rigorous analogue of these identities for the theory with external (background) field A_i and internal field $A_\rho - A_i = \sum_{j=i=1}^{\rho}$ integrated with the corresponding Gaussian measure. This theory has an ultraviolet cutoff at scale ρ and an infrared cutoff at scale i, $i = \rho - 1, \rho - 2, \ldots, 0$.

In each slice there is both a gauge breaking ultraviolet cutoff and a gauge breaking infrared cutoff. Therefore the relations (III.5.25) and (III.5.27–28) are no longer exact. Order by order in perturbation the violations can be studied in terms of Feynman diagrams (as is done at one loop in the previous section). These violations fall into two classes, the ones linked to the ultraviolet cutoff and the ones linked to the infrared cutoff. The counterterms CT in (III.5.21) have been fitted to cancel the violations of gauge invariance due to the ultraviolet cutoff at scale ρ. The violations in the ρ slice due to the infrared cutoff have an opposite sign; the corresponding difference generates the discrete flow for the counterterms CT. All violating effects which are not included in CT correspond to contributions which are convergent from the point of view of power counting. In this way apart from the flow of the two

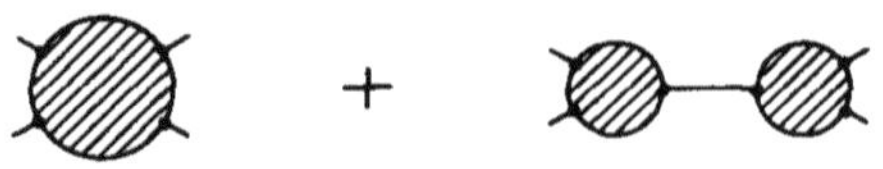

Figure III.5.4 The connected four point function in terms of proper (1PI) vertices

wave function constants corresponding to the A field and ghost field rescalings, there is a single flow for the effective coupling constant.

In the end at the zero slice we obtain a renormalized coupling constant λ_{ren}. As in section II.4 we can invert the relation between λ_ρ and λ_{ren}, reexpress everything in the theory as a formal power series in λ_{ren} and let ρ tend to infinity. In the case of a theory on a compact torus we can then check the Slavnov–Taylor identities (III.5.25) and (III.5.27–28) order by order in the renormalized coupling; in the limit $\rho \to \infty$ all violations disappear because they correspond to operators which are irrelevant, hence exponentially small in ρ. This means that at the perturbative level we have the right renormalized theory and that gauge invariance (in the form of Slavnov–Taylor identities) has been recovered in the limit $\rho \to \infty$, as expected.

Furthermore the analysis of section II.3 leads to uniform bounds of the type $c^n\sqrt{n!}$ at order n on the renormalized series in λ_{ren}. (The square root accounts for the fact that the quartic couplings have couplings λ^2 rather than λ like in φ^4.)

This derivation is perhaps too sketchy to be considered a full proof, but we do not see any basic difficulty in implementing this program in more details. In this way both the standard result on renormalization of non-Abelian gauge theories and a concrete bound on nth order contributions can be derived, and what is probably more important, an effective theory with only useful renormalization performed can be derived, together with a discrete flow equation for the running coupling constant. These are however only results in the sense of formal power series, and for instance although it is clear that the first term in this flow equation is asymptotically equal to the usual β_2 coefficient which has the correct sign for ultraviolet asymptotic freedom, it is impossible to check that a behavior like (III.5.12) corresponds really to a finite λ_{ren}.

E The positivity and domination problems

What happens if we try to apply directly the constructive method (a multiscale expansion) to the bare theory (III.5.20–21)?

We meet problems related to a lack of positivity of the functional integrals both for a direct bound on functional integration and when we try to dominate low momentum fields produced by the expansion.

The direct lack of positivity can be studied even in a single slice model in which there is no domination problem. Unfortunately the interaction terms in (III.5.20–21) are not all of the form of positive even monomials, and this is the main difference with a φ^4 theory such as the one of section III.3. The interaction term F_3 in particular is not positive, and the most natural positive bound recombines it with the Gaussian piece F_2 and the quartic piece F_4 to reconstruct F^2. Then only the gauge fixing term does remain, and it is obviously not positive definite, hence not sufficient for functional integration. In small field regions ($A \ll \lambda^{-1}$), we can bound the F_3 term by the small field condition. But then we need a non-perturbative bound in the regions $A \simeq \lambda^{-1}$ which tells us that the functional weight of these regions is small compared to the (Gaussian) weight of the region $A \simeq 0$. Although the A^4 term shows that the corresponding functional weight is bounded, we cannot prove solely with this term that this weight is small.

This is already in the author's opinion a serious problem, but presumably one can, in a single slice model, find some palliative solution, for instance using the fact that a typical slice cutoff of fixed finite width M enhances the Gaussian measure by a finite factor without enhancing similarly the interaction, so that it should be possible to prove that for such a fixed width cutoff the normalization of the theory tends to the Gaussian normalization as $\lambda \to 0$. Rather than further discussing this point we prefer to focus on the domination problem, which is perhaps more interesting because it is certainly intimately related to the decomposition of the theory into several scales.

A phase space expansion consists in both spatial cluster expansions and momentum decoupling expansions. The only important restriction to build these interpolations is to preserve positivity requirements for the interpolated theory. The covariance (III.5.5) in the Feynman gauge is the same as the one of infrared φ^4_4, hence for the horizontal expansion there is no particular positivity problem and we can use the tree cluster expansion scheme of the previous sections.

For the vertical t decoupling expansion, we need to interpolate in a way which preserves positivity and allows "domination" as much as possible. For an even positive monomial λA^{2n}, let us consider the field A to be the sum $H + L$, where H and L represent the high and low momentum parts of A. The interpolations $(H + tL)^{2n} + (1 - t^{2n})L^{2n}$ or $t(H + L)^{2n} + (1 - t)L^{2n}$ are both positive and suited for domination.

As remarked above, the interaction terms in (III.5.20–21) (in particular the F_3 term) are not of this form. Our first task is therefore to distinguish, in the exponential of the interaction, between two types of terms. In the first category are the ones which in French we call "dominable." These are the ones which come from the decomposition of even positive monomials like above, or which cannot really couple high and low momenta without violating momentum conservation, or which contain only anticommuting low momentum fields (ghosts); such fields can indeed be bounded using the Pauli principle, as in section III.4. In the second category we put the rest. Arguments based on conservation of momenta work only if there is a gap of at least one momentum slice. Therefore let us cut the propagator into the pth slice, the $p-1$ slice and the rest with a C_0^∞ momentum cutoff. Explicitly we write

$$A_p = A^p + A^{p-1} + A_{p-2}, \qquad \text{(III.5.29)}$$

$$\bar{\eta}_p = \bar{\eta}^p + \bar{\eta}^{p-1} + \bar{\eta}_{p-2}, \qquad \text{(III.5.30)}$$

$$\eta_p = \eta^p + \eta^{p-1} + \eta_{p-2}, \qquad \text{(III.5.31)}$$

so that (III.5.20) becomes:

$$\exp(J_{p-1} + I + K)d\mu_p(A_p)d\nu_p(\bar{\eta}_p, \eta_p) \qquad \text{(III.5.32)}$$

where I are the dominable terms, J_{p-1} is the part of the interaction which contains only low momentum fields, and K is the rest, or "non-dominable" terms. In I we want to put first all the coupling pieces coming from counterterms, because they are all dominable by the A^4 term, which is a positive monomial. Therefore we define:

$$I_1 = CT(A_p) - CT(A_{p-1}). \qquad \text{(III.5.33)}$$

We put also in I the following terms, for which the Pauli principle would provide an analogue of domination:

$$I_2 = \lambda\left\{\partial\bar{\eta}_p[A^p, \eta_p] + \partial\bar{\eta}^p[A^{p-1}, \eta_{p-1}] + \partial\bar{\eta}_{p-1}[A^{p-1}, \eta^p]\right\}. \qquad \text{(III.5.34)}$$

Then we have terms for which Gaussian integration will suffice:

$$\begin{aligned} I_3 = \lambda\{&\partial\wedge A_p[A^p,A^p] + \partial\wedge A_{p-1}([A^p,A^{p-1}] + [A^{p-1},A^p]) \\ &+ \partial\wedge A^p[A^{p-1},A^{p-1}]\}. \end{aligned} \qquad \text{(III.5.35)}$$

Finally we have dominable terms coming from the double commutator:

$$\begin{aligned} I_4 = \lambda^2\{&([A^p + A^{p-1},A^p + A^{p-1}][A^p + A^{p-1},A^p + A^{p-1}] \\ &- [A^{p-1},A^{p-1}][A^{p-1},A^{p-1}]) \\ &+ ([A_{p-2},A^p + A^{p-1}][A^p + A^{p-1},A^p + A^{p-1}] \\ &- [A_{p-2},A^{p-1}][A^{p-1},A^{p-1}] + \text{permutations of } A_{p-2}) \\ &+ 2[A_{p-2},A_{p-2}]([A^p + A^{p-1},A^p + A^{p-1}] \end{aligned}$$

$$- [A^{\rho-1}, A^{\rho-1}])\}. \qquad \text{(III.5.36)}$$

All other terms are non-dominable and put into K. Basically K contains the terms with commutators between high and low momentum fields and the coupling between two high momentum ghosts and a low momentum field. More explicitly we have:

$$\begin{aligned} K = {} & \lambda\partial \wedge A^{\rho}([A^{\rho}, A_{\rho-1}] + [A_{\rho-1}, A^{\rho}] + [A^{\rho-1}, A_{\rho-2}] + [A_{\rho-2}, A^{\rho-1}]) \\ & + \lambda\partial \wedge A^{\rho-1}([A^{\rho}, A_{\rho-2}] + [A_{\rho-2}, A^{\rho}]) \\ & + 2\lambda^2 \, ([A^{\rho} + A^{\rho-1}, A_{\rho-2}]([A^{\rho} + A^{\rho-1}, A_{\rho-2}] + [A_{\rho-2}, A^{\rho} + A^{\rho-1}]) \\ & - \, [A^{\rho-1}, A_{\rho-2}]([A^{\rho-1}, A_{\rho-2}] + [A_{\rho-2}, A^{\rho-1}])) \\ & + \lambda(\partial\bar{\eta}^{\rho}[A_{\rho-2}, \eta^{\rho} + \eta^{\rho-1}] + \partial\bar{\eta}^{\rho-1}[A_{\rho-2}, \eta^{\rho}]). \end{aligned} \qquad \text{(III.5.37)}$$

If the reader adds $I = I_1 + I_2 + I_3 + I_4$, $J_{\rho-1}$ and K algebraically he will not find all of J_ρ, but all the missing terms couple one single field of slice ρ to two or three fields of cutoff $\rho - 2$. Hence since the cutoff has compact support these terms are identically 0 when integrated over the torus Λ (provided M is large enough).

The non-dominable terms in K create a problem. However they have quite a regular structure. If one considers the low momentum field as a background field, they come from replacing the ordinary derivatives by covariant derivatives with respect to this background field. This suggests that one should adapt the Gaussian measure (and in particular the Feynman gauge condition) in the higher slices to the low momentum background field. We tried to follow this idea without much success until now. To implement this idea is particularly difficult if, as here, the Gaussian measure itself is used to create the momentum slices, but even with another slicing rule (such as Balaban's method of averaging over block-spins) the problem remains difficult.

This resistance of the gauge fixed non-Abelian functional integral to ordinary treatments is presumably related to the phenomenon of gauge ambiguity discovered by Gribov, which establishes that not every gauge orbit intersects the gauge condition once. Gribov discovery can be considered also as expressing a lack of positivity or of monotonicity of the gauge fixed functional measure. Our last section is devoted to a brief discussion of this problem. Its analysis points towards abandoning the simple ansatz (III.5.20–21) for a more sophisticated one in which additional gauge conditions resolve the Gribov ambiguities.

F The Gribov problem

Gribov discovered [Gri] that in the Landau gauge there can be different smooth field configurations which are nevertheless related by a gauge transformation. The Landau gauge is defined by the condition $\sum_\mu \partial_\mu A_\mu = 0$, and it gives rise to the same Faddeev–Popov determinant than the Feynman gauge [IZ]. A configuration $B \neq A$ such that $\sum_\mu \partial_\mu A_\mu = \sum_\mu \partial_\mu B_\mu = 0$ and such that there exists a gauge transformation g with $B = A^g$ (A^g being defined as in (III.5.3)) is called a Gribov copy of A. The weak Gribov phenomenon is that there are always some configurations which have copies; it is true even on a compact space and for configurations with the same topological properties, hence inside a given topological sector; it is also true for any regular gauge [Sin], not just the Landau gauge. What we call the strong Gribov phenomenon is when there are Gribov copies of the 0 configuration, hence pure gauges which satisfy the gauge condition. This is possible in an infinite space if weak decay at infinity is allowed, as shown in [Gri], but typically this strong Gribov phenomenon does not occur inside a given topological sector in a compact space, or under strong decay conditions at infinity if the space is not compact. For instance we have:

Lemma III.5.2

Let A be a pure gauge which is a smooth configuration vanishing outside some compact domain Δ. Then if A satisfies the Landau condition, A is identically zero.

Proof Without loss of generality (extending Δ if necessary, and using a translation) we can assume that Δ is a cube with the origin at the "lower left" corner. Then we write

$$0 = \sum_\nu \int_\Delta d^4x \int_0^{x^\nu} dy^\nu A_\nu \cdot \left(\sum_\mu \partial_\mu A_\mu)(x(y^\nu)) \right), \tag{III.5.38}$$

where by definition $x^\mu(y^\nu) = x^\mu$ for $\mu \neq \nu$, $x^\nu(y^\nu) = y^\nu$ (and the dot is a scalar product in su(2)). Integrating by parts, if $\nu \neq \mu$ we find

$$\sum_\mu \sum_{\nu\neq\mu} - \int_\Delta d^4x \int_0^{x^\nu} dy^\nu \partial_\mu A_\nu \cdot A_\mu(x(y^\nu))$$

$$= - \sum_\mu \sum_{\nu\neq\mu} \int_\Delta d^4x \int_0^{x^\nu} dy^\nu (1/2) \partial_\nu (A_\mu^2)(x(y^\nu))$$

$$= -\frac{3}{2} \sum_{\mu} \int_{\Delta} d^4x (A_\mu^2)(x), \tag{III.5.39}$$

where in the first equality we used that A is a pure gauge, hence that $F_{\mu\nu} = 0$, which gives $\partial_\mu A_\nu \cdot A_\mu = \partial_\nu A_\mu \cdot A_\mu$.

If $\nu = \mu$ we get instead:

$$\sum_{\mu} \int_{\Delta} d^4x \int_0^{x^\mu} dy^\mu A_\mu \cdot \partial_\mu A_\mu(x(y^\mu)) = \frac{1}{2} \sum_{\mu} \int_{\Delta} d^4x A_\mu^2(x). \tag{III.5.40}$$

Combining both equations we obtain $-\int_\Delta \sum_\mu A_\mu^2 = 0$, hence A, being smooth is identically zero.

The strong Gribov phenomenon happens when a change of topological sector or weak decay at infinity is allowed. Let us write first the gauge condition for a configuration which is smooth and a pure gauge. In the usual quaternionic representation, SU(2) is the unit sphere in $\mathbb{R}^4$, which is spanned by $1, i, j, k$ with the usual rules of multiplication $i^2 = j^2 = k^2 = -1$, $ij = -ji = k = i \wedge j, \ldots$ of quaternions (i, j, k may be thought as Pauli matrices but this is not necessary). The wedge product is the ordinary three dimensional wedge product in the imaginary quaternionic space $\mathbb{R}^3$ spanned by i, j and k. Scalar product in this space is shown by a dot. We can write

$$g = \cos a + \Omega \sin a; \qquad g^{-1} = \cos a - \Omega \sin a \tag{III.5.41}$$

where Ω is the vector piece of g (hence a unit three dimensional vector which is a combination of i, j and k). Then since $\Omega^2 = -1$ and $\partial_\mu \Omega \cdot \Omega = 0$:

$$A_\mu = (\partial_\mu a)\Omega + \sin a \cos a \partial_\mu \Omega - \sin^2 a \partial_\mu \Omega \wedge \Omega. \tag{III.5.42}$$

The Landau gauge condition written in terms of g gives:

$$\begin{aligned} 0 = \sum_\mu \partial_\mu A_\mu &= \Delta a \Omega + \cos a \left[\sin a \Delta\Omega + 2 \cos a \sum_\mu \partial_\mu a \partial_\mu \Omega \right] \\ &\quad - \sin a \left[\sin a \Delta\Omega + 2 \cos a \sum_\mu \partial_\mu a \partial_\mu \Omega \right] \wedge \Omega, \end{aligned} \tag{III.5.43}$$

Therefore the gauge condition is equivalent to two conditions, the longitudinal gauge condition obtained by projecting (III.5.43) on Ω:

$$\Delta a + \frac{\sin 2a}{2} \Omega \cdot \Delta\Omega = 0, \tag{III.5.44}$$

and the transverse gauge condition:

$$\left[\sin a(\Delta\Omega)_T + 2\cos a \sum_{\mu} \partial_\mu a \partial_\mu \Omega\right] = 0, \qquad \text{(III.5.45)}$$

where by definition $(\Delta\Omega)_T = \Delta\Omega - (\Delta\Omega \cdot \Omega)\Omega$. This follows from the straightforward analysis of equation (III.5.43) in the imaginary quaternionic space spanned by i, j, k, since sin and cos do not vanish simultaneously.

In the original example treated in [Gri], the slightly simpler case of a Coulomb gauge is considered (the extension to the Landau gauge is easy). Let us distinguish the time coordinate ($\mu = 0$) from the three spatial coordinates, for which we use roman letters like i or j. The Coulomb gauge is the condition $A_0 = 0$ plus the condition $\sum_j \partial_j A_j = 0$). In such a gauge a pure gauge configuration is constant in the time coordinate ($F_{0,j} = 0 \Rightarrow \partial_0 A_j = 0$). The problem reduces therefore to find a nontrivial three dimensional smooth configuration A_j with $\sum_j \partial_j A_j = 0$ which satisfies to $A_j = (\partial_j g)g^{-1}$ where g is a smooth function from $\mathbb{R}^3$ to SU(2).

To show that this problem has a solution, Gribov considers the case of spherical symmetry, which means that one assumes both $\Omega^i = x^i/r$ and a solely function of $r = \sqrt{(x^1)^2 + (x^2)^2 + (x^3)^2}$. It is easy to check that the transverse condition (III.5.45) is automatically satisfied; the only nontrivial equation is the longitudinal condition (III.5.44), which can be rewritten as

$$\frac{d^2 a}{dr^2} + \frac{2}{r}\frac{da}{dr} - \frac{1}{r^2}\sin 2a = 0, \qquad \text{(III.5.46)}$$

or using $u = 2a$ and $t = \log r$:

$$\frac{d^2 u}{dt^2} + \frac{du}{dt} - 2\sin u = 0, \qquad \text{(III.5.47)}$$

the equation of a damped pendulum. It is a well known fact that there is a solution of this equation, smooth at $r = 0$, which tends to $u = \pi$ or $a = \pi/2$ for $t \to \infty$, hence decreases like $1/r$. This solution starts at $t = -\infty$ from the unstable position at the top and takes an infinite amount of time to pass for the first time at the stable position $u = \pi$ at the bottom, say at $t = 0$. To prove that such a solution exists, one reverses time, namely one considers throwing at $t = 0$ the pendulum upwards with a certain initial velocity. If this velocity is too small, the pendulum will never get at the top; if it is too large it will pass over the top. Therefore at the supremum of the set of velocities with which it does not pass over the top, one can prove by some limit argument that it approaches asymptotically the top position at infinity. Running the corresponding solution backwards we obtain the desired solution; furthermore it must tend to the stable equilibrium $u = \pi$ at $t = +\infty$ because there is strict dissipation of energy in the system.

This nontrivial solution is a Gribov copy of the origin hence it is an example of the strong Gribov phenomenon. However the decrease at infinity as $1/r$ suggests that one can eliminate this phenomenon by an infrared or volume cutoff. This is not explicitly stated in this way in [Gri], but it is certainly well known to the experts. Lemma III.5.2 gives an explicit proof of this fact in the case of smooth configurations with compact support on $\mathbb{R}^4$. The case of a smooth configuration on a compact four dimensional Riemannian manifold is slightly more complicated; the Landau gauge condition is less natural, because the gauge connection transforms as a one form and to take its scalar product with ∇, we have to raise its indices, hence to consider in fact the dual vector field. This raising of indices is nontrivial if the metric is nontrivial. There are nevertheless two compact cases where this raising of indices is canonical because the metric itself is canonical: the case of a compact subset of $\mathbb{R}^4$, treated in Lemma III.5.2, and the case of the four dimensional torus T_4 with its flat canonical metric, which is explicitly treated in the next Lemma. (The proof of Lemma III.5.2 does not work directly for the torus because to get the second piece of equation (III.5.39) it was assumed that A vanishes on the border of the cube Δ.) We believe that there is presumably no strong Gribov phenomenon on any four dimensional compact Riemannian manifold, and this may perhaps be well known, but we just do not know where to find the appropriate literature on this question.

Lemma III.5.3

Let A be a pure gauge which is a smooth function (a topologically trivial connection) on the torus T_4. If A satisfies the Landau condition, A is identically zero.

Proof Let A be a pure gauge on the torus; there are corresponding smooth periodic functions a and Ω respectively to $\mathbb{R}$ and to the three dimensional sphere which define the corresponding gauge transformation g.

The Landau gauge condition implies that any linear combination of the longitudinal and transverse gauge conditions (III.5.44–5) is also zero. We introduce the convenient notation $f_\mu = \partial_\mu f$, and integrate a combination which is $x^\nu a_\nu$ times the longitudinal condition (III.5.44) plus the scalar product of $x^\nu \sin a\Omega_\nu$ with the transverse condition (III.5.45) (repeated indices are summed). Since Ω has unit norm, $\Omega \cdot \Delta\Omega = \Omega \cdot \Omega_{\mu\mu} = -\Omega_\mu \cdot \Omega_\mu$ and we obtain the equation

$$I = \int_{T_4} (x^\nu a_\nu a_{\mu\mu} - x^\nu a_\nu \sin a \cos a\Omega_\mu \cdot \Omega_\mu$$
$$+ \sin^2 a x^\nu \Omega_\nu \cdot \Omega_{\mu\mu} + 2a_\mu \cos a \sin a x^\nu \Omega_\nu \cdot \Omega_\mu) d^4x = 0. \qquad \text{(III.5.48)}$$

Integration by parts on a torus does not give boundary terms. Hence we obtain that the first term in the integral is equal to $+(d-2)/2 \int_{T_4} a_\mu^2$; similarly the last three combine to give

$$\int_{T_4} d^4x \left[-x^\nu (\frac{\sin^2 a}{2})_\nu \Omega_\mu \cdot \Omega_\mu + x^\nu \left(\sin^2 a\Omega_\mu\right)_\mu \cdot \Omega_\nu \right]$$
$$= \int_{T_4} d^4x \left[d \frac{\sin^2 a}{2} \Omega_\mu \cdot \Omega_\mu + x^\nu \sin^2 a\Omega_{\mu\nu} \cdot \Omega_\mu \right.$$
$$\left. - \; x^\nu \sin^2 a\Omega_{\mu\nu} \cdot \Omega_\mu - \sin^2 a\Omega_\mu \cdot \Omega_\mu \right]$$

hence we get finally

$$I = \int_{T_4} \frac{d-2}{2} (a_\mu^2 + \sin^2 a\Omega_\mu \cdot \Omega_\mu) d^4x. \qquad \text{(III.5.49)}$$

In this formulas $d = 4$, hence $(d - 2)/2 = 1$. If the integral (III.5.48) is zero we conclude that we must have $a_\mu = \sin a \cdot \Omega_\mu = 0$ for any μ. Therefore a is constant. If this constant is $k\pi$, then by (III.5.41) $g = 1$ and $A = 0$. If it is not equal to $k\pi$ we must have Ω constant, hence g constant and again $A = 0$. This concludes the proof.

In constructive theory we are interested in explicit proofs of absence of the strong Gribov phenomenon such as Lemmas III.5.2–3 because it might point the way to useful inequalities showing what is the exact amount of positivity which lies in the combination of the action F^2 and the Feynman gauge term $(\partial_\mu A_\mu)^2$. Unfortunately the positivity which comes from (III.5.39–40) or (III.5.49) seems too weak when the frequencies considered are much larger than the inverse size of the volume cutoff. In intuitive terms, the positivity which prevents the strong Gribov phenomenon from happening is linked to the way the infrared problem has been truncated; for instance in the situation of Lemma III.5.2 it is tied to the 0 boundary conditions on A. The corresponding positivity is useful for the analysis of the few last (physical) momentum slices in a phase space decomposition, at high momenta the distance from the boundary at the center of the cube, as measured in units of the relevant very small scale, becomes very large and the corresponding positivity becomes much too weak to be useful, i.e., to constrain the field to lie in the region where perturbative analysis around the Gaussian is valid.

Although in the constructive study of the ultraviolet limit of non-Abelian gauge theories we can avoid the strong Gribov phenomenon just as we avoid the infrared problem (i.e., by appropriate boundary conditions or compactification), we cannot avoid the weak Gribov phenomenon. This phenomenon indeed certainly occurs in the vicinity of null vectors of the Faddeev–Popov operator

$$K = -\partial_\mu D_\mu = -\Delta + \lambda\partial_\mu[A_\mu, .]. \tag{III.5.50}$$

For $\lambda = 0$ the Faddeev Popov operator reduces to minus the Laplacian and is positive definite once the 0-momentum mode (hence translation invariance) has been deleted. But for non-zero λ and A it is possible to show that rescaling $A \to kA$ one can always have negative eigenvalues of K for k large enough (which correspond physically to bound states of the ghost system). This fact is intimately linked to the Gribov phenomenon. The configurations where $\det K = 0$, hence where there exists null vectors for K are the so-called Gribov horizons. The region inside the first Gribov horizon, where K is positive, is called the first Gribov region Ω; similarly one can define the second Gribov region which lies in between the first and second Gribov horizons and so on. Then in [Gri] it is shown by a simple perturbative analysis that near a Gribov horizon there are typically Gribov copies, one on each side of the horizon. These copies can be rapidly decreasing at infinity (or smooth on a compact space) so this "weak Gribov phenomenon" is completely general and cannot be eliminated like the strong one by an infrared cutoff or by topological restrictions.

Remark however that in terms of functional integration the weak Gribov phenomenon is less dangerous than the strong one. Roughly speaking we can think to the strong Gribov phenomenon as a lack of strict positivity and to the weak Gribov phenomenon as a lack of monotonicity for the functional measure. The fact that the Faddeev–Popov operator is not always positive definite at large fields is nevertheless quite disturbing. Since the determinant of this operator occurs in the functional measure, we must conclude that the ordinary measure formula for functional integration is a signed measure.

It is argued in [Hi] that this signed measure, although derived in an incorrect way, is nevertheless the correct one; essentially the argument is that as we move away from the gauge orbit of the origin (which, by absence of the strong Gribov phenomenon, must cut the gauge condition only once) the Gribov copies of the weak Gribov phenomenon should by some sort of continuity argument occur in pairs with equal and opposite values, therefore in pairs which cancel out so that the ordinary prescription with the signed determinant is equivalent to a more correct prescription in which a single point on each gauge orbit is selected, and the absolute value of the determinant of the operator K is used. In any case even if this conjecture was true it seems difficult to prove it; in constructive theory one does not know well how to use signed measures and to check rigorously the corresponding cancellations, especially if they are not a marginal effect but are truly crucial for the existence of the limit.

Thinking again to the difficulties in positivity and domination that are met in the constructive analysis, and presumably their link to the existence of the weak Gribov phenomenon, it is tempting to conclude that the Gribov phenomenon should be eliminated at the beginning by use of a better gauge fixing process, hence a better ansatz than (III.5.20–21). We might hope that such a process will lead to a functional integral with only positive weights and with better large field decay, two things of enormous practical value for constructive theory. However we know that no simple smooth gauge condition is free from Gribov problems [Sin], hence we expect that if it exists, a prescription which bypasses the Gribov problem will be a limit process which is not captured by a single analytic expression.

This direction has been followed over many years by Zwanziger and collaborators. We will limit ourselves to a brief discussion of the approach of [Zw1–4][DeZw1–2], in which one tries to eliminate Gribov copies by additional gauge conditions and to define in this way a correct configuration space for the functional measure. The simplest condition which comes to mind is to choose on gauge orbits the point closest to the origin. The simplest norm to measure distance to the origin is the L^2 norm on each component, which is with our previous notations:

$$\|A\| = \left(\int A^2\right)^{1/2} = \left[\int (1/2)\sum_{\mu,a}(A_\mu^a)^2\right]^{1/2} \qquad \text{(III.5.51)}$$

In [Zw1] it is shown that on a gauge orbit the condition that A is a critical point for the function $A \to \|A\|$ is the Landau gauge condition and the condition that it is a second order local minimum is the condition that at A the Faddeev–Popov operator $K(A)$ in (III.5.50) is positive. This remarkable fact is obtained by a very simple computation, which we would like to reproduce.

At second order we have, if $g = e^{\varepsilon} = 1 + \varepsilon + \varepsilon^2/2$:

$$A^g \equiv gAg^{-1} + (\partial g)g^{-1} = A + D\varepsilon + \frac{1}{2}[\varepsilon, D\varepsilon], \tag{III.5.52}$$

where D is the covariant derivative. Using the cyclicity of the trace which lies in the definition of our scalar product we obtain:

$$(A^g)^2 = \left(A + D\varepsilon + \frac{1}{2}[\varepsilon, D\varepsilon]\right)^2 = A^2 + 2A \cdot D\varepsilon + D\varepsilon \cdot D\varepsilon + A \cdot [\varepsilon, D\varepsilon]. \tag{III.5.53}$$

At first order, using integration by parts and the orthogonality of A and $[A, \varepsilon]$:

$$\int (A^\varepsilon)^2 = \int A^2 - 2\int (\partial_\mu A_\mu \cdot \varepsilon), \tag{III.5.54}$$

which shows that the condition of criticality for the L^2 norm on a gauge orbit is the Landau condition. The second order term in ε gives rise to a nice cancellation due again to cyclicity of the trace:

$$+D\varepsilon \cdot D\varepsilon + A \cdot [\varepsilon, D\varepsilon] = \partial\varepsilon \cdot \partial\varepsilon - \partial\varepsilon \cdot [A, \varepsilon]. \tag{III.5.55}$$

and when integrated by parts we find that the second order term is therefore

$$\int \partial\varepsilon \cdot (\partial\varepsilon - [A, \partial\varepsilon]) = \int \varepsilon \cdot K\varepsilon. \tag{III.5.56}$$

More recently it was proved that on any gauge orbit obtained by applying gauge transformations with finite L^2 norm (in the sense that $\|\partial g g^{-1}\|$ is finite) to a given field with finite L^2 norm, the norm function (III.5.51) achieves its minimum [DeZw2]. Let us consider the configuration space Λ obtained by selecting this absolute minimum of the norm $\|\ \|$ on each gauge orbit (plus additional conditions if there are several absolute minima, which hopefully is not a generic case). We conclude that Λ should be a correct classical configuration space (it seems reasonable to select a single point on each gauge orbit to form this space). What do we know about Λ? Its most important property is to lie entirely inside the first Gribov region Ω, defined as the region in which the operator K is positive; the corresponding determinant being positive we no longer have to deal with a signed measure, hence it seems that we have a better constructive starting point. Recently Λ has been shown to be convex [Zw4] and *not* equal to the first Gribov region (i.e., it is a strict subset of Ω).

Once the correct classical configuration space Λ has been identified, functional quantization should lead to an everywhere positive measure supported on Λ which is formally the exponential of the action times a Lebesgue measure, at least for the bare theory. This point of view is certainly more reasonable than to use the naive Landau gauge condition, which selects the whole class of all stationary points in every gauge orbit, even if one hopes for cancellations as in [Hi]. Nevertheless we think that one of the main unsolved problem in this approach is to give the definition of Λ in the presence of an ultraviolet (gauge breaking) cutoff and to construct a corresponding ansatz for the bare theory in which gauge invariance is recovered in the limit of removing the cutoff. This would be the analogue of what was done in the previous sections for the naive Landau gauge condition. Before that is done, we cannot reach rigorous and definitive conclusions.

With this restriction in mind, the fact that the bare functional integral should be a positive measure supported by a domain $\Lambda \subset \Omega$ has nevertheless surprising consequences which seem at odds with the conventional wisdom about the high energy behavior of non-Abelian gauge theory [DeZw1]. Indeed the first Gribov region Ω (which is in fact better known than Λ itself) is itself included into an ellipsoid in the Hilbert space of L^2 connections (hence it is in particular "bounded in every direction"). This ellipsoidal bound found in [DeZw1] is the source of curious effects for the bare expectation values of field operators, and we explain it now briefly.

In the Landau gauge the Faddeev–Popov operator K reduces to

$$-\Delta + \lambda[A_\mu, \partial_\mu \cdot] = -\Delta + L(A). \qquad \text{(III.5.57)}$$

Following [DeZw1] we rewrite the condition that K is positive as

$$M(A) \equiv (-\Delta)^{-1/2} L(A) (-\Delta)^{-1/2} \geq -1. \qquad \text{(III.5.58)}$$

But $M(A)$ is linear in A (as is $L(A)$). It is therefore (in the case of $SU(2)$) the sum of three components M^a, $a = 1, 2, 3$, obtained by decomposing $A = \sum_a A^a \sigma_a$ in an $su(2)$ basis. More explicitly:

$$(M^a(A) \cdot f)^c = (-\Delta)^{-1/2} \varepsilon_{abc} \sum_\mu A^a_\mu (\partial_\mu f)^b (-\Delta)^{-1/2}. \qquad \text{(III.5.59)}$$

Each M^a is a densely defined operator on the sum of three copies of $H = L^2(T)$, the Hilbert space of ordinary functions on our base manifold T (typically the torus T_4). This sum $H^1 \oplus H^2 \oplus H^3$ is by definition the same as the tensor product $H \otimes \mathbb{C}^3$ and M^a is in fact the tensor product of the real antisymmetric (because of the derivative) x-space operator

$$f \to (-\Delta)^{-1/2} \sum_{\mu} A_\mu^a(\partial_\mu f)(-\Delta)^{-1/2} \tag{III.5.60}$$

by the real antisymmetric operator ε^{abc} in color space. A real antisymmetric operator has a purely imaginary spectrum which is symmetric around zero, and the spectrum and eigenvectors of a tensor product of two such operators is obtained by taking the product of their spectra and the tensor products of their eigenvectors. Therefore the spectrum of each M^a must be real and symmetric around zero. Furthermore each M^a satisfies individually the bound $M^a \geq -1$ similar to (III.5.58), because in color space ε^{abc} is i times the angular momentum in a spin one basis, hence has eigenvalues $-i$, 0 and i; it is a fact familiar to every physicist that the eigenvectors of the momentum operator in one direction have zero expected momenta in the two other orthogonal directions. In other words, if ψ is a normalized eigenvector of M^a, it has to be $\varphi(x) \otimes u$, with u an eigenvector of the three by three matrix m^a, $(m^a)_{bc} = \varepsilon^{abc}$, and one can easily check that:

$$\langle \psi, M^a \psi \rangle = \langle \psi, M \psi \rangle \geq -1. \tag{III.5.61}$$

Combining this information with the fact that the spectrum of M^a is real symmetric we obtain that $-1 \leq M^a \leq +1$ for each a, hence we conclude:

Lemma III.5.4 ([DeZw1])

In the first Gribov region where $K \geq 0$, we have:

$$-1 \leq M(A) \leq 3. \tag{III.5.62}$$

that is when an operator like $M(A)$ is bigger than -1 it is also smaller than 3 (similar bounds of course also exist in the case of a group larger than SU(2)).

The inequality $-1 \leq M(A) \leq 3$ on an operator which is linear in A strongly suggests that A itself should be bounded in terms of a suitable norm. Indeed this is the case and one can prove that e.g., on the torus for SU(2) one has:

Lemma III.5.5 ([DeZw1])

For any A in the first Gribov region

$$N(A)^2 = \sum_k \frac{\operatorname{Tr} A_\mu A_\mu(k)}{k^2} \leq 240 \qquad \text{(III.5.63)}$$

where the sum $\sum_k$ stands for a sum over the frequencies on the lattice dual to the torus, with the zero frequency deleted.

The factor k^2 in the denominator of $N(A)$ can be traced back easily to the derivative factors $(-\Delta)^{-1/2}\partial(-\Delta)^{-1/2}$ in the definition of $M(A)$; however for the detailed derivation of the bound, we refer the reader to [DeZw1] (it is called an ellipsoidal bound because the norm is not the standard one (III.5.51)).

Averaging the pointwise inequality (III.5.63) on configuration space with a positive measure leads to a behavior of the Fourier transform of the two point function which is

$$\langle A^a_\mu(x) A^b_\nu(0)\rangle = \sum_k \left(\delta_{\mu\nu} - \frac{k_\mu k_\nu}{k^2}\right)\delta^{ab} g(k) e^{ikx} \qquad \text{(III.5.64)}$$

with

$$\sum_k \frac{g(k)}{k^2} \leq 80/3, \qquad \text{(III.5.65)}$$

where in contrast the standard renormalization group analysis [IZ] predicts for the renormalized two point function:

$$g_r(k) \simeq_{k\to\infty} \frac{1}{k^2(\log k)^{13/22}} \qquad \text{hence} \quad \sum_k \frac{g_r(k)}{k^2} = +\infty. \qquad \text{(III.5.66)}$$

The bound (III.5.64–65) applies however to the geometrical (bare) quantities whether (III.5.66) applies to the renormalized two point function, which is an expectation values of renormalized fields which have no geometrical interpretation because of the wave function and coupling constant renormalizations. Therefore it is not clear whether there is really a contradiction between the behaviors (III.5.64–65) and (III.5.66). However in some gauges there is no field strength renormalization (hence (III.5.64–65) and (III.5.66) should coincide) and the contradiction, although weaker, still persists [Zw4].

Although it may be too early to derive conclusions, it seems that further interesting and surprising results may await us in this direction. From the physical point of view we conclude that the Gribov phenomenon appears as a non-perturbative effect which seems to work against the coexistence of too many energy scales in the non-Abelian theory. On the infrared side, this was already the content of the initial paper of Gribov [Gri], in which it was argued that this effect leads to an effective infrared cutoff related to confinement. The inequalities (III.5.64–65) show that in the ultraviolet direction as well there is a truncation on frequencies which seems unexpected on the basis of standard perturbative computations. It might be interesting to investigate the high energy behavior of gauge invariant quantities and search for corresponding analogues of (III.5.64–65); a departure of their high energy behavior from standard asymptotic freedom would be even more surprising because it would have physical consequences and because the high energy behavior of objects such as Wilson loops seems well established through Monte-Carlo simulations apparently in good agreement with the standard perturbative wisdom.

In conclusion constructive theory has still no solution to offer to those who are not completely satisfied with lattice results such as those of Balaban [Ba2–9] and who would like to construct the ultraviolet limit of non-Abelian gauge theories in the sense of expectation values of products of fields. We can only hope that for this future task the examples and ideas discussed above will be of some help; we must however add that there are several interesting other approaches, such as stochastic quantization, that we have not taken the time to discuss. Tentatively we think at the moment that the most promising program is to search for a correct ansatz for the bare theory with a gauge breaking ultraviolet cutoff such that gauge invariance in the form of Slavnov identities is recovered in a convincing way when the cutoff is removed, and such that the regularized measure corresponds in some sense to a classical configuration space without Gribov copies, perhaps in the style of [Zw1–4]. If such an ansatz can be found, hopefully the technical details of controlling the limit might be not worse than in the case of the other renormalizable models treated in this book.

References and Bibliography

[Aiz] M. Aizenman, Geometric analysis of φ^4 fields and Ising models, Comm. Math. Phys. **86**, 1 (1982).

[Am] D. Amit, Field Theory, the Renormalization Group and Critical Phenomena, McGraw-Hill.

[AM] G. Auberson and G. Menessier, Some properties of Borel summable functions, Journ. Math. Phys. **22**, 2472 (1981).

[Aub] T. Aubin, CR Acad. Sci. Paris, **280A**, 279 (1975).

[Ba1] T. Balaban, $(\text{Higgs})_{2,3}$ Quantum fields in a finite volume, I, II and III, Comm. Math. Phys. **85**, 603; **86**, 555 (1982) and **88**, 411 (1983).

[Ba2] T. Balaban, Propagators and renormalization transformations for lattice gauge theories I and II, Comm. Math. Phys. **95**, 17 and **96**, 223 (1984).

[Ba3] T. Balaban, Averaging operations for lattice gauge theories, Comm. Math. Phys. **98**, 17 (1985).

[Ba4] T. Balaban, Spaces of regular gauge field configurations on a lattice and gauge fixing conditions, Comm. Math. Phys. **99**, 75 (1985).

[Ba5] T. Balaban, Propagators for lattice gauge theories in a background field, Comm. Math. Phys. **99**, 389 (1985).

[Ba6] T. Balaban, The variational problem and background fields in renormalization group method for lattice gauge theories, Comm. Math. Phys. **102**, 277 (1985).

[Ba7] T. Balaban, Renormalization group approach to lattice gauge field theories, I: Generation of effective actions in a small fields approximation and a coupling constant renormalization in four dimensions, Comm. Math. Phys. **109**, 249 (1987).

[Ba8] T. Balaban, Renormalization group approach to lattice gauge field theories, II: Cluster expansions, Comm. Math. Phys. **116**, 1 (1988); Convergent Renormalization Expansions for lattice gauge theories, Comm. Math. Phys. **119**, 243 (1988).

[Ba9] T. Balaban, Large Field Renormalization I: The basic step of the $\mathbb{R}$ Operation, Comm. Math. Phys. **122**, 175 (1989); II Localization, Exponentiation and bounds for the $\mathbb{R}$ Operation, Comm. Math. Phys. **122**, 355 (1989).

[BaF1] G. A. Battle and P. Federbush, A phase cell cluster expansion for Euclidean field theories, Ann. Phys. **142**, 95 (1982).

[BaF2] G. A. Battle and P. Federbush, A phase cell cluster expansion for a Hierarchical φ_3^4 Model, Comm. Math. Phys **88**, 263 (1983).

[BaF3] G. A. Battle and P. Federbush, A note on cluster expansions, tree graph identities, extra 1/N! factors!!! Lett. Math. Phys. **8**, 55 (1984).

[BaF4] G. A. Battle and P. Federbush, Comm. Math. Phys. **109**, 417 (1987).

[Bat1] G. Battle, A new combinatoric estimate for cluster expansions, Comm. Math. Phys. **94**, 133 (1984).

[Bat2] G. Battle, Comm. Math. Phys. **110**, 601 (1987).

[Bat3] G. Battle, Comm. Math. Phys. **114**, 93 (1988).

[BCGNOPS] G. Benfatto, M. Cassandro, G. Gallavotti, F. Nicolò, E. Olivieri, E. Presutti and E. Scacciatelli, Comm. Math. Phys. **59**, 1433 (1978); Comm. Math. Phys. **71**, 95 (1980).

[BCS] W. Boenkost, F. Constantinescu and U. Schaffenberger, The inverse of a Borel summable function, Journ. Math. Phys. **29**, 1118 (1988).

[BD] J. Bjorken and S. Drell, "Relativistic Quantum Mechanics" and "Relativistic Quantum Fields," McGraw-Hill, New York, 1964 and 1965.

[Be] D. Bessis, A new method in the combinatoric of the topological expansion, Comm. Math. Phys. **69**, 147 (1979).

[BDZ] E. Brézin, F. David, J. Zinn-Justin, private communications.

[BeDa1] M. Bergère and F. David, Ambiguities of renormalized φ_4^4 field theory and the singularities of its Borel transform, Phys. Lett. **135B**, 412 (1984).

[BeDa2] M. Bergère and F. David, Journ. Math. Phys. **20**, 2144 (1979).

[BFS] D. Brydges, J. Fröhlich and T. Spencer, The random walk representation of classical spin systems and correlations inequalities, Comm. Math. Phys. **83**, 123 (1982).

[BGZ] E. Brézin, J. C. Le Guillou and J. Zinn-Justin, Perturbation theory at large order, I: The φ^{2N} interaction and II: Role of the vacuum instability, Phys. Rev **D15**, 1544, 1558 (1977).

[BK] E. Brézin and V. A. Kazakov, ENS preprint, October 1989.

[BIPZ] E. Brézin, C. Itzykson, G. Parisi and J. B. Zuber, Planar diagrams, Comm. Math. Phys. **59**, 35 (1978).

[BL] M. Bergère and Y. M. P. Lam, Bogoliubov-Parasiuk theorem in the α-parametric representation, Journ. Math. Phys. **17**, 1546 (1976).

[BP] N. Bogoliubov and Parasiuk, Acta Math. **97**, 227 (1957).

[Bre] S. Breen, PhD thesis and Leading large order asymptotics for φ_2^4 perturbation theory, Comm. Math. Phys. **92**, 197 (1983).

[BrFe] D. Brydges and P. Federbush, A new form of the Mayer expansion in classical statistical mechanics, Journ. Math. Phys. **19**, 2064 (1978).

[BRS] C. Becchi, A. Rouet, and R. Stora, Ann. of Phys. **98**, 287 (1976).

[Bry] D. Brydges, A short course on cluster expansions, in Critical phenomena, random systems, gauge theories, Les Houches session XLIII, 1984, Elsevier Science Publishers, 1986.

[BS] N. Bogoliubov and D. Shirkov, Introduction to the theory of quantized fields, Wiley Interscience, New York, 1959.

[BZ] M. Bergère and J. B. Zuber, Renormalization of Feynman amplitudes and parametric integral representation, Comm. Math. Phys. **35**, 113 (1974).

[Ca1] C. Callan, Phys. Rev. **D2**, 1541 (1970).

[Ca2] C. Callan, in "Méthodes en théorie des champs," Proceedings of Les Houches summer school, 1975, North Holland ed. Amsterdam, 1976.

[CFR] A. Cooper, J. Feldman, L. Rosen, Legendre transforms and r-particle irreducibility in quantum field theory: the formal power series framework. Ann. Phys. **137**, 213 (1981).

[CL] P. Colella and O. Lanford, contribution to [Er1].

[Co] J. Collins, "Renormalization," Cambridge University Press, 1984.

[dC] C. de Calan, Cours de troisième cycle, Université d'Orsay et preprint Ecole Polytechnique (1982).

[dCdVMS] C. de Calan, P. Faria da Veiga, J. Magnen and R. Sénéor, The Gross-Neveu model in three dimensions, in preparation.

[dCM] C. de Calan, A. Malbouisson, Ann. Inst. Henri Poincaré **32**, 91 (1980).

[dCPR] C. de Calan, D. Petritis and V. Rivasseau, Local existence of the Borel transform in Euclidean massless φ_4^4, Comm. Math. Phys. **101**, 559 (1985).

[dCR1] C. de Calan and V. Rivasseau, Local existence of the Borel transform in Euclidean φ_4^4, Comm. Math. Phys. **82**, 69 (1981); A comment on the local existence of the Borel transform in Euclidean φ_4^4, Comm. Math. Phys. **91**, 265 (1983).

[dCR2] C. de Calan and V. Rivasseau, The perturbation series for φ_3^4 field theory is divergent, Comm. Math. Phys. **83**, 77 (1982).

[DeZw1] G. Dell'Antonio and D. Zwanziger, Ellipsoidal bound on the Gribov horizon contradicts the perturbative renormalization group, Nucl. Phys. **B326**, 333 (1989).

[DeZw2] G. Dell'Antonio and D. Zwanziger, in preparation.

[DFR] F. David, J. Feldman and V. Rivasseau, On the large order behavior of φ_4^4, Comm. Math. Phys. **116**, 215 (1988).

[DS] M. Douglas and S. Shenker, Rutgers University preprint, October 1989.

[Dy1] F. Dyson, Phys. Rev. **75**, 1736 (1949).

[Dy2] F. Dyson, Divergence of perturbation theory in quantum electrodynamics, Phys Rev. **85**, 631 (1952).

[Ec] J. Ecalle, Les fonctions résurgentes et leurs applications, Tomes I et II, Publications mathématiques de l'Université d'Orsay 81.05-06 (1981).

[EG] H. Epstein and V. Glaser, Ann. Inst. Henri Poincaré, **XIX**, 2111 (1973).

[EMS] J. P. Eckmann, J. Magnen and R. Sénéor, Decay properties and Borel summability for the Schwinger functions in $P(\varphi)_2$ theories, Comm. Math. Phys. **39**, 251 (1975).

[Er1] Constructive Quantum field theory, Proceedings of the 1973 Erice Summer School, ed. by G. Velo and A. Wightman, Lecture Notes in Physics, Vol. 25, Springer, 1973.

[Er2] Renormalization theory, Proceedings of the 1975 Erice Summer School, G. Velo and A. Wightman eds.

[FaPo] L. D. Faddeev and V. Popov, Phys. Letters **25B**, 29 (1967).

[Fed1] P. Federbush, A mass Zero cluster expansion I and II, Comm. Math. Phys. **81**, 327 and 341 (1981).

[Fed2] P. Federbush, A phase cell approach to Yang-Mills theory. I Modes, lattice-continuum duality. Comm. Math. Phys. **107**, 3319 (1986).

[Fed3] P. Federbush, A phase cell approach to Yang-Mills theory. III. Local stability, modified renormalization group transformation. Comm. Math. Phys. **110**, 293 (1987).

[Fed4] P. Federbush, A phase cell approach to Yang-Mills theory. IV The choice of variables, Comm. Math. Phys. **114**, 317 (1988).

[Fed5] P. Federbush, A phase cell approach to Yang-Mills theory. VI Non Abelian lattice-continuum duality. Ann. Inst. Henri Poincaré **47**, 17 (1987).

[Fed6] P. Federbush, A phase cell approach to Yang-Mills theory. V Analysis of a chunk, University of Michigan preprint (1987).

[Fed7] P. Federbush, A phase cell approach to Yang-Mills theory. VII Stability: Basics and small field result, University of Michigan preprint (1988).

[FedW] P. Federbush and C. Williamson, A phase cell approach to Yang-Mills theory. II Analysis of a mode, Journ. Math. Phys. **28**, 1416 (1987).

[Fel] J. Feldman, The $\lambda\varphi^4_3$ field theory in a finite volume, Comm. Math. Phys. **37**, 93 (1974).

[Fey1] R. P. Feynman, Space time approach to non relativistic quantum mechanics, Rev. Mod. Phys. **20**, 367 (1948).

[Fey2] R. P. Feynman, Acta Phys. Polon. **24**, 697 (1963).

[FG] G. Felder and G. Gallavotti, Perturbation theory and non renormalizable scalar fields, Comm. Math. Phys. **102**, 549 (1985).

[FH] R. Feynman and A. Hibbs, "Quantum Mechanics and Path Integrals," McGraw-Hill, New York, 1965.

[FHRW] J. Feldman, T. Hurd, L. Rosen and J. Wright, QED: a proof of renormalizability, Lecture Notes in Physics, Vol. 312, Springer Verlag, 1988.

[FMRS1] J. Feldman, J. Magnen, V. Rivasseau and R. Sénéor, Bounds on completely convergent Euclidean Feynman graphs, Comm. Math. Phys. **98**, 273 (1985).

[FMRS2] J. Feldman, J. Magnen, V. Rivasseau and R. Sénéor, Bounds on renormalized graphs, Comm. Math. Phys. **100**, 23 (1985).

[FMRS3] J. Feldman, J. Magnen, V. Rivasseau and R. Sénéor, Infrared φ_4^4, Proceedings of Les Houches Summer school on Critical phenomena, Random systems, Gauge theories, 1984, North Holland.

[FMRS4] J. Feldman, J. Magnen, V. Rivasseau and R. Sénéor, A renormalizable field theory: the massive Gross-Neveu model in two dimensions, Comm. Math. Phys. **103**, 67 (1986).

[FMRS5] J. Feldman, J. Magnen, V. Rivasseau and R. Sénéor, Construction of infrared φ_4^4 by a phase space expansion, Comm. Math. Phys. **109**, 437 (1987).

[FO] J. Feldman and K. Osterwalder, The Wightman axioms and the mass gap for weakly coupled φ_3^4 quantum field theories, Ann. Phys. **97**, 80 (1976).

[FR] J. Feldman and V. Rivasseau, Existence of an instanton singularity in φ_3^4 Euclidean field theory, Ann. Inst. H. Poincaré, **44**, 427 (1986).

[Frö] J. Fröhlich, On the triviality of $\lambda\varphi_d^4$ theories and the approach to the critical point in $d \geq 4$ dimensions, Nucl. Phys. **B200** (FS4), 281 (1982).

[Gal] G. Gallavotti, Renormalization theory and ultraviolet stability for scalar fields via renormalization group methods, Rev. Mod. Phys. **57**, 471 (1985).

[GaNi] G. Gallavotti and F. Nicoló, Renormalization theory in four dimensional scalar fields I and II, Comm. Math. Phys. **100**, 545 and **101**, 247 (1985).

[GaRi] G. Gallavotti and V. Rivasseau, Φ^4 field theory in dimension 4; a modern introduction to its unsolved problems, Ann. Inst. H. Poincaré, **40**, 185 (1984).

[GGS] S. Graffi, V. Grecchi and B. Simon, Borel summability, application to the anharmonic oscillator, Phys. Lett. **B32**, 631 (1970).

[GJ1] J. Glimm and A. Jaffe, Positivity of the φ_3^4 Hamiltonian, Fortschr. Phys. **21**, 327 (1973).

[GJ2] J. Glimm and A. Jaffe, "Quantum Physics. A functional integral point of view." McGraw-Hill, New York, 1981.

[GJS] J. Glimm, A. Jaffe and T. Spencer, The particle structure of the weakly coupled $P(\varphi)_2$ model and other applications of high temperature expansions, Part II: The cluster expansion, in [Er1] above.

[GK1] K. Gawedzki and A. Kupiainen, A rigorous block spin approach to massless lattice theories, Comm. Math. Phys. **77**, 31 (1980).

[GK2] K. Gawedzki and A. Kupiainen, Massless lattice φ_4^4 theory: Rigorous control of a renormalizable asymptotically free model, Comm. Math. Phys. **99**, 197 (1985).

[GK3] K. Gawedzki and A. Kupiainen, Asymptotic freedom beyond perturbation theory, Proceedings of Les Houches Summer school on Critical phenomena, Random systems, Gauge theories, 1984, North Holland eds.

[GK4] K. Gawedzki and A. Kupiainen, Gross-Neveu model through convergent perturbation expansions, Comm. Math. Phys. **102**, 1 (1985).

[GK5] K. Gawedzki and A. Kupiainen, Non-trivial continuum limit of a φ_4^4 model with negative coupling constant, Nucl. Phys. **B257**, 474 (1985).

[GKT] K. Gawedzki, A. Kupiainen and Tirozzi, Renormalons, a dynamical system approach, Nucl. Phys. **B257**, 610 (1985).

[Gl] J. Glimm, Comm. Math. Phys. **10**, 1 (1968).

[GL] M. Gell-Mann and F. Low, Phys. Rev. **95**, 1300 (1954).

[GM] D. Gross and A. A. Migdal, Princeton University preprint, October 1989.

[GrNe] D. Gross and A. Neveu, Dynamical symmetry breaking in asymptotically free field theories, Phys Rev. **D10**, 3235 (1974).

[Gri] V. Gribov, Quantization of non-Abelian gauge theories, Nucl. Phys. **B139**, 1 (1978).

[GrK] D. Gross and Kitazawa, A quenched momentum prescription for large N theories, Nucl. Phys. **B206**, 440 (1982).

[GW] D. Gross and F. Wilczek, Phys. Rev. Lett. **30**, 1343 (1973).

['tH1] G. 't Hooft, Nucl. Phys. **B33**, 173 and **B35**, 167 (1971).

['tH2] G. 't Hooft, A planar diagram theory for strong interactions, Nucl. Phys. **B72**, 461 (1974).

['tH3] G. 't Hooft, Lectures given at Ettore Majorana School, Erice Sicily (1977).

['tH4] G. 't Hooft, Is asymptotic freedom enough? Phys. Lett. **109B**, 474 (1982).

['tH5] G. 't Hooft, On the convergence of planar diagram expansion, Comm. Math. Phys. **86**, 449 (1982).

['tH6] G. 't Hooft, Rigorous construction of planar diagram field theories in four dimensional Euclidean space, Comm. Math. Phys. **88**, 1 (1983).

['tH7] G. 't Hooft, Borel summability of a four dimensional field theory, Phys. Lett. **119B**, 369 (1982).

[Ha] G. Hardy, Divergent series, Oxford University Press, London, 1949.

[He1] K. Hepp, Proof of the Bogoliubov-Parasiuk theorem on renormalization, Comm. Math. Phys. **2**, 301 (1966).

[He2] K. Hepp, Théorie de la renormalisation, Berlin, Springer Verlag, 1969.

[Hi] P. Hirschfeld, Strong evidence that Gribov copying does not affect the gauge theory functional integral, Nucl. Phys. **B157**, 37 (1979).

[Hid] T. Hida, Stationary stochastic processes, Mathematical Notes, Princeton University Press, 1970.

[HS] E. Harrell and B. Simon, Duke Math. Journ. **47**, 845 (1980).

[Hu] T. Hurd, A renormalization group proof of perturbative renormalizability, Comm. Math. Phys. **124**, 153 (1989).

['tHV] G. 't Hooft and M. Veltman, Nucl. Phys. **B44**, 189 and **B50**, 318 (1972).

[Ia] D. Iagolnitzer, Book in preparation on Asymptotic completeness and multiparticle structure in field theories.

[IM1] D. Iagolnitzer and J. Magnen, Asymptotic Completeness and multiparticle structure in field theories, Comm. Math. Phys. **110**, 511 (1987).

[IM2] D. Iagolnitzer and J. Magnen, Asymptotic Completeness and multiparticle structure in field theories, II, Theories with renormalization, Comm. Math. Phys. **111**, 81 (1987).

[IM3] D. Iagolnitzer and J. Magnen, Large momentum properties and Wilson short distance expansion in non-perturbative field theory, Comm. Math. Phys. **119**, 609 (1988).

[IM4] D. Iagolnitzer and J. Magnen, Bethe-Salpeter kernel and short distance expansion in the Massive Gross-Neveu model, Comm. Math. Phys. **119**, 567 (1988).

[IZ] C. Itzykson and J. B. Zuber, Quantum field theory, McGraw-Hill, New York, 1980.

[Ja] A. Jaffe, Divergence of perturbation theory for bosons, Comm. Math. Phys. **1**, 127 (1965).

[Ka] L. Kadanoff, Physics, **2**, 263 (1966) and Phys. Rev. Lett. **23**, 1430 (1969).

[Kac1] M. Kac, On some connection between probability theory and differential and integral equations, *in* Proceedings 2nd Berkeley symposium on mathematics and probability, Univ. of Calif. Press, Berkeley pp189, (1950).

[Kac2] M. Kac, Probability and related topics in the Physical sciences, Interscience, New York (1950).

[KNN] J. Koplik, A. Neveu and S. Nussinov, Some aspects of the planar perturbation series, Nucl. Phys. **B123**, 109 (1977).

[Kop] C. Kopper, preprint Max Planck Institut, 1989.

[KopRi] C. Kopper and V. Rivasseau, unpublished.

[KW] K. Wilson and J. Kogut, Phys Rep. **12C**, 77 (1974); J. Kogut and K. Wilson, The renormalization group and the ε expansion, Phys. Rep. **2C**, 75 (1975).

[La] Collected papers of L. D. Landau, New York, Gordon and Breach, 1965.

[Lau] B. Lautrup, On high order estimates in QED, Phys. Lett. **69B**, 109 (1977).

[Lip] L. N. Lipatov, Divergence of the perturbation series and the quasi classical theory, JETP **45**, 216 (1977).

[LSZ] H. Lehmann, K. Symanzik and W. Zimmermann, Nuovo Cimento, **1**, 205 (1955).

[LZ] B. W. Lee and J. Zinn Justin, Spontaneously broken gauge symmetries, Phys. Rev. **D5**, 3121, 3137, 3155 (1972) and **D7**, 1049 (1973).

[MaS] P. T. Matthews and A. Salam, The Green functions of Quantized Fields, Nuovo Cimento, Vol. **12**, 563 (1954).

[Me] M. L. Mehta, A method of integration over Matrix variables, Comm. Math. Phys. **79**, 327 (1981).

[MNRS] J. Magnen, F. Nicolò, V. Rivasseau and R. Sénéor, A Lipatov bound for convergent graphs of φ_4^4, Comm. Math. Phys. **108**, 257 (1987).

[MP] G. Mack and A. Pordt, Convergent perturbation expansions for Euclidean quantum field theory, Comm. Math. Phys. **97**, 267 (1985).

[MR] J. Magnen and V. Rivasseau, The Lipatov argument for φ_3^4 perturbation theory, Comm. Math. Phys. **102**, 59 (1985).

[MS1] J. Magnen and R. Sénéor, The infinite volume limit of the φ_3^4 model, Ann. Inst. Henri Poincaré, **24**, 95 (1976).

[MS2] J. Magnen and R. Sénéor, Phase space cell expansion and Borel summability for the Euclidean φ_3^4 theory, Comm Math. Phys. **56**, 237 (1977).

[MS3] J. Magnen and R. Sénéor, The infrared behavior of $\nabla\varphi_3^4$, Ann. Phys. **152**, 30 (1984).

[MW] P. Mitter and P. Weisz, Asymptotic scale invariance in a massive Thirring model with U(n) symmetry, Phys. Rev **D8**, 4410 (1973).

[Nak] N. Nakanishi, Graph Theory and Feynman integrals, Gordon and Breach, New York, 1971.

[Ne] F. Nevanlinna, Ann. Acad. Sci. Fenn. Ser. **A12**, 3 (1919).

[Nel] E. Nelson, Construction of Quantum Fields from Markoff Fields, Journ. Funct. Anal. **12**, 97 (1973).

[OS] K. Osterwalder and R. Schrader, Axioms for Euclidean Green's functions, Comm. Math. Phys. **31**, 83 (1973).

[Pa1] G. Parisi, The Borel Transform and the renormalization group, Phys. Rep. **49**, 215 (1979).

[Pa2] G. Parisi, Singularities of the Borel transform in renormalizable theories, Phys. Lett. **76B**, 65 (1978).

[Po] H. D. Politzer, Phys. Rev. Lett. **30**, 1346 (1973).

[Ree] M. Reed, contribution to [Er1].

[Ri1] V. Rivasseau, Construction and Borel summability of planar 4 dimensional Euclidean field theory, Comm. Math. Phys. **95**, 445 (1984).

[Ri2] V. Rivasseau, Constructive Renormalization, in VIIIth International Congress on Mathematical Physics, Juillet 1986, World Scientific Publishing Co, Singapore.

[Ros] L. Rosen, in Canadian Mathematical Society, Conference Proceedings Montréal, Volume 9 (1988).

[RS] V. Rivasseau and E. Speer, The Borel Transform in Euclidean φ^4_ν, Local existence for $\mathrm{Re}\,\nu < 4$, Comm. Math. Phys. **72**, 293 (1980).

[Se] E. Seiler, Gauge theories as a problem of constructive quantum field theory and statistical mechanics, Lect. Notes in Physics, Vol. 159, Springer Verlag, 1982.

[Sen] R. Sénéor, Renormalization group, in Proceedings of the 1989 Erice summer school.

[Si1] B. Simon, The $P(\varphi)_2$ (Euclidean) Quantum field theory, Princeton University Press, 1974.

[Si2] B. Simon, Large orders and summability of eigenvalue perturbation theory; a mathematical overview, International Journal of Quantum Chemistry **XXI**, 3 (1982).

[Sin] I. Singer, Some remarks on the Gribov Ambiguity, Comm. Math. Phys. **60**, 7 (1978).

[Sla] A. Slavnov, Theor. Math. Phys. **10**, 99 (English translation) (1972).

[Sok1] A. Sokal, An improvement of Watson's theorem on Borel summability, Journ. Math. Phys. **21**, 261 (1980).

[Sok2] A. Sokal, An alternative constructive approach to the φ_3^4 quantum field theory and a possible destructive approach to φ_4^4, Ann. Inst. Henri Poincaré **137**, 317 (1982).

[SP] E. Stueckelberg and A. Petermann, Helv. Phys. Acta **26**, 499 (1953).

[Sp1] E. Speer, Generalized Feynman amplitudes, Princeton University Press, 1969.

[Sp2] T. Spencer, The decay of the Bethe-Salpeter kernel in $P(\varphi)_2$ quantum field models, Comm. Math. Phys. **44**, 143 (1975).

[Sp3] T. Spencer, The Lipatov argument, Comm. Math. Phys. **74**, 273 (1980).

[Sy1] K. Symanzik, Small distance behavior in field theory and power counting, Comm. Math. Phys. **18**, 227 (1970).

[Sy2] K. Symanzik, Lett. al Nuovo Cimento, **6**, n°2 (1973).

[SW] R. Streater and A. Wightman, PCT, Spin statistics and all that, Benjamin eds, New York, 1964.

[Tay] J. C. Taylor, Nucl. Phys. **B33**, 173 (1971).

[We] S. Weinberg, Phys. Rev. **118**, 838 (1960).

[Wet] W. Wetzel, Two loop β-function for the Gross-Neveu model, Phys. Lett. **153B**, 297 (1985).

[Wig1] A. Wightman, Quantum field theory in terms of vacuum expectation values, Phys. Rev. **101**, 860 (1956).

[Wig2] A. Wightman, φ_v^4 and generalized Borel summability, in Canadian Mathematical Society, Conference Proceedings, Volume 9 (1988).

[Wil] K. Wilson, Renormalization group and critical phenomena, II Phase space cell analysis of critical behavior, Phys. Rev. **B4**, 3184 (1974).

[Wit] E. Witten, Topological quantum field theory, Comm. Math. Phys. **117**, 353 (1988).

[Zim] W. Zimmermann, Convergence of Bogoliubov's method for renormalization in momentum space, Comm. Math. Phys. **15**, 208 (1969).

[Zin] J. Zinn-Justin, The principles of instanton calculus, in Proceedings of Les Houches 1982 summer school, Elsevier Science Publishers, 1984.

[Zw1] D. Zwanziger, Non-perturbative modification of the Faddeev-Popov formula and banishment of the naive vacuum, Nucl. Phys. **B209**, 336 (1982).

[Zw2] D. Zwanziger, Action from the Gribov horizon, Nucl. Phys. **B321**, 591 (1989).

[Zw3] D. Zwanziger, Local and renormalizable action from the Gribov horizon, Nucl. Phys. **B323**, 513 (1989).

[Zw4] D. Zwanziger, preprints to appear and private communication.

Index

GPSR Authorized Representative: Easy Access System Europe - Mustamäe tee 50, 10621 Tallinn, Estonia, gpsr.requests@easproject.com

www.ingramcontent.com/pod-product-compliance
Ingram Content Group UK Ltd.
Pitfield, Milton Keynes, MK11 3LW, UK
UKHW021457290926
475585UK00012B/16
9780691636825